**Bioprocessing Technology for Production of
Biopharmaceuticals and Bioproducts**

Wiley Series in Biotechnology and Bioengineering

Significant advancements in the fields of biology, chemistry, and related disciplines have led to a barrage of major accomplishments in the field of biotechnology. **Wiley Series in Biotechnology and Bioengineering** focuses on showcasing these advances in the form of timely, cutting-edge textbooks and reference books that provide a thorough treatment of each respective topic.

Topics of interest to this series include, but are not limited to, protein expression and processing; nanotechnology; molecular engineering and computational biology; environmental sciences; food biotechnology, genomics, proteomics and metabolomics; large-scale manufacturing and commercialization of human therapeutics; biomaterials and biosensors; and regenerative medicine. We expect these publications to be of significant interest to the practitioners both in academia and industry. Authors and editors were carefully selected for their recognized expertise and their contributions to the various and far-reaching fields of biotechnology.

Bioprocessing Technology for Production of Biopharmaceuticals and Bioproducts
by Claire Komives (Editor), Weichang Zhou (Editor)

Preparative Chromatography for Separation of Proteins
by Arne Staby, Anurag S. Rathore, Satinder (Sut) Ahuja

Vaccine Development and Manufacturing
by Emily P. Wen (Editor), Ronald Ellis (Editor), Narahari S. Pujar (Editor)

Risk Management Applications in Pharmaceutical and Biopharmaceutical Manufacturing
by Hamid Mollah (Editor), Harold Baseman (Editor), Mike Long (Editor)

Emerging Cancer Therapy: Microbial Approaches and Biotechnological Tools
by Arsenio Fialho (Editor), Ananda Chakrabarty (Editor)

Quality by Design for Biopharmaceuticals: Principles and Case Studies
by Anurag S. Rathore (Editor), Rohin Mhatre (Editor)

Bioprocessing Technology for Production of Biopharmaceuticals and Bioproducts

Edited by

Claire Komives
San Jose State University
San Jose, CA, US

Weichang Zhou
WuXi Biologics
Shanghai, China

The right of Claire Komives and Weichang Zhou to be identified as the editors of this work has been asserted in accordance with law.

Registered Office
John Wiley & Sons, Inc., 111 River Street, Hoboken, NJ 07030, USA

Editorial Office
111 River Street, Hoboken, NJ 07030, USA

For details of our global editorial offices, customer services, and more information about Wiley products visit us at www.wiley.com.

Wiley also publishes its books in a variety of electronic formats and by print-on-demand. Some content that appears in standard print versions of this book may not be available in other formats.

Library of Congress Cataloging-in-Publication Data

Names: Komives, Claire, editor. | Zhou, Weichang, 1963- editor.
Title: Bioprocessing technology for production of biopharmaceuticals and
   bioproducts / edited by Claire Komives, Weichang Zhou.
Description: First edition. | Hoboken, NJ : John Wiley & Sons, Inc., 2019. |
   Series: Wiley series in biotechnology and bioengineering | Includes
   bibliographical references and index. |
Identifiers: LCCN 2018036496 (print) | LCCN 2018037472 (ebook) | ISBN
   9781119378280 (Adobe PDF) | ISBN 9781119378303 (ePub) | ISBN 9781118361986
   (hardback)
Subjects: | MESH: Biological Products–chemistry | Drug Compounding |
   Technology, Pharmaceutical | Bioreactors
Classification: LCC RM301.25 (ebook) | LCC RM301.25 (print) | NLM QV 241 |
   DDC 615.1/9–dc23
LC record available at https://lccn.loc.gov/2018036496

Cover design by Wiley
Cover image: © iStock.com/MiljaPhoto

Set in 10/12pt WarnockPro by SPi Global, Chennai, India

Printed in the United States of America

V10006281_112818

Contents

List of Contributors   *xi*

Part I   Case Study   *1*

1   ***Bacillus* and the Story of Protein Secretion and Production**   *3*
          Giulia Barbieri, Anthony Calabria, Gopal Chotani, and Eugenio Ferrari
1.1   *Bacillus* as a Production Host: Introduction and Historical
          Account   *3*
1.2   The Building of a Production Strain: Genetic Tools for *B. subtilis*
          Manipulation   *5*
1.2.1   Promoters   *5*
1.2.2   Vectors for Building a Production Strain   *6*
1.2.3   *B. subtilis* Competent Cell Transformation   *7*
1.2.4   Protoplasts-Mediated Manipulations   *9*
1.2.5   Genetics by Electroporation   *9*
1.3   *B. subtilis* Secretion System and Heterologous Protein Production   *9*
1.3.1   *Bacillus* Fermentation and Recovery of Industrial Enzyme   *11*
1.3.2   Fermentation Stoichiometry   *12*
1.3.3   Fermentor Kinetics and Outputs   *14*
1.3.4   Downstream Processing   *17*
1.4   Summary   *21*
          References   *21*

2   **New Expression Systems for GPCRs**   *29*
          Dimitra Gialama, Fragiskos N. Kolisis, and Georgios Skretas
2.1   Introduction   *29*
2.2   Recombinant GPCR Production – Traditional Approaches for
          Achieving High-Level Production   *39*
2.3   Engineered Expression Systems for GPCR Production   *42*
2.3.1   Bacteria   *42*

2.3.2   Yeasts   *48*
2.3.3   Insect Cells   *51*
2.3.4   Mammalian Cells   *54*
2.3.5   Transgenic Animals   *54*
2.3.6   Cell-Free Systems   *56*
2.4   Conclusion   *57*
         References   *58*

3   Glycosylation   *71*
         Maureen Spearman, Erika Lattová, Hélène Perreault, and Michael Butler
3.1   Introduction   *71*
3.2   Types of Glycosylation   *72*
3.2.1   N-linked Glycans   *72*
3.2.2   O-linked Glycans   *74*
3.3   Factors Affecting Glycosylation   *76*
3.3.1   Nutrient Depletion   *76*
3.3.2   Fed-batch Cultures and Supplements   *79*
3.3.3   Specific Culture Supplements   *80*
3.3.4   Ammonia   *82*
3.3.5   pH   *82*
3.3.6   Oxygen   *83*
3.3.7   Host Cell Systems   *83*
3.3.8   Other Factors   *85*
3.4   Modification of Glycosylation   *86*
3.4.1   siRNA and Gene Knockout/Knockin   *86*
3.4.2   Glycoprotein Processing Inhibitors and *In Vitro* Modification of
         Glycans   *88*
3.5   Glycosylation Analysis   *89*
3.5.1   Release of Glycans from Glycoproteins   *90*
3.5.2   Derivatization of Glycans   *91*
3.6   Methods of Analysis   *91*
3.6.1   Lectin Arrays   *91*
3.6.2   Liquid Chromatography   *93*
3.6.2.1   HILIC Analysis   *93*
3.6.2.2   Reversed Phase (RP) and Porous Graphitic Carbon (PGC)
         Chromatography   *95*
3.6.2.3   Weak Anion Exchange (WAX) HPLC Analysis   *96*
3.6.2.4   High pH Anion Exchange Chromatography with Pulsed
         Amperometric Detection (HPAEC-PAD)   *96*
3.6.3   Capillary Electrophoresis (CE)   *97*
3.6.4   Fluorophore-assisted Carbohydrate Electrophoresis (FACE) and
         CGE-LIF   *99*

3.6.5 Mass Spectrometry (MS) *100*
3.6.5.1 Ionization *100*
3.6.5.2 Derivatization Techniques Used for MS Analysis of Glycans *102*
3.6.5.3 Fragmentation of Carbohydrates *103*
3.7 Conclusion *109*
References *109*

Part II Bioreactors *131*

4 Bioreactors for Stem Cell and Mammalian Cell Cultivation *133*
Ana Fernandes-Platzgummer, Sara M. Badenes, Cláudia L. da Silva, and Joaquim M. S. Cabral
4.1 Overview of (Mammalian and Stem) Cell Culture Engineering *133*
4.1.1 Cell Products for Therapeutics *134*
4.1.2 Cell as a Product: Stem Cells *136*
4.2 Bioprocess Characterization *140*
4.2.1 Cell Cultivation Methods *140*
4.2.2 Cell Metabolism *141*
4.2.3 Culture Medium Design *143*
4.2.4 Culture Parameters *144*
4.2.5 Culture Modes *145*
4.3 Cell Culture Systems *147*
4.3.1 Static Culture Systems *147*
4.3.2 Roller Bottles *150*
4.3.3 Spinner Flask *150*
4.3.4 Airlift Bioreactor *151*
4.3.5 Fixed/Fluidized-Bed Bioreactor *152*
4.3.6 Wave Bioreactor *152*
4.3.7 Rotating-Wall Vessel Bioreactor *154*
4.3.8 Stirred Tank Bioreactor *155*
4.3.8.1 Agitation/Shear Stress *156*
4.4 Cell Culture Modeling *157*
4.5 Case Studies *159*
4.5.1 Antibody Production in Bioreactor Systems *159*
4.5.2 mESC Expansion on Microcarriers in a Stirred Tank Bioreactor *161*
4.6 Concluding Remarks *162*
List of Symbols *163*
References *164*

5 **Model-Based Technologies Enabling Optimal Bioreactor Performance** *175*
Rimvydas Simutis, Marco Jenzsch, and Andreas Lübbert
5.1 Introduction *175*
5.2 Basics *176*
5.2.1 Balances *176*
5.2.2 Model Identification *177*
5.2.3 Model-Based Process Optimization *178*
5.3 Examples *180*
5.3.1 Model-Based State Estimation *180*
5.3.1.1 Static Model Approach *180*
5.3.1.2 Dynamic Alternatives *183*
5.3.2 Optimizing Open Loop-Controlled Cultivations *184*
5.3.2.1 Robust Cultivation Profiles *184*
5.3.2.2 Evolutionary Modeling Approach *188*
5.3.3 Optimization by Model-Aided Feedback Control *190*
5.3.3.1 Improving the Basic Control *190*
5.3.3.2 Optimizing the Amount of Soluble Product *190*
5.3.4 CO_2-Removal in Large-Scale Cell Cultures *194*
5.4 Conclusion *197*
References *198*

6 **Monitoring and Control of Bioreactor: Basic Concepts and Recent Advances** *201*
James Gomes, Viki Chopda, and Anurag S. Rathore
6.1 Introduction *201*
6.2 Challenges in Bioprocess Control *202*
6.2.1 Process Dynamics and Modeling *202*
6.2.2 Limits of Hardware and Software and Their Integration *203*
6.2.3 Regulatory Aspects *204*
6.3 Basic Elements of Bioprocess Control *205*
6.3.1 Bioprocess Monitoring *205*
6.3.2 Parameter Estimators *205*
6.3.3 Bioprocess Modeling *206*
6.4 Current Practices in Bioprocess Control *208*
6.4.1 PID Control *208*
6.4.2 Model-Based Control *209*
6.4.3 Adaptive Control *211*
6.4.4 Nonlinear Control *214*
6.5 Intelligent Control Systems *217*
6.5.1 Fuzzy Control *217*
6.5.2 Neural Control *219*

6.5.3   Statistical Process Control   *222*
6.5.4   Integrated and Plant-Wide Bioprocess Control   *224*
6.5.5   Metabolic Control   *225*
6.6   Summary   *226*
6.7   Future Perspectives   *227*
        Acknowledgments   *227*
        References   *227*

Part III   Host Strain Technologies   *239*

7   Metabolic Engineering for Biocatalyst Robustness to Organic Inhibitors   *241*
        Liam Royce and Laura R. Jarboe
7.1   Introduction   *241*
7.2   Mechanisms of Inhibition   *243*
7.3   Mechanisms of Tolerance   *245*
7.4   Membrane Engineering   *246*
7.5   Evolutionary and Metagenomic Strategies for Increasing Tolerance   *251*
7.6   Reverse Engineering of Improved Strains   *254*
7.7   Concluding Remarks   *255*
        Acknowledgments   *255*
        References   *255*

Index   *267*

List of Contributors

Sara M. Badenes
Department of Bioengineering and
Institute for Bioengineering and
Biosciences
Insituto Superior Técnico
Universidade de Lisboa
Lisboa, Portugal

Giulia Barbieri
Dipartimento de Genetica
Molecolare Batterica
Universidad di Pavia
Pavia PV, Italy

Michael Butler
Department of Microbiology
University of Manitoba
Winnipeg, Manitoba, Canada

Joaquim M. S. Cabral
Department of Bioengineering and
Institute for Bioengineering and
Biosciences
Insituto Superior Técnico
Universidade de Lisboa
Lisboa, Portugal

Anthony Calabria
DuPont Industrial Biosciences
Palo Alto, CA, USA

Viki Chopda
Department of Chemical Engineering
Indian Institute of Technology Delhi
New Delhi, India

Gopal Chotani
DuPont Industrial Biosciences
Palo Alto, CA, USA

Ana Fernandes-Platzgummer
Department of Bioengineering and
Institute for Bioengineering and
Biosciences
Insituto Superior Técnico
Universidade de Lisboa
Lisboa, Portugal

Eugenio Ferrari
Dipartimento de Genetica
Molecolare Batterica
Universidad di Pavia
Pavia PV, Italy

Dimitra Gialama
Institute of Biology, Medicinal
Chemistry and Biotechnology
National Hellenic Research
Foundation
Athens, Greece

and

Biotechnology Laboratory, School of
Chemical Engineering
National Technical University of
Athens, Zografou Campus
Athens, Greece

James Gomes
Kusuma School of Biological
Sciences
Indian Institute of Technology Delhi
New Delhi, India

Laura R. Jarboe
4134 Biorenewables Research
Laboratory
Department of Chemical and
Biological Engineering
Iowa State University
Ames, IA, USA

Marco Jenzsch
Roche Diagnostics Operations
Nonnenwald 2
Penzberg, Germany

Fragiskos N. Kolisis
Biotechnology Laboratory
School of Chemical Engineering
National Technical University of
Athens, Zografou Campus
Athens, Greece

Erika Lattová
Department of Chemistry
University of Manitoba
Winnipeg, Manitoba, Canada

Andreas Lübbert
Martin Luther University
Halle-Wittenberg, Halle/Salle
Germany

Hélène Perreault
Department of Chemistry
University of Manitoba
Winnipeg, Manitoba, Canada

Anurag S. Rathore
Kusuma School of Biological
Sciences
Indian Institute of Technology Delhi
New Delhi, India

Liam Royce
4134 Biorenewables Research
Laboratory
Department of Chemical and
Biological Engineering
Iowa State University
Ames, IA, USA

Cláudia L. da Silva
Department of Bioengineering and
Institute for Bioengineering and
Biosciences
Insituto Superior Técnico
Universidade de Lisboa
Lisboa, Portugal

Rimvydas Simutis
Martin Luther University
Halle-Wittenberg, Halle/Salle
Germany

Georgios Skretas
Institute of Biology, Medicinal
Chemistry and Biotechnology
National Hellenic Research
Foundation
Athens, Greece

Maureen Spearman
Department of Microbiology
University of Manitoba
Winnipeg, Manitoba, Canada

Part I

Case Study

1

Bacillus and the Story of Protein Secretion and Production

Giulia Barbieri[1], Anthony Calabria[2], Gopal Chotani[2], and Eugenio Ferrari[1]

[1] *Dipartimento de Genetica Molecolare Batterica, Universidad di Pavia, Pavia PV, Italy*
[2] *DuPont Industrial Biosciences, Palo Alto, CA, USA*

1.1 *Bacillus* as a Production Host: Introduction and Historical Account

Contrary to logical thinking, the use of enzymes in daily activities may actually predate the development of modern agricultural societies. Nomad populations of hunters and gatherers exploited rennin produced by the stomach of ruminants for the cheese-making process; while the development of fermentation processes for alcohol can be traced back to more than 7000 years (McGovern et al. 2004; Alba-Lois and Segal-Kischinevzky 2010). However, it is only in the nineteenth-century that enzymes were identified as responsible factors for century old processes such as leather tanning and conversion of starch to sugar (Payen and Persoz 1833).

At the beginning of the twentieth-century, thanks to the work of Otto Rohm, enzymes started playing a wider role in industrial processes as well as in household applications (Wallerstein 1939; Maurer 2010). Two US patents were granted on the use of enzymes for the conversion of starch to sugar; one filed by Schultz et al. (1939) describing the use of a *Bacillus mesentericus* "extract," and the other by Dale and Langlois (1940) claiming the use of fungal saccharifying enzymes.

In the mid-1950s, microbial enzymes started being used extensively in several applications. Large-scale enzyme preparations, obtained via microbial fermentation thus prominently entered the industrial world (Underkofler et al. 1958). The 1960s saw the dawn of *Bacillus* as a production workhorse. Toward the end of the decade, *Bacillus*-derived proteases took hold as essential components of laundry detergents (Roald and De Tieme 1969). At about the same time, high temperature-resistant amylases, useful in the

Bioprocessing Technology for Production of Biopharmaceuticals and Bioproducts, First Edition.
Edited by Claire Komives and Weichang Zhou.

saccharification process, were identified in *Bacillus licheniformis* and *Bacillus amyloliquefaciens*. At first, due to the insufficient genetic characterization, the strains used in large-scale fermentation were isolated via a labor-intensive and time-consuming approach of mutagenesis and screening. Most likely, several thousand mutants were tested for improved production characteristics, such as relief of catabolite repression, antibiotic resistance (most likely mutation in one or more ribosomal components), and sporulation deficiency (Ingle and Boyer 1976). The choice of sporulation mutations is particularly important since it allows extending production time in fermentors and, due to poor survival of nonsporulating cells in the environment, precludes isolation of production strains by competitors. The advent of genetic engineering allowed making rapid targeted changes in enzymes and accelerated construction of *ad hoc* production strains starting from laboratory strains, allowing budding industrial biotechnology companies, such as Genencor, to introduce the first detergent alkaline protease produced by a recombinant microorganism in 1984 (E. Ferrari, unpublished).

For the reasons mentioned above, and for their ease of growth in large-scale submerged fermentation, members of the genus *Bacillus* play a very important role in the manufacture of a number of industrially important products. While the use of *Bacilli* has been explored for the synthesis of pharmaceutical products, their most important commercial role is in the production of industrial enzymes (Aehle 2007). It is estimated that in the current greater than $4-billion industrial enzyme market, *Bacilli* produce about 50% of the enzymes (G. Nedwin, personal communication). These products are employed in a variety of important commercial applications such as laundry, dishwashing, starch-derived ethanol and sweeteners, baking, animal feed, textile, and leather (for review see Aehle 2007).

Several traits make the genus *Bacillus* attractive for protein production especially since *B. subtilis* has a long history of safe use. *Bacillus natto*, a very close relative of the laboratory strain *B. subtilis*, has been used to obtain natto, a staple of Japanese cuisine from soybean fermentation, for over a thousand years (Nishito et al. 2010). Furthermore, what makes the use of *Bacillus* for the production of industrial enzymes particularly attractive is its ability to secrete proteins in the culture fluid. This is a necessary feature to keep the cost of the enzymes low, an essential aspect for this class of product. In fact, in most cases, the cost of enzyme production has to be below the $500 kg^{-1} mark, hence the necessity to have low recovery-associated costs. Over the years, a number of tools have been developed to ease and speed up *Bacillus* genetic manipulation. The availability of the sequenced genomes of both *B. subtilis* (Kunst et al. 1997) and *B. licheniformis* (Rey et al. 2004; Veith et al. 2004) has allowed studies aimed at better understanding their behavior during growth and production (Buescher et al. 2012; Nicolas et al. 2012). Moreover, the well-characterized fermentation, its relatively short time, and the possibility to use cheap feedstock add to the appeal of using these bacteria for the production of industrial enzymes.

This chapter is divided in two main sections: the first section focuses on the genetic tools and strategies useful for the efficient cloning and expression of proteins in *Bacillus*, while the second section provides an up-to-date status on fermentation and recovery of heterologous enzymes from *Bacillus*.

1.2 The Building of a Production Strain: Genetic Tools for *B. subtilis* Manipulation

Numerous genetic manipulation techniques of *B. subtilis* laboratory strains have been established over the years. These tools have helped in refining the genetic and biochemical characterization of this microbe. Hence, even if *B. subtilis* had never played a major role in the historical development of the genus *Bacillus* as an industrial workhorse, the availability of new or genetically engineered enzymes with new properties, and the need to express them at very high levels, has placed this laboratory microbe as a frontrunner in the expression of industrial enzymes. In fact, the only tool available to carry out the needed manipulations in the traditional industrial strains, namely *B. licheniformis* and *B. amyloliquefaciens*, was and is to a large extent protoplast transformation. But this approach is very time-consuming and not always reliable. Recently, however, in some instances, it has become possible to develop competence in *B. licheniformis* strains using the *comK* induction system described below (Diaz-Torres et al. 2003; Hoffmann et al. 2010). Given that the tools and ways to transform *B. subtilis* are simple, the building of a *B. subtilis* production strain for a secreted protein is relatively straightforward when the transcriptional and translational determinants for the synthesis of the protein of interest as well as a signal sequence to direct its secretion are available.

Some of the genetic techniques and tools currently available for building a *B. subtilis* production strain are briefly outlined in the next section.

1.2.1 Promoters

There are a number of promoters that can be used to direct transcription of any target protein. One of the best-characterized *B. subtilis* promoters is *aprE*, which is responsible for the transcription of the alkaline protease. It is yet difficult to explain why a promoter responsible for the expression of one of several scavenging enzymes is so complexly regulated (Ferrari et al. 1993). The transcription of *aprE* is controlled by at least two different repressors, AbrB and ScoC, and by a pleiotropic transcriptional activator, DegU (Henner et al. 1988), which can boost the transcription of the *aprE* mRNA by about 100-fold. The presence of both AbrB and ScoC assure that AprE is not synthesized before the transition phase, e.g. before the culture enters the stationary phase. One advantage of using the *aprE* promoter for heterologous expression is the presence of a transcriptional leader sequence responsible for extending the half-life of its mRNA to about 25–30 minutes (Hambraeus et al. 2002). The mRNA stability is transferred to most genes hooked to this transcriptional leader, allowing robust expression.

Another widely used promoter is the amylase promoter, in its different versions, *amyE*, *amyQ*, and *amyL*, which are derived from *B. subtilis*, *B. amyloliquefaciens*, and *B. licheniformis*, respectively. The amylase promoter, albeit not under strict sporulation control, is temporally regulated and its transcription is turned on at the end of the vegetative growth, just before the cells enter the stationary phase. This is most likely due to the control exerted by catabolite repression in both *B. subtilis* and *B. licheniformis* (Nicholson et al. 1987; Laoide et al. 1989).

Two other promoters worth mentioning are the *sacB* and *sacC* promoters. The *sacB* promoter directs the transcription of a gene responsible for the conversion of sucrose to glucose and fructose and for the production of levans (Gay et al. 1983; Steinmetz 1993). The expression of the *sacB* gene is transcriptionally boosted by certain *degU* mutations, similar to the *aprE* promoter. However, *sacB* is not under sporulation control and is transcribed during vegetative growth. A useful synthetic promoter, widely used in research studies in *B. subtilis* is the *spac* promoter and all its derivatives. It is a hybrid promoter built by fusing a promoter taken from the *B. subtilis* phage SP01 and the lac operator sequence from *Escherichia coli* (Yansura and Henner 1984). This promoter is very useful to understand the possible toxicity of an expressed gene because it allows, to a certain extent, to modulate the transcription of any gene fused to it.

1.2.2 Vectors for Building a Production Strain

There are three types of plasmid vectors that can be used for carrying out genetic manipulations in *B. subtilis*: replicating, temperature sensitive (Ts), and integrative. We will limit the description of the vectors to their use in cloning/expression experiments and will refer the curious reader to three extensive reviews on this subject (Bron 1990; Janniere et al.; 1993; Perego 1993).

Most replicating plasmids are inadequate when building an expression strain due to their inherent instability: in some cases these plasmids are lost during an overnight incubation, even when the strain expressing them is grown under selective pressure conditions. Their usefulness is, therefore, limited to initial cloning and expression testing tasks. Furthermore, since only multimers are effective in transformation of competent *B. subtilis* cells (see section on transformation), "shuttle" vectors carrying both an origin of replication for *B. subtilis* as well as for *E. coli* were developed. One drawback of using shuttle vectors is that often *Bacillus* genes, especially the secreted ones, are toxic to *E. coli*, therefore the outcomes of this approach need to be monitored carefully.

To build expression strains it is, therefore, preferable to work with Ts or integrative vectors that allow stable integration of the desired expression construct into the chromosome. A temperature-sensitive origin of replication, such as the one from pE194Ts (Bron 1990), forces the plasmid to integrate into the chromosome, upon raising the temperature under selective pressure, provided that the vector carries a region of homology with the host DNA. The use of these Ts vectors is a necessary step when transforming protoplasts, which are incapable of integrating incoming DNA directly in their chromosome in contrast to competent cells. Furthermore, when working with a *Bacillus* strain that can be made competent, it is sufficient to create a circular expression cassette carrying: (i) a fragment of DNA homologous to the chromosome of the recipient host; (ii) an antibiotic-resistance locus for selection; and; and (iii) the construct with the gene to be expressed. In this case, because of the multimeric requirement, one must resort to the *in vitro* amplification steps described in Figure 1.1.

1.2.3 *B. subtilis* Competent Cell Transformation

The ability of *B. subtilis* to differentiate in a physiological state, known as "competence," associated with the ability to take up exogenous DNA in response to the exhaustion of nutrients in the environment, has been widely exploited for the genetic manipulation of this industrially relevant microorganism. Under laboratory growth conditions, when nutrients become limiting and cells enter stationary phase, a limited fraction of the cell population switches to the competent state for DNA transformation (Hamoen et al. 2003). Since the pioneering work of Anagnostopolous and Spizizen (1961), different protocols for the preparation of competent *B. subtilis* cells have been developed. The majority of them relies on a two-step procedure: growing cells to an early stationary phase, at 37 °C, in minimal medium with 0.5% glucose as the sole carbon source, then, diluting the cells in a poorer minimal medium containing, in most cases, a lower concentration of amino acids. After 90 minutes of incubation, the cells are highly competent and ready to be transformed by addition of purified chromosomal or plasmid DNA. In optimal conditions, only about 10–20% of the cells in the culture will develop competence (Somma and Polsinelli 1970).

Remarkably, only multimeric plasmids, either produced by rec^+, $recB\text{-}C^-$, or $recF^-$ *E. coli* hosts strains (Bedbrook and Ausubel 1976), or generated *in vitro* can be used to successfully transform *B. subtilis* competent cells at high frequency. Though the requirement for multimerization does not allow the use of a ligation mixture to directly transform *Bacillus* competent cells; multimers created via PCR (Shafikhani et al. 1997) or the use of commercial kits (e.g. Templiphi, sold by GE Healthcare Lifesciences) are sufficient to transform

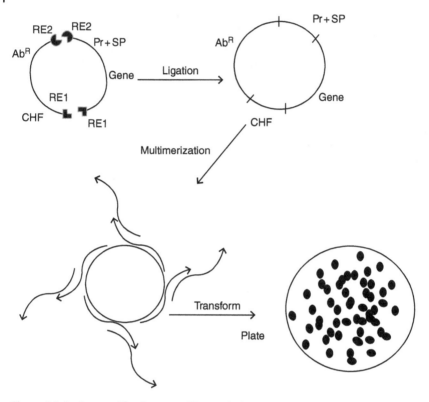

Figure 1.1 *In vitro* amplification steps. REs, restriction enzymes; Pr + SP, promoter and signal peptide; CHF, chromosome homology fragment; and Ab^R, antibiotic resistance or any other selectable marker.

Bacillus competent cells with ligation mixtures and thereby skip the initial *E. coli* cloning step.

The selection of cells reaching competency within a population is random and dependent on the variable expression of a single protein, the transcriptional regulator of competence ComK (van Sinderen et al. 1995). The expression of this master regulator of competence is controlled, via a quorum-sensing-associated mechanism, both at the transcriptional and posttranslational level and is subject to an auto-regulatory positive feedback loop (van Sinderen and Venema 1994).

The role of ComK as the master regulator of competence has been exploited for the induction of competence and the development of "super-competent" *B. subtilis* cells. An extra copy of the *comK* gene, placed under the control of the xylose-inducible promoter PxylA, was integrated in the *lacA* locus of the chromosome (Hahn et al. 1996, Zhang and Zhang 2011). Xylose-induced competent cells can be transformed with efficiencies greater than 1×10^7 transformants μg^{-1} of multimeric plasmid DNA.

1.2.4   Protoplasts-Mediated Manipulations

Use of protoplast-mediated techniques requires methods that allow conversion of cells to protoplasts and subsequent reversion to the bacillary form (regeneration) (Bourne and Dancer 1986). *B. subtilis* cells growing exponentially in rich medium can be easily converted to protoplasts by lysozyme treatment at 37 °C in an osmotic stabilizing medium. Protoplast formation can be monitored by microscopic observation and generally begins within 30 minutes after addition of lysozyme. The incubation period in *B. subtilis* can generally be prolonged to 60–90 minutes to ensure complete protoplast formation (Chang and Cohen 1979). However, for some bacilli, such as *B. licheniformis*, a long incubation period in the presence of lysozyme can ultimately affect the efficiency of regeneration.

1.2.5   Genetics by Electroporation

High transformation efficiencies in *B. subtilis* have also been obtained by electroporation, a technique in which the application of an electric pulse to the cells alters the membrane potential causing a temporary breakdown of the cell membrane permeation barrier that allows the entry of DNA into the cells (Tsong 1992). The procedure varies depending on the strain employed and is very sensitive to different parameters, such as field strength, growth and electroporation medium composition, concentration of competent cells, plasmid variety, and so forth (Lu et al. 2012). The addition of DL-threonine and glycine to the growth medium made it possible to obtain reliable electro-competent *B. subtilis* str. 168 cells that could be efficiently electro-transformed (McDonald et al. 1995). A hyper-osmolarity electroporation method developed by Xue and coworkers (1999) using high concentrations of sorbitol and mannitol gave 1.4×10^6 transformants μg^{-1} of plasmid DNA. In addition to the introduction of circular replicating DNA, methods for successful electroporation with linear integrative DNA have been recently described (Yang et al. 2010; Cao et al. 2011; Meddeb-Mouelhi et al. 2012; Wang et al. 2012).

1.3   *B. subtilis* Secretion System and Heterologous Protein Production

There is a large body of excellent reviews dealing with the *Bacillus* Sec-dependent secretion machinery (Simonen and Palva 1993; van Wely et al. 2001; Tjalsma et al. 2004; Harwood and Cranenburgh 2008) as well as the TAT secretion pathway (van Dijl et al. 2002); hence, we will not review the subject here. We will only point out that the translocase complex of *B. subtilis* is homologous to the system found in *E. coli* (de Keyzer et al. 2003). SecY, SecE, and SecG proteins form the core of a heterotrimeric integral membrane pore that interacts with SecA, an ATPase that drives translocation (Meyer et al. 1999). However, in Gram-positive bacteria, unlike the Gram-negative ones, proteins exported through the cytoplasmic membrane are released directly into the external environment. This ability to export high amounts ($>20\,g\,l^{-1}$) of proteins into the growth medium (Schallmey et al. 2004) renders *B. subtilis* an ideal host for the production of industrial enzymes. In fact, from a commercial point of view, the purification of proteins from the culture supernatant rather than from the cytoplasm is considerably more cost-effective, less time-consuming, and often leads to improved structural authenticity (Pohl and Harwood 2010). Nevertheless, the secretion of heterologous or mutagenized proteins is frequently inefficient because of a variety of secretion bottlenecks, not fully elucidated. These could include poor membrane targeting, inefficient membrane translocation, slow or incorrect polypeptide chain folding, and degradation by extracellular proteases (Li et al. 2004). While the degradation of the protein of interest by the extracellular proteases is, at least in part, under control thanks to the availability of strains deleted for genes encoding multiple proteases (Wu et al. 2002), there is no definite solution to the export blocks. However, the current understanding of the secretion machinery does not allow making rational changes that would result in solving this problem. It is likely that changes in the Sec components would affect the fitness of the cell, given their important role in targeting a large number of vital components to the membrane/cell wall apparatus. It is difficult to imagine that such a weakened cell could perform well under stressful fermentation conditions; hence, most approaches should focus on introducing changes in the protein to be exported. One such approach would be the testing of different signal peptides that are necessary to efficiently direct the protein to the translocase (Kakeshita et al. 2011). As recently revealed by a systematic study in which all *B. subtilis* Sec-type signal peptides were screened for their ability to direct heterologous protein secretion, an optimal fit between the signal peptide and the mature protein is required for efficient secretion (Brockmeier et al. 2006). Target proteins must fold rapidly as they emerge from the Sec translocase. Proteins that fold slowly or partially in fact expose protease-sensitive sites that are recognized and cleaved by WprA, HtrA, and HtrB, the "quality control" proteases expressed by *B. subtilis* to monitor proteins at the membrane and wall interface (Jensen et al. 2000).

Another solution devised in the case of mutagenized commercial proteases was the work in which the proregion of the subtilisin in question was mutagenized via a site evaluation library (Estell and Ferrari 2009; Ferrari et al. 2010). Each of the 84 triplets of the pro-region was mutagenized using 84 libraries of *in vitro*-generated primers (one library for each residue). The anticipated outcome was to obtain pro-region libraries in which each targeted codon contained the triplets encoding for all the 20 natural amino acids. *Bacillus*-competent cells were transformed with the mutagenized constructs and protease expression in the variant cells was tested. Every mutation introduced in some of the residues close to the autoproteolytic site (Power et al. 1986) was deleterious to protease expression, probably due to the interference with the self-maturation process. However, changes at other sites boosted expression up to 100% of the initial titer. The rationale behind this outcome or whether this approach can be applied in a broader context is not yet obvious.

In conclusion, while achieving the secretion of heterologous proteins appears, at present, to have a hit or miss outcome, some recent studies seem to suggest that there are ways to overcome this problem.

1.3.1 *Bacillus* Fermentation and Recovery of Industrial Enzyme

This section examines an industrial process used to produce an engineered thermostable α-amylase for anti-staling in baking applications. A description of this process was chosen because it delivers critical enzymes for the baking industry, e.g. α-amylases that maintain bread freshness for a longer time that reduces bread waste. However, the baking process involves raising the internal bread temperatures to at least 100 °C, and typical enzymes are inactivated around 85 °C. Therefore, thermostable amylases are desired for this application. A production process begins with the engineering of an α-amylase for thermostability and concludes with the enzyme product in the form of a spray-dried powder. Various examples from the literature are highlighted to provide specific process details. Rather than describing an optimal production process, the goal of this example is to provide the reader with a flavor of the options to consider when choosing between various types of processes for the production of a protein product from *B. subtilis*.

A thermostable amylase may be discovered as a novel enzyme from nature, selected from existing enzyme databases, or engineered by using either rational or combinatorial approaches. An example of enzyme engineering is described in the European patent application from Genencor International, Inc. entitled, "Thermostable amylase polypeptides, nucleic acids encoding those polypeptides and uses thereof," EP2292745A1 (Gernot et al. 2011). A plasmid containing the variant gene encoding the desired thermostable amylase is then transformed into a protease-deficient sporulation mutant of *B. subtilis*, according to the methods of Wells et al. (1983) as described in detail in the earlier section of this chapter.

The engineered *B. subtilis* strain is exposed to a representative production environment after preliminary screening and characterization. Optimization of fermentation conditions includes the development of an appropriate medium. Ideally, the organism's requirements for the basic elements, such as carbon, nitrogen, phosphorus, sulfur, magnesium, potassium, and trace elements, are met in a cost-effective manner that allow for optimal growth and product formation rates. These nutrients can be provided using complex (plant, microbial, or animal derived) or chemically defined sources that also include reducing and oxidizing agents (Theil 1998). Medium optimization can be performed using various stochastic and statistical design techniques (Kennedy and Krouse 1999; Weuster-Botz 2000). Considerations to include when evaluating a medium recipe (as well as other fermentation conditions) entail examination of the potential trade-offs between fermentation production metrics and the ease of downstream processing. For the purpose of this α-amylase application, a defined media is chosen for the fermentation step. Specific fermentation details can be found in the research by Huang et al. (2004).

1.3.2 Fermentation Stoichiometry

Knowledge of the chemical composition of the production microorganism can assist fermentation process development toward several objectives, namely, optimizing media composition, rationalizing carbon and nitrogen sources for generating cells and products, and estimating respiration (oxygen, carbon dioxide) and volumetric rate parameters. The composition of a typical microbial cell is shown in Table 1.1 (although ash/minerals content is expected to be much higher) and has a molecular formula of $CH_{1.8}O_{0.5}N_{0.2}$, molecular weight (MW) = 24.6 (Ingram et al. 1983). A similar composition has been observed for *B. subtilis* ($CH_{1.7}O_{0.5}N_{0.2}$, MW = 24.5) (Sauer et al. 1996). General equations for aerobic cell growth, product formation, and cell maintenance have been described in the literature (Hong 1989). Briefly, the equation for cell mass formation is

$$S + n_1 N + o_1 O_2 \rightarrow x_1 X + e_1 CO_2 + w_1 H_2 O \tag{1.1}$$

Table 1.1 Typical composition of a microbial cell (approximately 70% water).

Molecule	Dry cell weight (%)
Protein	55
RNA	20
Lipids	9
Glycogen	3
DNA	3
Liposaccharide	3
Peptidoglycan	3
Metabolites	3
Metal ions	1

where S is the key (limiting) substrate, typically a carbon source; N is the nitrogen source; X is cell mass; and n, o, x, e, and w are stoichiometric coefficients. For cell mass formation from glucose in defined media, the equation is

$$C_6H_{12}O_6 + 0.76NH_3 + 2.03O_2 = 0.63C_6H_{10.8}O_3N_{1.2} + 2.22CO_2$$
$$+ 3.7H_2O \quad (1.2)$$

where the stoichiometric coefficient 0.63 is calculated from metabolic flux experiments. A similar equation can be written for product formation, e.g.

$$S + n_2N + o_2O_2 \rightarrow p_2P + e_2CO_2 + w_2H_2O \quad (1.3)$$

where P is product of interest and p is a stoichiometric coefficient. This equation can be written as follows for an α-amylase molecule such as the thermostable version from *Geobacillus stearothermophilus* with the GenBank Accession Number AAA22227.1 (Suominen et al. 1987).

$$C_6H_{12}O_6 + 1.1NH_3 + 1.4O_2 = 0.73C_6H_{8.9}O_{1.7}N_{1.5} + 1.6CO_2 + 4.4H_2O \quad (1.4)$$

where the stoichiometric coefficient 0.73 is calculated from metabolic flux experiments. Finally, an equation for cell maintenance can be used to describe the consumption of substrates for purposes not resulting in cell mass or product:

$$S + n_3N + o_3O_2 \rightarrow \text{maintenance} + e_3CO_2 + w_3H_2O \quad (1.5)$$

This equation becomes the following when adapted to the α-amylase example:

$$C_6H_{12}O_6 + 6O_2 = 6CO_2 + 6H_2O \quad (1.6)$$

These equations assume that the only fermentation products containing carbon are the cell mass, α-amylase, and carbon dioxide, though there are often additional side products that divert carbon and energy away and can be taken into account to improve the accuracy of the model.

The theoretical cell mass and product yield coefficients derived from Eqs. (1.2) and (1.4) are

$$Y_{x/s} = \frac{x_1M_x}{M_s} = \frac{0.63 \times 147.8 \, \text{g mol}^{-1}}{180.2 \, \text{g mol}^{-1}} = 0.52 \quad (1.7)$$

and

$$Y_{p/s} = \frac{p_2M_p}{M_s} = \frac{0.73 \times 129.1 \, \text{g mol}^{-1}}{180.2 \, \text{g mol}^{-1}} = 0.52 \quad (1.8)$$

respectively, where M_x is the MW of cell mass, M_p is the MW of product, and M_s is the MW of substrate.

These equations have uses throughout fermentation development, such as a comparison of the fermentor's product yield to its theoretical value (see the following section), and the calculation of the cumulative heat of reaction to provide an understanding of the fermentation cooling requirements.

1.3.3 Fermentor Kinetics and Outputs

The proper selection of a fermentation process involves a consideration of a number of factors. The microorganism may prefer aerobic or anaerobic conditions, the product may be extracellular or intracellular (or sometimes the cell mass itself), and it may be produced in a growth or nongrowth associated manner (Asenjo and Merchuk 1991; Van't Riet and Tramper 1991; Arbige et al. 1993). In addition, fermentation process selection must be considered in the context of the entire production process since this upstream step can significantly affect downstream processing.

The following section will focus on the *fed-batch* fermentation process due to its extensive use in industry and its ability to generate high-production rates. The reader can find a detailed description of this and other fermentor processes in the literature where the equations for total mass and species balances around the fermentor are presented (Chotani et al. 2007). Additional equations of interest include the specific substrate consumption rate, q_s, which can be defined as follows:

$$q_s = \frac{\mu}{Y_{x/s}^{max}} + \frac{q_p}{Y_{p/s}^{max}} + m \tag{1.9}$$

where μ is the growth rate of the cell mass, $Y_{x/s}^{max}$ is the maximum growth yield, q_p is the specific product formation rate, $Y_{p/s}^{max}$ is the maximum product yield, and m is the maintenance coefficient. The specific product formation rate, q_p, can be expressed by the Leudeking–Piret equation (Atkinson and Mavituna 1983):

$$q_p = \alpha \times \mu + \beta \tag{1.10}$$

where α is a growth rate-associated coefficient and β is a nongrowth rate-associated coefficient of product formation.

In our specific example, the engineered α-amylase strain (biocatalyst) is fermented for 27 hours at which time the dissolved oxygen level decreases below 20%, and the fermentation is ended. The cell density, α-amylase, and broth glucose concentration profiles are shown in Figure 1.2a. The fermentation finished with a cell density of 17.6 g l^{-1} and an α-amylase concentration of 7.3 g l^{-1}. No residual glucose was detected in the final fermentation broth, and the total glucose added to the fermentor (including batched and fed) was equivalent to

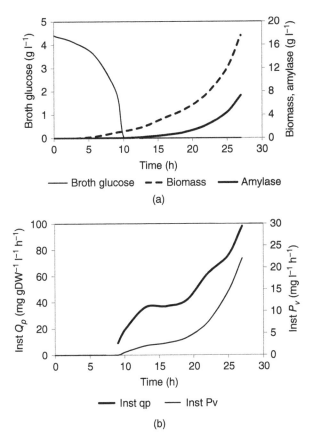

Figure 1.2 Selected fermentation trends illustrating (a) broth glucose, biomass, and α-amylase concentrations over time. (b) Oxygen uptake rate (OUR), instantaneous volumetric productivity (P_v), and hinstantaneous specific productivity (Q_p).

68.3 g l^{-1}. Titer, fermentor productivity, and yield are fermentation metrics used to evaluate the economics of this unit operation and the performance of the biocatalyst. The product concentration in the broth (titer), C_p, at time t, is

$$C_p = \frac{\int_0^t (q_p \times X \times V) dt}{V_t} = \frac{\text{cumulative product mass}}{\text{cumulative volume}}$$

$$= 7.3 \text{ g l}^{-1} \text{ at 27 hours} \tag{1.11}$$

where X is the cell mass concentration in the fermentor broth, V is the volume of the culture, and V_t is the final fermentor broth volume. The average volumetric productivity, P_v, is

$$P_v = \frac{\int_0^t (q_p \times X \times V)dt}{V_t \times t} = \frac{C_p}{t} = \frac{\text{cumulative product mass}}{\text{cumulative volume} \times \text{time}}$$

$$= \frac{7.3 \, \text{g} \, \text{l}^{-1}}{27 \, \text{hours}} = 270 \, \text{mg} \, (\text{l} \, \text{h})^{-1} \tag{1.12}$$

Equation (1.12) can be modified to estimate instantaneous productivity by replacing V_t with the average working volume, $\overline{V}$, or nominal productivity by replacing V_t with nominal fermentor volume. The cumulative observed cell mass yield, $Y_{x/s}^{obs}$, at time t is

$$Y_{x/s}^{obs} = \frac{X_t \times V_t}{\left[S_0 \cdot V_0 + S_f \int_0^t F(t)dt \right]} = \frac{\text{biomass}}{\text{substrate mass consumed}}$$

$$= \frac{17.6 \, \text{g} \, \text{l}^{-1}}{68.3 \, \text{g} \, \text{l}^{-1}} = 0.26 \tag{1.13}$$

The cumulative observed product yield, $Y_{p/s}^{obs}$, at time t is

$$Y_{p/s}^{obs} = \frac{C_p \times V_t}{\left[S_0 \times V_0 + S_f \int_0^t F(t)dt \right]} = \frac{\text{product mass}}{\text{substrate mass consumed}}$$

$$= \frac{7.3 \, \text{g} \, \text{l}^{-1}}{68.3 \, \text{g} \, \text{l}^{-1}} = 0.11 \tag{1.14}$$

where S_0 is the substrate concentration at time zero, and V_0 is the volume at time zero. These observed cell mass and product yields were 50% and 21% of the theoretical maximum values (Eqs. (1.7) and (1.8)), respectively.

Additional calculations provide information about the biocatalyst performance. The ability of the biocatalyst to generate product can be quantified by the specific productivity metric, $\overline{q}_p$:

$$\overline{q}_p = \frac{P_v}{\overline{X}}$$

$$= \frac{\text{cumulative product mass}}{\text{average working volume} \times \text{average grams dry cell weight} \times \text{time}} \tag{1.15}$$

Specific productivity can also be quantified on cumulative cell mass basis or instantaneous product formation rate basis by changing numerator and denominator of Eq. (1.15). The instantaneous fermentor productivity $(\overline{P}_v)$, and biocatalyst specific productivity $(\overline{q}_p)$ profiles are shown in Figure 1.2b. Other measurements provide information of the metabolic activity and include the oxygen uptake rate (OUR) and carbon dioxide evolution rate (CER) and are described in detail elsewhere in the literature (Chotani et al. 2007). These metrics help characterize biocatalyst performance by quantifying the interplay between metabolic activity and product formation over the course of the fermentation.

Additional end-point calculations provide important metrics for an overall fermentation process evaluation. A mass balance over the fermentor indicates a significant fraction of the glucose provided for cell mass and product was also used for maintenance.

$$\text{Maintenance glucose} = 68.3 \ \text{g l}^{-1} - \frac{17.6 \, \text{g l}^{-1} \times 180.2 \ \text{g mol}^{-1}}{0.63 \times 147.8 \ \text{g mol}^{-1}}$$
$$- \frac{7.3 \ \text{g l}^{-1} \times 180.2 \ \text{g mol}^{-1}}{0.73 \times 129.1 \ \text{g mol}^{-1}} = 20.3 \, \text{g l}^{-1}$$

Thus, $\frac{20.3 \, \text{g l}^{-1}}{68.3 \, \text{g l}^{-1}} = 0.3$ is the fraction of glucose used for maintenance (0.5 for cell mass and 0.2 for α-amylase). This value can range between 0.05 and 0.9 for biological systems producing heterologous protein products. This information can also be presented as the glucose carbon fate (Figure 1.3), which indicates the overall efficiency of product generation. These data indicate that fermentation costs could be lowered by genetic changes to the biocatalyst and/or process changes that improve cell mass and/or α-amylase yield on carbon.

1.3.4 Downstream Processing

Downstream processing typically refers to the separation, purification, and formulation of fermentation products. The overall objective is to (i) recover the product in an efficient manner and (ii) formulate it in a cost-effective manner for the properties required by the application. Numerous recovery processes exist since the required characteristics of the products can vary widely. However, they typically include some modular combination of the following procedures: fermentor harvest, cell separation from the broth, cell

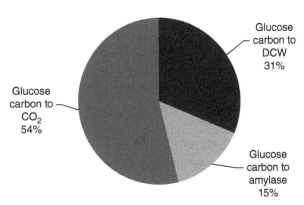

Figure 1.3 The fraction of glucose carbon residing in dry cell weight (DCW), α-amylase, or carbon dioxide (CO_2) after 27 hours of fermentation.

Glucose carbon to DCW 31%

Glucose carbon to CO_2 54%

Glucose carbon to amylase 15%

lysis, product extraction, crude and refined separation steps, concentration, purification, formulation, and/or drying. Factors taken into consideration include the type and physiological state of microorganism used for production, the fermentation media, and the physical and chemical properties of the product. The final product form can be as simple as fermentation broth or require more sophisticated procedures such as those utilized in making an enzyme granule.

The recovery of proteins and enzymes requires particular care since they can become inactivated by physical or chemical denaturation. They can be produced as either intracellular or extracellular products. Intracellular product recovery typically involves some type of cell disruption technique prior to a separation process. For extracellular products, cell disruption is avoided since it can liberate intracellular components that hinder separation processes and/or induce issues related to regulatory compliance. In any of these cases, cell separation is typically one of the initial processes and is accomplished in industry using centrifugation and/or filtration techniques. What follows is a discussion of filtration methods. The reader is referred to equipment manufacturer's manual for an in-depth discussion of centrifugation theory and practice.

Filtration processes include various types of equipment that are operated in batch, continuous, or semicontinuous mode, such as membrane modules, rotary vacuum drum filters (RVDFs), filter presses, and belt filters. The membrane's mode of operation can be described as dead-end filtration since the only fluid flow is through the membrane. A membrane acts as a separating boundary for pressure and concentration-driven mass transfer between feed and filtrate. Darcy's law relates the physical properties within this process to the fluid flow:

$$J = \frac{dV}{A\,dt} = \frac{\Delta p}{v_0(R_m + R_c)} = \frac{\Delta p}{v_0\left(R_m + \alpha \rho_c \frac{V}{A}\right)} = \frac{\text{fluid volume}}{\text{membrane area} \times \text{time}}$$

$$(1.16)$$

where J is the solvent flux, V is the fluid volume, A is the membrane area, Δp is the transmembrane pressure (TMP), v_0 is the permeate viscosity, R_m is the resistance of the membrane, and R_c is the resistance of the filter cake. The resistance of the filter cake is usually the dominating resistance to permeability and can be cast in terms of the specific cake resistance, α, the mass of dry filter cake per unit volume of permeate, ρ_c, as shown in Eq. (1.16). Integration of both sides of Eq. (1.16) gives

$$\frac{tA}{V} = \frac{\rho_c \alpha v_0}{2\Delta p}\left(\frac{V}{A}\right) + \frac{v_0 R_m}{\Delta p}$$

$$(1.17)$$

where α can be determined from the slope of a plot of $\frac{tA}{V}$ versus $\frac{V}{A}$. Filter aids are used for RVDF and filter press systems to improve permeability through the cell cake and filtrate clarity.

In addition to permeate flux rate, other important metrics include product yield and purity. Recovery process product can be improved by increasing the separation selection for the product (physical) and by controlling degradation losses (chemical or biochemical) within the unit operation. For example, proteases from *Bacillus* can undergo self-degradation during recovery processes, and so concentration, temperature, and pH must be appropriately controlled. Product purity is typically dictated by the end-user application and can be affected by numerous upstream conditions including the fermentation media composition and physiological state of the cells.

Following fermentation, the α-amylase-containing broth is subjected to tangential flow microfiltration (MF) for cell separation (Caridis and Papathanasiou 1997; Keefe and Dubbin 2005). While MF equipment costs can be relatively high compared to other cell separation technologies, advantages include the potential for high throughputs and purity as well as a better contained and cleaner process. MF also fits well into the theme of integrated process development. For example, the replacement of centrifugation and rotary drum vacuum filtration steps with a MF step can consolidate a two-step process to one, eliminate the need for adding filter aids that can cause broth disposal issues, improve downstream ultrafiltration (UF) fluxes, and provide higher product yields by improving the selectivity of the separation and by maintaining biological activity. MF membranes are typically defined by pore sizes ranging between 0.01 and 0 μm and operated at TMP values of 0.1–3 bar. They can be constructed from numerous materials such as synthetic polymers, ceramics, carbon and stainless steel. As opposed to polymeric spiral or hollow fiber membranes, ceramic and steel membranes provide open, tubular channels that avoid clogging and facilitate cleaning and sterilization. Their use is commonplace in industry as a result of their durability (years of use before replacement), though fouling must be minimized to maintain flux. However, unlike dead-end filtration processes, turbulent crossflow rates (≥ 5 m s^{-1}) can assist in minimizing the membrane cake layer formation and thereby reduce fouling. For illustrative purposes, it will be assumed that an average MF permeate flux rate of 20 l m^{-2} h^{-1} is observed over a 36-hour period with a single pass α-amylase transmission rate of 65% over the range of concentrations in question. Thus, a MF unit containing three membrane stages can be used to recover α-amylase over a 12-hour period with the following membrane load:

$$20 \, \frac{\text{l broth}}{\text{m}^2 \, \text{h}^{-1}} \times 12 \text{ hours} = 240 \, \text{l m}^{-2},$$

and yield:

$$1 - (1 - 0.65)^{3 \text{ stages}} = 0.96$$

leading to a 96% product recovery. Purity is observed at 90%, where the impurities consist primarily of salts, cell-derived polysaccharides, and some proteins and peptides.

Next, the permeate stream from the MF stage is concentrated using UF. The considerations for UF systems are similar to those described for MF systems since they utilize similar materials and configurations. The filtration unit consists of plate-and-frame style cassettes containing polyethersulfone membranes with a nominal molecular weight cut-off value of 30 kDa. UF concentration results in comparable flow rates, yields (98%), and purity levels (90%) to the MF process, concentrating the MF-permeate α-amylase concentration from 7.3 to $65 \, g \, l^{-1}$ (for other products there can be up to a 20-fold increase from the final fermentor concentration). Finally, the same UF equipment is used to perform diafiltration (DF) to reduce the solution conductivity to $4 \, mS \, cm^{-1}$ while improving purity to 95% by removing salts (98% yield).

Purification is usually achieved by less expensive techniques for industrial products, if it is performed at all. Crystallization of the α-amylase was chosen for this example though precipitation and extraction also meet this criteria. Crystallization is initiated by adding solid sodium chloride to the DF retentate to a final concentration of 400 mM. This solution is stirred at 25 °C for 16 hours before separating the crystallized enzyme using a filter press (Becker and Lawlis, Jr. 1991). The crystallization step increases the purity to 99% at the expense of a relatively low yield (80%).

The final product form can be a liquid, granule, or powder. The choice is typically dictated by the requirements of the application, and each has its own considerations to take into account. For the α-amylase example, a spray-dried powder (Neubeck 1980; Becker and Crowley 1998) was chosen since this form is commonly used for baking applications. A spray dryer feed solution is fed into the chamber at a rate of $1.7 \, cm^3 \, s^{-1}$. Inlet air at a constant flow rate of $2.3 \, m \, s^{-1}$ and 220 °C is used as the drying media. Process measurements include the powder moisture content and the remaining α-amylase activity (Nath and Satpathy 1998; Samborska et al. 2005). The spray-drier is able to produce $1.6 \, kg \, h^{-1}$ of powder with a residual moisture level of 6% and 90% of the initial α-amylase activity remaining.

Table 1.2 Example data for an α-amylase downstream production process illustrating yield and purity values for individual process unit operations.

Process step	Yield (%)	Purity (%)
Microfiltration	96	90
Ultrafiltration	96	90
Diafiltration	98	95
Crystallization	80	99
Spray drying	90	–
Total	65	99

Table 1.2 shows individual and cumulative process step yield values (65%), as well as purity levels (99%). Further integrated process development work could be directed toward improving the entire production system yield as well as product quality characteristics.

1.4 Summary

The systematic approach for developing *Bacillus* as host of choice for expression and secretion is a useful cell factory design model. The advances in *Bacillus* molecular genetics and cell engineering in the last three decades have reshaped enzyme production. It has become possible to clone engineered gene sequences encoding efficient enzymes and express them in *Bacilli* suitable for large-scale industrial fermentation processes. Enzyme productivity has steadily been increased through efficient promoters, regulatory gene mutations, deletions, and multiple copy insertions. Industrial protein engineering continues to tailor enzyme properties such as optimum temperature, pH, stability, and substrate interaction. The development of nonpathogenic and nontoxigenic microbial production strains, such as *B. subtilis* and *B. licheniformis*, well characterized by generally accepted safety evaluation procedures (e.g. Pariza and Johnson 2001) has found wide acceptance by regulatory agencies such as the US FDA (Olempska-Beer et al. 2006) and US EPA (http://www.epa.gov/opptintr/biotech/pubs/rulesupc.htm). The general acceptance of the safety of microbial enzymes tested against international standards (JECFA 2006) has allowed increased enzyme manufacture and use worldwide over the last few decades. Industrial bioprocessing will expand as gene sequencing and engineering of production microorganisms continue to become faster.

References

Aehle, W. (2007). *Enzymes in Industry: Production and Applications*. Weinheim: Wiley-VCH.

Alba-Lois, L. and Segal-Kischinevzky, C. (2010). Beer and wine makers. *Nat. Educ.* 3: 17.

Anagnostopoulos, C. and Spizizen, J. (1961). Requirements for transformation in *Bacillus subtilis*. *J. Bacteriol.* 81: 741–746.

Arbige, M.V., Bulthuis, B.A., Schultz, J., and Crabb, D. (1993). Fermentation of *Bacillus*. In: *Bacillus subtilis and Other Gram-Positive Bacteria: Biochemisty, Physiology and Molecular Genetics* (ed. A.L. Sonenshein, J.A. Hoch and R. Losick), 871–895. Washington, DC: American Society for Microbiology.

Asenjo, J.A. and Merchuk, J.C. (1991). *Bioreactor System Design* (ed. J.A. Asenjo and J.C. Merchuk). New York, NY, USA: Marcel Dekker.

Atkinson, B. and Mavituna, F. (1983). *Biochemical Engineering and Biotechnology Handbook*. New York, NY, USA: Nature.

Becker, N.T. and Crowley, R.P. (1998). Process for making dust-free enzyme-containing particles from an enzyme-containing fermentation broth. US Patent 5,814,501.

Becker, T. and Lawlis, Jr., V.B. (1991). Subtilisin crystallization process. US Patent 5,041,377.

Bedbrook, J.R. and Ausubel, F.M. (1976). Recombination between bacterial plasmids leading to the formation of plasmid multimers. *Cell* 9: 707–716.

Bourne, N. and Dancer, B.N. (1986). Regeneration of protoplasts of *Bacillus subtilis* 168 and closely related strains. *J. Gen. Microbiol.* 132: 251–255.

Brockmeier, U., Caspers, M., Freudl, R. et al. (2006). Systematic screening of all signal peptides from *Bacillus subtilis*: a powerful strategy in optimizing heterologous protein secretion in Gram-positive bacteria. *J. Mol. Biol.* 362: 393–402.

Bron, S. (1990). *Plasmids. Molecular Biological Methods for Bacillus* (ed. C.R. Harwood and S.M. Cutting), 75–174. Chichester, New York: Wiley.

Buescher, J.M., Liebermeister, W., Jules, M. et al. (2012). Global network reorganization during dynamic adaptations of *Bacillus subtilis* metabolism. *Science* 335: 1099–1103.

Cao, G., Zhang, X., Zhong, L., and Lu, Z. (2011). A modified electro-transformation method for *Bacillus subtilis* and its application in the production of antimicrobial lipopeptides. *Biotechnol. Lett* 33: 1047–1051.

Caridis, K.A. and Papathanasiou, T.D. (1997). Pressure effects in cross-flow microfiltration of suspensions of whole bacterial cells. *Bioprocess. Biosyst. Eng.* 16 (4): 199–208.

Chang, S. and Cohen, S.N. (1979). High frequency transformation of *Bacillus subtilis* protoplasts by plasmid DNA. *Mol. Gen. Genet.* 168: 111–115.

Chotani, G.K., Dodge, T.C., Gaertner, A.L., and Arbige, M.V. (2007). Industrial biotechnology: discovery to delivery. In: *Kent and Riegel's Handbook of Industrial Chemistry and Biotechnology*, 11e, vol. 1 (ed. J.A. Kent), 1311–1374. New York: Springer Science & Business Media.

Dale, J.K. and Langlois, D.P. (1940). Sirup and method of making the same. US Patent 2,201,609 (to Staley Mfg Co A E.).

Diaz-Torres, M.R., Lee, E.W., Morrison, T.B. et al. (2003). Bacillus transformation, transformants and mutant libraries. EP1309677 A2.

van Dijl, J.M., Braun, P.G., Robinson, C. et al. (2002). Functional genomic analysis of the *Bacillus subtilis* Tat pathway for protein secretion. *J. Biotechnol.* 98: 243–254.

Estell, D.A. and Ferrari, E. (2009). Modified proteases. US Patent application 2009075332.

Ferrari, E., Jarnagin, A.S., and Schmidt, B.F. (1993). Commercial production of extracellular enzymes. In: *Bacillus subtilis and Other Gram-Positive Bacteria:*

Biochemistry, Physiology, and Molecular Genetics (ed. A.L. Sonenhein, J.A. Hoch and R. Losick), 917–937. Washington, DC: American Society for Microbiology.

Ferrari, E., Fioresi, C., and van Kimmenade, A. (2010). Proteases with modified pro regions. US Patent 8,530,218 (to Danisco US Inc.).

Gay, P., Le Coq, D., Steinmetz, M. et al. (1983). Cloning structural gene *sacB*, which codes for exoenzyme levansucrase of *Bacillus subtilis*: expression of the gene in *Escherichia coli. J. Bacteriol.* 153: 1424–1431.

Gernot, A., Berg, C.T., Derkx, P.M. et al. (2011). Thermostable amylase polypeptides, nucleic acids encoding those polypeptides and uses thereof. EP2292745A1. European Union.

Hahn, J., Luttinger, A., and Dubnau, D. (1996). Regulatory inputs for the synthesis of ComK, the competence transcription factor of *Bacillus subtilis. Mol. Microbiol.* 21: 763–775.

Hambraeus, G., Karhumaa, K., and Rutberg, B. (2002). A 5′ stem-loop and ribosome binding but not translation are important for the stability of *Bacillus subtilis aprE* leader mRNA. *Microbiology* 148: 1795–1803.

Hamoen, L.W., Venema, G., and Kuipers, O.P. (2003). Controlling competence in *Bacillus subtilis*: shared use of regulators. *Microbiology* 149: 9–17.

Harwood, C.R. and Cranenburgh, R. (2008). Bacillus protein secretion: an unfolding story. *Trends Microbiol.* 16: 73–79.

Henner, D.J., Yang, M., and Ferrari, E. (1988). Localization of *Bacillus subtilis sacU*(Hy) mutations to two linked genes with similarities to the conserved procaryotic family of two-component signalling systems. *J. Bacteriol.* 170: 5102–5109.

Hoffmann, K., Wollherr, A., Larsen, M. et al. (2010). Facilitation of direct conditional knockout of essential genes in *Bacillus licheniformis* DSM13 by comparative genetic analysis and manipulation of genetic competence. *Appl. Environ. Microbiol.* 76: 5046–5057.

Hong, J. (1989). Communications to the editor: yield coefficients for cell mass and product formation. *Biotechnol. Bioeng.* 33: 506–507.

Huang, H., Ridgway, D., Gu, T., and Moo-Young, M. (2004). Enhanced amylase production by *Bacillus subtilis* useing a dual exponential feeding strategy. *Bioprocess. Biosyst. Eng.* 27: 63–69.

Ingle, M.B. and Boyer, E.W. (1976). Production of industrial enzymes. In: *Microbiology 1976* (ed. D. Schlessinger), 420–426. Washington, DC: Am. Soc. Microbiol.

Ingram, J.L., Maaloe, O., and Neidhart, F.C. (1983). *Growth of the Bacterial Cell.* Sunderland, MA: Sinauer Associates.

Janniere, L., Gruss, A., and Ehrlich, S.D. (1993). Plasmids. In: *Bacillus subtilis and Other Gram-Positive Bacteria: Biochemistry, Physiology, and Molecular Genetics* (ed. A.L. Sonenhein, J.A. Hoch and R. Losick), 625–644. Washington, DC: American Society for Microbiology Press.

JECFA (Joint FAO/WHO Expert Committee on Food Additives) (2006). *Combined Compendium of Food Additive Specifications*, Analytical Methods, Test Procedures and Laboratory Solutions Used by and Referenced in the Food Additive Specifications, vol. 4. Rome: Food and Agriculture Organization of the United Nations. (FAO JECFA Monograph No. 1) http://www.fao.org/docrep/009/a0691e/A0691E00.htm.

Jensen, C.L., Stephenson, K., Jorgensen, S.T., and Harwood, C. (2000). Cell-associated degradation affects the yield of secreted engineered and heterologous proteins in the *Bacillus subtilis* expression system. *Microbiology* 146 (Pt 10): 2583–2594.

Kakeshita, H., Kageyama, Y., Endo, K. et al. (2011). Secretion of biologically-active human interferon-beta by *Bacillus subtilis*. *Biotechnol. Lett* 33: 1847–1852.

Keefe, R.J. and Dubbin, D.M. (2005). Specifying microfiltration systems. *Chem. Eng (New York, NY, United States)* 112 (8): 48–51.

Kennedy, M. and Krouse, D. (1999). Strategies for improving fermentation medium performance: a review. *J. Ind. Microbiol. Biotechnol.* 23: 456–475.

de Keyzer, J., van der Does, C., and Driessen, A.J. (2003). The bacterial translocase: a dynamic protein channel complex. *Cell. Mol. Life Sci.* 60: 2034–2052.

Kunst, F., Ogasawara, N., Moszer, I. et al. (1997). The complete genome sequence of the gram-positive bacterium *Bacillus subtilis*. *Nature* 390: 249–256.

Laoide, B.M., Chambliss, G.H., and McConnell, D.J. (1989). *Bacillus licheniformis* alpha-amylase gene, *amyL*, is subject to promoter-independent catabolite repression in *Bacillus subtilis*. *J. Bacteriol.* 171: 2435–2442.

Li, W., Zhou, X., and Lu, P. (2004). Bottlenecks in the expression and secretion of heterologous proteins in *Bacillus subtilis*. *Res. Microbiol.* 155: 605–610.

Lu, Y.P., Zhang, C., Lv, F.X. et al. (2012). Study on the electro-transformation conditions of improving transformation efficiency for *Bacillus subtilis*. *Lett. Appl. Microbiol.* 55: 9–14.

Maurer, K.H. (2010). *Enzymes, Detergent. Encyclopedia of Industrial Biotechnology* (ed. M.C. Flickinger), 1–16. New York: Wiley.

McDonald, I.R., Riley, P.W., Sharp, R.J., and McCarthy, A.J. (1995). Factors affecting the electroporation of *Bacillus subtilis*. *J. Appl. Bacteriol.* 79: 213–218.

McGovern, P.E., Zhang, J., Tang, J. et al. (2004). Fermented beverages of pre- and proto-historic China. *Proc. Natl. Acad. Sci. U.S.A* 101: 17593–17598.

Meddeb-Mouelhi, F., Dulcey, C., and Beauregard, M. (2012). High transformation efficiency of *Bacillus subtilis* with integrative DNA using glycine betaine as osmoprotectant. *Anal. Biochem.* 424: 127–129.

Meyer, T.H., Menetret, J.F., Breitling, R. et al. (1999). The bacterial SecY/E translocation complex forms channel-like structures similar to those of the eukaryotic Sec61p complex. *J. Mol. Biol.* 285: 1789–1800.

Nath, S. and Satpathy, G.R. (1998). A systematic approach for investigation of spray drying processes. *Drying Technol.* 16 (6): 1173–1193.

Neubeck, C.E. (1980). Process for spray drying enzymes. US Patent 4233405.

Nicholson, W.L., Park, Y.K., Henkin, T.M. et al. (1987). Catabolite repression-resistant mutations of the *Bacillus subtilis* alpha-amylase promoter affect transcription levels and are in an operator-like sequence. *J. Mol. Biol.* 198: 609–618.

Nicolas, P., Mäder, U., Dervyn, E. et al. (2012). Condition-dependent transcriptome reveals high-level regulatory architecture in *Bacillus subtilis*. *Science* 335: 1103–1106.

Nishito, Y., Osana, Y., Hachiya, T. et al. (2010). Whole genome assembly of a natto production strain *Bacillus subtilis* natto from very short read data. *BMC Genomics* 11: 243.

Olempska-Beer, Z.S., Merker, R.L., Ditto, M.D., and DiNovi, M.J. (2006). Food-processing enzymes from recombinant microorganisms–a review. *Regul. Toxicol. Pharm.* 45: 144–158.

Pariza, M.W. and Johnson, E.A. (2001). Evaluating the safety of microbial enzyme preparations used in food processing: update for a new century. *Regul. Toxicol. Pharm.* 33: 173–186.

Payen, A. and Persoz, J.F. (1833). Mémoire sur la diastase, les principaux produits de ses réactions et leurs applications aux arts industriels. *Annales de chimie et de physique* 53: 73–92.

Perego, M. (1993). Integrational vectors for genetic manipulation in *Bacillus subtilis*. In: *Bacillus subtilis and Other Gram-Positive Bacteria: Biochemistry, Physiology, and Molecular Genetics* (ed. A.L. Sonenhein, J.A. Hoch and R. Losick), 615–624. Washington, DC: American Society for Microbiology Press.

Pohl, S. and Harwood, C.R. (2010). Heterologous protein secretion by *Bacillus* species from the cradle to the grave. *Adv. Appl. Microbiol.* 73: 1–25.

Power, S.D., Adams, R.M., and Wells, J.A. (1986). Secretion and autoproteolytic maturation of subtilisin. *Proc. Natl. Acad. Sci. U.S.A* 83: 3096–3100.

Rey, M.W., Ramaiya, P., Nelson, B.A. et al. (2004). Complete genome sequence of the industrial bacterium *Bacillus licheniformis* and comparisons with closely related *Bacillus* species. *Genome Biol.* 5: R77.

Roald, A.S. and De Tieme, O.N. (1969). Granular enzyme-containing laundry composition. US Patent 3451935 (to Procter & Gamble).

Samborska, K., Witrowa-Rajchert, D., and Goncalves, A. (2005). Spray-drying of alpha-amylase: the effect of process variable on the enzyme inactivation. *Drying Technol.* 23: 941–953.

Sauer, U., Hatzimanikatis, V., Hohmann, H.-P. et al. (1996). Physiology and metabolic fluxes of wild-type and riboflavin-producing *Bacillus subtilis*. *Appl. Environ. Microbiol.* 62 (10): 3687–3696.

Schallmey, M., Singh, A., and Ward, O.P. (2004). Developments in the use of *Bacillus* species for industrial production. *Can. J. Microbiol.* 50: 1–17.

Schultz, A., Atkin, L., and Frey, C.N. (1939). Preparation of an enzymic material. US Patent 2159678 A (to Standard Brands Inc).

Shafikhani, S., Siegel, R.A., Ferrari, E., and Schellenberger, V. (1997). Generation of large libraries of random mutants in *Bacillus subtilis* by PCR-based plasmid multimerization. *Biotechniques* 23: 304–310.

Simonen, M. and Palva, I. (1993). Protein secretion in *Bacillus* species. *Microbiol Rev.* 57: 109–137.

van Sinderen, D. and Venema, G. (1994). *comK* acts as an autoregulatory control switch in the signal transduction route to competence in *Bacillus subtilis*. *J. Bacteriol.* 176: 5762–5770.

van Sinderen, D., Luttinger, A., Kong, L. et al. (1995). *comK* encodes the competence transcription factor, the key regulatory protein for competence development in *Bacillus subtilis*. *Mol. Microbiol.* 15: 455–462.

Somma, S. and Polsinelli, M. (1970). Quantitative autoradiographic study of competence and deoxyribonucleic acid incorporation in *Bacillus subtilis*. *J. Bacteriol.* 101: 851–855.

Steinmetz, M. (1993). Carbohydrate catabolism: enzymes, pathways and evolution. In: *Bacillus subtilis and Other Gram-Positive Bacteria: Biochemistry, Physiology, and Molecular Genetics* (ed. A.L. Sonenhein, J.A. Hoch and R. Losick), 157–170. Washington, DC: American Society for Microbiology Press.

Suominen, I., Karp, M., Lautamo, J. et al. (1987). Thermostable alpha amylase of *Bacillus stearothermophilus*: cloning, expression, and secretion by *Escherichia coli*. In: *Extracellular Enzymes of Microorganisms* (ed. J. Chaloupka, V. Krumphanzl, J. Chaloupka and V. Krumphanzl), 129–137. New York: Plenum Press.

Theil, E.C. (1998). *Principles of Chemistry in Biology* (ed. E.C. Theil). Washington, DC: American Chemical Society.

Tjalsma, H., Antelmann, H., Jongbloed, J.D. et al. (2004). Proteomics of protein secretion by *Bacillus subtilis*: separating the "secrets" of the secretome. *Microbiol. Mol. Biol. Rev.* 68: 207–233.

Tsong, T.Y. (1992). Molecular recognition and processing of periodic signals in cells: study of activation of membrane ATPases by alternating electric fields. *Biochim. Biophys. Acta* 1113: 53–70.

Underkofler, L.A., Barton, R.R., and Rennert, S.S. (1958). Production of microbial enzymes and their applications. *Appl. Microbiol.* 6: 212–221.

Van't Riet, K. and Tramper, J. (1991). *Basic Bioreactor Design* (ed. K. Van't Riet and J. Tramper). New York, USA: Marcel Dekker.

Veith, B., Herzberg, C., Steckel, S. et al. (2004). The complete genome sequence of *Bacillus licheniformis* DSM13, an organism with great industrial potential. *J. Mol. Microbiol. Biotechnol.* 7: 204–211.

Wallerstein, L. (1939). Enzyme preparations from microorganisms: commercial production and industrial application. *Ind. Eng. Chem.* 31: 1218–1224.

Wang, Y., Weng, J., Waseem, R. et al. (2012). *Bacillus subtilis* genome editing using ssDNA with short homology regions. *Nucleic Acids Res.* 40: e91.

Wells, J.A., Ferrari, E., Henner, D.J. et al. (1983). Cloning, sequencing, and secretion of *Bacillus amyloliquefaciens* subtilisin in *Bacillus subtilis*. *Nucleic Acids Res.* 11 (22): 7911–7925.

van Wely, K.H., Swaving, J., Freudl, R., and Driessen, A.J. (2001). Translocation of proteins across the cell envelope of Gram-positive bacteria. *FEMS Microbiol. Rev.* 25: 437–454.

Weuster-Botz, D. (2000). Experimental design for fermentation media development: statistical design or global random searches? *J. Biosci. Bioeng.* 90 (5): 473–483.

Wu, S.C., Yeung, J.C., Duan, Y. et al. (2002). Functional production and characterization of a fibrin-specific single-chain antibody fragment from *Bacillus subtilis*: effects of molecular chaperones and a wall-bound protease on antibody fragment production. *Appl. Environ. Microbiol.* 68: 3261–3269.

Xue, G., Johnson, J.S., and Dalrumple, B.P. (1999). High osmolarity improves the electrotransformation efficiency of the gram-positive bacteria *Bacillus subtilis* and *Bacillus licheniformis*. *J. Microbiol. Methods* 34: 183–191.

Yang, M.M., Zhang, W.W., Bai, X.T. et al. (2010). Electroporation is a feasible method to introduce circularized or linearized DNA into *B. subtilis* chromosome. *Mol. Biol. Rep.* 37: 2207–2213.

Yansura, D.G. and Henner, D.J. (1984). Use of the *Escherichia coli* lac repressor and operator to control gene expression in *Bacillus subtilis*. *Proc. Natl. Acad. Sci. U.S.A* 81: 439–443.

Zhang, X.Z. and Zhang, Y.H. (2011). Simple, fast and high-efficiency transformation system for directed evolution of cellulase in *Bacillus subtilis*. *Microb. Biotechnol.* 4: 98–105.

2

New Expression Systems for GPCRs

Dimitra Gialama[1,2], Fragiskos N. Kolisis[2], and Georgios Skretas[1]

[1] Institute of Biology, Medicinal Chemistry and Biotechnology, National Hellenic Research Foundation, Athens, Greece
[2] Biotechnology Laboratory, School of Chemical Engineering, National Technical University of Athens, Zografou Campus, Athens, Greece

2.1 Introduction

The G-protein coupled receptors (GPCRs) are proteins that reside in the plasma membranes of most eukaryotic cells, from yeasts to humans. They consist of seven trans-membrane α helices, connected to one another by three cytosolic and three extracellular loops of variable lengths (Figure 2.1). Their N-terminal tails are always facing the extracellular space, while the C-terminal, the cytosol. According to a recent classification, GPCRs can be divided into five families that share sequence and structural similarities: the rhodopsin, secretin, glutamate, adhesion, and Frizzled/Taste2 families (Fredriksson et al. 2003). The rhodopsin family is by far the largest and its members contain, in general, much shorter N-terminal tails compared to the other classes (Fredriksson et al. 2003).

GPCRs are the major mediators of intercellular signaling. They have the ability to bind to a remarkably wide array of ligands, varying from amines, ions, and photons, to lipids, peptides, and glycoproteins (Kristiansen 2004). Binding of these ligands causes a conformational change, which enables the receptor to interact with and activate the GTPase activity of intracellular guanine nucleotide-binding proteins (G proteins). These, in turn, interact with effector proteins, such as enzymes and ion channels, which modify the intracellular concentration of various signaling molecules, such as cAMP, cGMP, inositol phosphates, diacylglycerol, ions (Cabrera-Vera et al. 2003; Kristiansen 2004). Altered concentrations of these molecules can induce a plethora of specific physiological responses, which are responsible for the senses of vision and smell, inflammatory responses, behavioral changes, and other vital functions in animals and humans, as well as the mating behavior of yeasts.

Bioprocessing Technology for Production of Biopharmaceuticals and Bioproducts, First Edition.
Edited by Claire Komives and Weichang Zhou.
© 2019 John Wiley & Sons, Inc. Published 2019 by John Wiley & Sons, Inc.

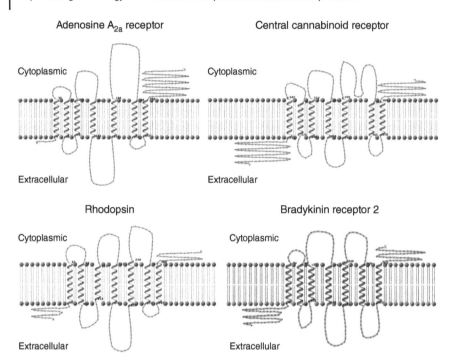

Figure 2.1 Schematics of the predicted/actual topologies of four human GPCRs: the adenosine A_{2a} receptor (uniprot P29274), the central cannabinoid receptor (uniprot P21554), rhodopsin (uniprot P08100), and the bradykinin receptor 2 (uniprot P30411). Figures were generated using TMRPres2D (Spyropoulos et al. 2004).

Apart from this canonical (G protein-dependent) signaling pathway, GPCRs can mediate signal transduction through interactions with other proteins as well, such as β-arrestins, Homer proteins, PDZ domain proteins, GPCR kinases, protein phosphatases (Rajagopal et al. 2010; Magalhaes et al. 2012). In these cases, signaling can be directed to other routes, which are distinct from the ones involving G proteins and can lead to receptor desensitization, receptor trafficking and endocytosis, regulation of chemotaxis and apoptosis, and modulation of receptor-ligand specificity among others (Rajagopal et al. 2010; Magalhaes et al. 2012).

GPCRs constitute one of the largest protein families in the human body, with more than 800 distinct genes encoding for this type of receptors (Fredriksson et al. 2003). About half of them are dedicated to our sense of smell (olfactory receptors). Among the nonolfactory GPCRs, approximately 50 receptors constitute the targets of about 40% of all currently marketed drugs (Zhao and Wu 2012). Notable examples include the blockbuster drugs Zyprexa®, which functions as a mixed antagonist for three receptors (dopamine D1, D2 receptors, and 5-hydroxytryptamine receptor 2) and is used for the treatment of schizophrenia; and Telfast® (Allegra), a modulator of the histamine H1 receptor and one of the most frequently used anti-allergens (Jacoby et al. 2006). The "druggability," i.e. their potential as targets for pharmaceutical discovery, of the rest of the more than 300 GPCRs needs to be investigated more, but according to predictions, the majority of them are potential therapeutic targets for future, yet undeveloped drugs (Russ and Lampel 2005). Very interesting and promising from a drug discovery standpoint, is the group of approximately 150 GPCRs for which endogenous ligands have not been discovered yet, the so-called orphan receptors.

Potentially therapeutic agents that target GPCRs include compounds that aim to modulate the function of specific receptors and regulate their signaling in processes involved in health and disease, such as receptor agonists (full, partial, inverse, etc.) and antagonists, i.e. molecules which can act as activators or inhibitors of the native signaling pathways, respectively. Examples of such GPCR modulators include TAK-875, an agonist of G protein-coupled receptor 40 (GPR40), which completed successfully phase II; however, failed to pass phase III clinical trials for the treatment of type II diabetes (Araki et al. 2012; Kaku et al. 2015); ponesimod, a potential therapeutic compound against multiple sclerosis with selective antagonistic effects toward sphingosine-1-phosphate receptor 1, which has completed phase II trials and is proceeding to the last stage of clinical development (Piali et al. 2011; Subei and Cohen 2015); and atrasentan, an endothelin A subtype receptor antagonist at various stages of clinical development against ovarian, prostate, and other types of cancer (Lappano and Maggiolini 2011), which has reached phase III clinical trials. Recently, atrasentan failed a phase III trial for prostate cancer in patients unresponsive to hormone therapy.

Another class of therapeutically valuable GPCR-targeting compounds is receptor-specific pharmacological chaperones. It has been shown that a number of serious human diseases, such as nephrogenic diabetes insipidus (NDI) and hypogonadotropic hypogonadism (HH), occur due to missense mutations in the genes encoding for GPCRs (the V_2 vasopressin receptor and the gonadotropin-releasing hormone receptor in the cases of NDI and HH, respectively), which result in receptor misfolding, enhanced retention, or degradation in the endoplasmic reticulum (ER), with concomitant reduced surface expression, and activity of the protein (Conn et al. 2007). Pharmacological chaperones are compounds that bind to these mutant forms specifically, correct receptor misfolding, and restore cell surface accumulation and functionality back to the levels of the wild-type sequence (Bernier et al. 2004). Finally, compounds that can act as modulators of receptor oligomerization have emerged as potentially important factors in GPCR pharmacology. This is based on the discovery that a variety of GPCRs can form homo- and/or heteroligomers and that depending on the type and state of oligomerization, the physiological signaling pathways of a receptor can be dramatically affected (George et al. 2002).

In the postgenomic era, one would expect that rational methods of drug discovery would be major drivers of innovation in GPCR pharmacology, similarly to what has been achieved for receptor tyrosine kinases, proteases, metabolic enzymes, and other proteins (Congreve and Marshall 2010). However, currently marketed drugs and compounds in late stages of clinical development that target GPCRs have been discovered either serendipitously or by cell-based screening (Schlyer and Horuk 2006; Salon et al. 2011). The application of rational approaches in GPCR drug discovery requires detailed understanding of GPCR folding and misfolding of the modes of receptor binding and activation/inactivation by agonists/antagonists, of the possible oligomerization state(s) of a receptor, and of the interactions with partnering proteins that mediate GPCR signaling. These, in turn, require high-resolution crystal structures that will provide atomic-level snapshots of receptor conformations and interactions.

GPCRs, however, are notoriously difficult to crystallize. Due to the fact that they reside in the hydrophobic environment of the plasma membrane, they first have to be extracted and purified in the presence of appropriate detergents. Under these conditions, membrane proteins lose their stability and tend to aggregate or degrade over time, making the formation of high-quality protein crystals very challenging (Bill et al. 2011). Another factor contributing to the great difficulty of GPCR crystallization is the highly dynamic character of their trans-membrane α helices (Zhao and Wu 2012). Until quite recently, there was only one high-resolution crystal structure available, that of bovine rhodopsin (Palczewski et al. 2000). In 2007, a big breakthrough was achieved when the three-dimensional structure of the human β_2 adrenergic receptor was determined in the presence and absence of its partial inverse agonist carazolol (Cherezov et al. 2007; Rasmussen et al. 2007). Since then, there has been an explosion in the number of reported structures, which include different members of the GPCR superfamily, from different organisms (human, mouse, rat, squid, etc.), in the absence and presence of receptor agonists, inverse agonists, antagonists, and interacting proteins (Venkatakrishnan et al. 2013) (Table 2.1). Very importantly, structures have been solved not only for rhodopsin-like receptors but also for other classes of GPCRs (Table 2.1). In 2012, Lefkowitz and Kobilka were awarded the Nobel Prize in Chemistry for studies of GPCRs that included the development of generalizable approaches with which GPCRs can be crystallized and probed structurally. Examples of determined high-resolution structures of three GPCRs are shown in Figure 2.2. The solved structures have already been utilized for structure-guided design of GPCR modulators with potentially therapeutic effects (Congreve et al. 2011).

Table 2.1 Solved structures of GPCRs.

GPCR	Origin	Ligand	R	Expression host	PDB file
Rhodopsin	Bos taurus	—	No or yes	N/A or COS cells or HEK293S-GnTI⁻ cells	1F88, 1L9H, 1GZM, 1U19, 2J4Y, 2I35, 2I36, 2I37, 2X72, 3C9L, 3C9M, 3CAP, 3PQR, 3PXO, 3DQB, 4A4M, 4J4Q, 4X1H, 4PXF
	Todarodes pacificus	—	No	N/A	2Z73, 2ZIY, 3AYN, 3AYM
	Homo sapiens	—	Yes	HEK293S cells	4ZWJ
β₁ Adrenergic receptor	Meleagris gallopavo	Full and partial agonists, partial inverse agonists, antagonists	Yes	Trichoplusia ni	2VT4, 2Y01, 2Y02, 2Y03, 2Y04, 2Y00, 2YCW, 2YCX, 2YCY, 2YCZ, 4AMI, 4AMJ, 4GPO
Dopamine D3 receptor	Homo sapiens	Antagonist	Yes	Spodoptera frugiperda	3PBL
M2 muscarinic acetylcholine receptor	H. sapiens	Antagonist, agonist	Yes	S. frugiperda	3UON, 4MQS
Neurotensin receptor 1	Rattus norvegicus	Agonist	Yes	T. ni or Escherichia coli	4GRV, 4BUO, 4XEE
β₂ Adrenergic receptor	H. sapiens	Agonist, inverse agonists, partial inverse agonist, antagonist	Yes	S. frugiperda	2R4R, 3KJ6, 2RH1, 3D4S, 3NY8, 3NY9, 3NYA, 3P0G, 3PDS, 3SN6, 4GBR
CXC chemokine receptor type 1	H. sapiens	—	Yes	E. coli	2LNL
CXC chemokine receptor type 4	H. sapiens	Antagonists	Yes	S. frugiperda	3ODU, 3OE8, 3OE9, 3OE6, 3OE0, 4RWS
CXC chemokine receptor type 5	H. sapiens	Antagonist	Yes	S. frugiperda	4MBS
Histamine H₁ receptor	H. sapiens	Antagonist	Yes	Pichia pastoris	3RZE

(Continued)

Table 2.1 (Continued)

GPCR	Origin	Ligand	R	Expression host	PDB file
Sphingosine-1-phosphate receptor	H. sapiens	Antagonist	Yes	S. frugiperda	3V2W, 3V2Y
Adenosine A_{2a} receptor	H. sapiens	Agonists, inverse-agonist, antagonists	Yes	S. frugiperda, T. ni, P. pastoris	3EML, 3QAK, 2YDO, 2YDV, 3RFM, 3PWH, 3REY, 3VG9, 3VGA, 4EIY, 4UHR
M3 muscarinic acetylcholine receptor	R. norvegicus	Inverse agonist	Yes	S. frugiperda	4DAJ
κ-Opioid receptor	H. sapiens	Antagonist	Yes	S. frugiperda	4DJH
μ-Opioid receptor	Mus musculus	Antagonist, agonist	Yes	S. frugiperda	4DKL, 5C1M
Glucagon receptor	H. sapiens	Antagonist	Yes	S. frugiperda	4L6R
Nociceptin/orphanin FQ receptor	H. sapiens	Antagonists	Yes	S. frugiperda	4EA3, 5DHG
δ-Opioid receptor	M. musculus	Antagonist	Yes	S. frugiperda	4EJ4
	H. sapiens	Antagonist, antagonist/agonist	Yes	S. frugiperda	4N6H, 4RWD
Protease-activated receptor 1	H. sapiens	Antagonist	Yes	S. frugiperda	3VW7
Apelin receptor (helix 1)	H. sapiens	—	Yes	E. coli	2LOU
OX2 orexin receptor	H. sapiens	Antagonist	Yes	S. frugiperda	4S0V
US28 with human cytokine CX3CL1	Cytomegalovirus	—	Yes	HEK293s GntI− cells	4XT1, 4XT3
5-HT_{1B} serotonin receptor	H. sapiens	Agonist	Yes	S. frugiperda	4IAR

Receptor	Species	Ligand type	R	Expression system	PDB
5-HT$_{2B}$ serotonin receptor	H. sapiens	Agonist	Yes	S. frugiperda	4IB4, 4NC3
Corticotropin-releasing factor receptor type 1	H. sapiens	Antagonist	Yes	T. ni	4K5Y
Smoothened receptor	H. sapiens	Antagonist	Yes	S. frugiperda	4JKV
P2Y$_{12}$ receptor	H. sapiens	Agonist, antagonist	Yes	S. frugiperda	4NTJ, 4PXZ
P2Y$_1$ receptor	H. sapiens	Antagonist	Yes	S. frugiperda	4XNV
Gamma-aminobutyric acid (GABA)$_B$ receptor	H. sapiens	Agonists, antagonists	Yes	S. frugiperda	4MQE, 4MQF, 4MR7, 4MR8, 4MR9, 4MRM, 4MS1, 4MS3, 4MS4
Metabotropic glutamate receptor 1	H. sapiens	— (Bound allosteric modulator)	Yes	S. frugiperda	4OR2
Metabotropic glutamate receptor 5	H. sapiens	— (Bound allosteric modulator)	Yes	S. frugiperda	4OO9, 5CGC
GPR40 receptor (free fatty-acid receptor 1)	H. sapiens	Allosteric agonist	Yes	S. frugiperda	4PHU
LPAR1 (lysophosphatidic acid receptor 1)	H. sapiens	Antagonists	Yes	S. frugiperda	4Z34
Angiotensin type I receptor	H. sapiens	Antagonist, inverse agonist	Yes	S. frugiperda	4ZUD, 4YAY

R, recombinant; N/A, not applicable.

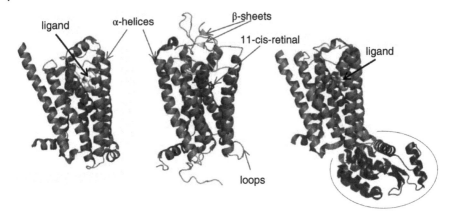

Figure 2.2 High-resolution structures of three GPCRs: the turkey β_1 adrenergic receptor in complex with its antagonist carazolol (PDB 2YCW) (a), bovine rhodopsin in an inactive state with bound ground-state chromophore, 11-cis-retinal (PDB 1F88) (b), and the rat M3 muscarinic acetylcholine receptor fused to the bacteriophage T4 lysozyme in complex with its inverse agonist tiotropium (PDB 4DAJ) (c). The α helices of the receptors, the β-sheets and their loops are identified by arrows. T4 lysozyme is indicated by the partial oval on figure (c). 11-*cis*-retinal is also indicated by an arrow. Structure schematics were generated using PyMOL (Schrodinger 2010).

Despite these tremendous successes, however, the different receptor structures, which have been solved until today, represent only a very small percentage of the about 400 nonolfactory GPCRs that constitute targets for current or potential drugs, while some of the crystallized receptors are closely related (e.g. the β_1 and β_2 adrenergic receptors) (Venkatakrishnan et al. 2013). Furthermore, most of the solved structures correspond to nonnative engineered sequences that contain missense mutations imparting enhanced thermostability (Dore et al. 2011; White et al. 2012), N- or C-terminal truncations for enhanced expression (Wu et al. 2010; Shimamura et al. 2011), or foreign protein domain insertions (e.g. T4 lysozyme) at the N terminus or in the third cytosolic loop between trans-membrane helices 5 and 6 that assist with the initiation of crystal contacts (Cherezov et al. 2007; Rasmussen et al. 2011a) (Table 2.1). The question arises if such stabilizing modifications of GPCRs allow the good representation of the ligand-binding pockets, the residues involved in the interactions with G proteins and, in general, the overall structures of the wild-type receptors. Thermodynamic stabilization of a GPCR by such modifications can have distinct effects on the affinities of different ligands and therefore affect the representation of ligand-binding pockets. Ligand-binding affinities of the modified receptors can remain identical or very similar to that corresponding to the original sequence (Rosenbaum et al. 2007; Jaakola et al. 2008; Warne et al. 2008; Chien et al. 2010; Wu et al. 2010; Dore et al. 2011; Lebon et al. 2011; Shimamura et al. 2011; Rasmussen et al. 2011b; Haga et al. 2012; Kruse et al. 2012; Manglik et al. 2012; Thompson et al. 2012; White et al. 2012; Hollenstein et al. 2013; Siu et al. 2013). In other cases, they can be altered significantly (Rosenbaum et al. 2007; Jaakola et al. 2008; Warne et al. 2008, 2011; Wu et al. 2010; Dore et al. 2011; Lebon et al. 2011; Granier et al. 2012), and this may reflect slight differences in the structure of the modified GPCR compared to wild-type. As far as G protein-coupling ability is concerned, most crystal structures of GPCRs have been solved in complex with ligands that normally favor dissociation of the receptors from G proteins (antagonists or inverse agonists) and therefore stabilize inactive conformations (Table 2.1). More importantly, the third intracellular loop of GPCRs is critical for interactions with G proteins and, therefore, the generation of fusion proteins by foreign domain insertions (T4 lysozyme or others) in this loop abrogates the ability of the receptors to couple to G proteins (Rosenbaum et al. 2007; Wu et al. 2010; Liu et al. 2012). Furthermore, the presence of T4 lysozyme within the sequence of a GPCR can result in the appearance of new interactions which may affect the local or overall structure of the receptor. For example, it has been shown that insertion of T4 lysozyme in the β2 adrenergic receptor led to the formation of a new salt bridge between an arginine residue contained in lysozyme with a glutamate of the receptor, a residue which has been found to be important for the stabilization of the inactive state in the case of rhodopsin (Cherezov et al. 2007). It must be mentioned, however, that the number of these new interactions are very limited (Cherezov et al. 2007; Jaakola et al. 2008) and that biophysical and biochemical assays have provided evidence showing that agonist-induced conformational changes are not prevented by fusion of GPCR to T4 lysozyme (Rosenbaum et al. 2007).

An additional way of evaluating the impact of stabilizing sequence modifications on receptor structures is to compare the effect(s) of a number of different such alterations on the structure of a single receptor. Such a comparison of structures of the A_{2a} adenosine receptor modified by different methods (fusion with T4 lysozyme or mutagenesis) revealed no significant differences between the two structures (Dore et al. 2011). On the contrary, A_{2a} adenosine receptor-T4 lysozyme structures bound to either agonist or inverse agonist show obvious differences, thus providing indications that ligand-induced conformational changes could be reliably captured in GPCR-T4 lysozyme structures (Xu et al. 2011). All this information suggests that the structures of modified/stabilized variants can be key in beginning to understand GPCR signaling and structure–function relationships but, in many cases, these modifications provide structures with nonnative properties, and it is important to be cautious when dealing with detailed interpretation of these structures. For the reasons mentioned above, there is a clear need for a large number of additional structures of GPCRs, especially ones corresponding to wild-type sequences, before the full pharmacological potential of GPCRs can be realized.

Difficulties in GPCR structure determination are not only related to the protein crystallization procedure per se, but arise from the very first step of the process: finding a source of a reasonable amount of properly folded protein to initiate purification and crystallization trials. Like the vast majority of integral membrane proteins, GPCRs are found in their native cells and tissues only in miniscule quantities (Salom et al. 2012). One of the few exceptions to this rule is rhodopsins, which are found in the eyes of animals in relative abundance, a characteristic that allows the isolation of a sufficient amount of properly folded protein from its native environment. This is one of the most important reasons why rhodopsin was the first GPCR whose crystal structure was determined and, for a long time, the only one (Palczewski et al. 2000). For the rest of the hundreds of GPCRs of pharmacological interest, production of sufficient amounts of protein relies on recombinant production in heterologous hosts, such as bacteria, yeasts, mammalian, and insect cell cultures. It has been proposed that, in general, a yield of at least 1 mg of ligand-binding-competent receptor per 5 l of cell culture is required before GPCR structural studies can be initiated (Grisshammer and Tate 1995; Sarramegna et al. 2003), although this number may vary a little across different expression systems. Even though a number of GPCRs can be produced in high-enough amounts for potential structural characterization, the great majority of GPCR overexpression attempts result in poor protein production (Sarramegna et al. 2003).

There are three main problems associated with recombinant GPCR production: (i) there is usually very little membrane-incorporated protein per cell, (ii) in cases where accumulation at the cell membrane occurs at appreciable levels, there is typically a very small amount of protein that is produced in a well folded and functional (ligand-binding competent) form, and (iii) overexpression is very frequently associated with severe cell toxicity that further limits the volumetric protein yields. Although, in general, prokaryotic integral membrane proteins express better in prokaryotic hosts and eukaryotic proteins in eukaryotic hosts, there is really no way of predicting which GPCR will accumulate in higher amounts in which host (Sarramegna et al. 2003). Thus, recombinant GPCR expression is carried out on a trial and error basis, and optimization of the production yields is routinely pursued by varying simple expression parameters, such as strains, expression vectors, media composition, incubation temperature (wherever possible). These approaches, however, have met only limited success, especially with the harder-to-express members of the superfamily. Furthermore, the optimized expression parameters differ depending on the target receptor and, thus, have to be optimized every time a different GPCR needs to be studied. Due to these reasons, there is an emerging trend to attempt to achieve enhanced GPCR production by genetically modifying the expression host in order to optimize its cellular machinery for recombinant membrane protein production (Bill et al. 2011). The great advantage of this approach is that it can potentially provide specialized "membrane protein-producing cell factories" with the ability to accumulate increased yields of properly folded protein for a variety of different receptors, without having to re-optimize the host each time. Furthermore, characterization of the modifications that lead to enhanced GPCR production will potentially provide clues about the bottlenecks associated with high-level recombinant membrane protein production and an improved level of understanding of recombinant protein production in general. In this chapter, we will describe advances in GPCR expression systems that have resulted from applying this concept of "cell host engineering."

2.2   Recombinant GPCR Production – Traditional Approaches for Achieving High-Level Production

Recombinant GPCRs have traditionally been expressed in four different types of cell hosts: bacteria, yeasts, insect, and mammalian cells (Sarramegna et al. 2003). There are no general guidelines for selecting an appropriate host for successful production of a particular GPCR. In the cases of receptors where posttranslational modifications are expected to be critical for the structure and/or function of the particular GPCR, an expression host that can perform such posttranslational protein processing should be selected. In general, prokaryotic integral membrane proteins express better in prokaryotic hosts and eukaryotic proteins in eukaryotic hosts (Bill et al. 2011). According to this, expression of mammalian GPCRs would be unsuccessful in simple bacterial cell hosts, such as *Escherichia coli*. However, a number of mammalian GPCRs have already been produced at sufficient amounts for structural determination in *E. coli*, such as the human adenosine A_{2a} receptor and the human thyroid-stimulating hormone receptor (Busuttil et al. 2001; Weiss and Grisshammer 2002). Very importantly, a human GPCR structure, that of the CXC chemokine receptor type 1 (CXCR1) has been determined by nuclear magnetic resonance (NMR) using protein overexpressed in *E. coli* (Park et al. 2012). Furthermore, a recent report that compared a number of different prokaryotic and eukaryotic organisms as expression hosts for membrane protein production revealed that both of the GPCRs included in the study, the human CXC chemokine receptor type 4 (CXCR4) and the human CC chemokine receptor type 5 (CCR5) could be produced at high levels in *E. coli* but were undetectable when expressed in insect or plant cell cultures (Bernaudat et al. 2011). Of course, these conclusions could be a result of the particular expression conditions that were tested, as CXCR4 has already been produced in a ligand-binding-competent form in insect cell cultures at high enough levels for purification, crystallization, and structural determination (Wu et al. 2010). The study by Bernaudat et al. also showed that heterologous membrane protein production was in a number of cases much more efficient, in terms of total protein accumulation, than overexpression in the homologous host. Clearly, the *a priori* selection of a suitable expression host for a particular GPCR is currently impossible and is usually a result of personal preference, expertise, and intuition or a trial-and-error procedure.

Once the expression host has been determined, a vector type needs to be selected (high- or low-copy number plasmid, recombinant virus) and a mode of expression (episomal expression or chromosomal integration, transient or stable transfection, etc.) applied. The expression of the gene encoding the GPCR of interest is then placed under the control of a suitable promoter (constitutive or inducible, strong, or weak). The use of high-copy number vectors is often beneficial as observed, for example, with the production of the rat neurotensin receptor in *E. coli* (Tucker and Grisshammer 1996); however, low-copy number vectors can be advantageous in the same host for the expression of other receptors, such as in the case of a number of human chemokine receptors (Ren et al. 2009). The recombinant gene can have the native sequence of the receptor's cDNA from the originating organism or be "codon optimized," i.e. contain substitutions for certain codons with synonymous ones that are found with higher frequencies in the heterologous host. Synonymous codon substitution can have a profound effect on membrane protein accumulation levels and folding (Norholm et al. 2012), and codon optimization has been reported to result in enhanced GPCR accumulation in a number of cases, such as for the rat β_2 adrenergic receptor expressed in stably transfected human embryonic kidney 293 (HEK293) cells (Chelikani et al. 2006), and for the human leukotriene B_4 receptor produced in *E. coli* (Baneres et al. 2003). However, codon optimization is not always beneficial for GPCR production, as exemplified by the higher accumulation of the nonoptimized sequence of the human adenosine A_{2a} receptor compared to the optimized one when expressed in *Caenorhabditis elegans* (Salom et al. 2012). Furthermore, full synonymous randomization of the entire amino acid sequence of a variant of the rat neurotensin receptor 1 with all 64 codons showed that the presence of rare codons within the receptor sequence did not lower its expression levels (Schlinkmann et al. 2012a), thus showing that codon optimization is not always effective.

Modifications of the amino acid sequence of a GPCR can also have a dramatic effect on its production yields. To this end, a very popular strategy has been the expression of the target GPCR in the form of a single or double end-to-end fusion with well-expressed fusion partners, such as the maltose-binding protein (MBP), thioredoxin (TrxA), Mistic, Rho tag, and glutathione-S-transferase (GST), which stabilize the produced receptor, assist its folding, or direct its proper embedding in the membrane (Young et al. 2012). For example, sandwiching the rat neurotensin receptor between MBP and TrxA led to a more than 200-fold increase in the number of active sites per cell when the receptor was expressed in *E. coli*, compared to the fusion-free receptor (Grisshammer et al. 1993; Tucker and Grisshammer 1996), and in enhanced production of a number of olfactory receptors in transiently transfected HEK293 cells (Kajiya et al. 2001). Truncations of the N- or C-terminal tails can also lead to enhanced GPCR expression and stability, and such modifications have been implemented in a number of the receptors whose three-dimensional structure has been determined (Wu et al. 2010; Shimamura et al. 2011). Finally, point mutations can markedly affect GPCR accumulation levels. This characteristic has been exploited by Plückthun and coworkers, who have applied directed evolution techniques in order to identify highly expressed variants for a number of different receptors (Sarkar et al. 2008; Dodevski and Pluckthun 2011; Schlinkmann et al. 2012a,b).

Once the cell host has been transformed/transfected with the constructed expression vector, lab-scale or large-scale culturing can be used for the production of the required protein amounts. GPCR production for biochemical and structural studies is usually carried out in small scales, but a few studies on large-scale productions have been reported (see below). Apart from the scale that determines levels of biomass production, culturing conditions can also have a significant impact on the productivity of the host – on a per cell basis – in membrane-incorporated receptor. Incubation temperature, composition of media (nutrients, supplements, ligands, etc.), protein production inducer concentrations, incubation times, and other factors can greatly affect the final yields. For example, ligand-binding-competent GPCR production in yeast was found to be generally enhanced at low incubation temperatures (20 °C instead of the optimal growth temperature of 30 °C) and by the addition of histidine, dimethylsulfoxide (DMSO), and of certain GPCR ligands (Andre et al. 2006), while *E. coli* culturing in terrific broth (TB) medium resulted in significantly increased cell growth and volumetric yields for human CCR3 and CXCR1, compared to less rich growth media (Ren et al. 2009). Similarly, supplementation of dipeptides containing branched-chain amino acids was found to enhance significantly the accumulation of recombinant membrane proteins in *Lactococcus lactis* (Marreddy et al. 2010). "Optimized" expression conditions, however, are typically found to be effective only for certain GPCRs, while for others, they do not offer any improvements in protein accumulations and, in some cases, they can even have deleterious effects. In the study of Andre et al. for example, both the incubation temperature and added ligand concentration, which were found to be optimal for the accumulation of the majority of the studied GPCRs, had a negative impact on the accumulation of ligand-binding-competent human serotonin receptor 1B (Andre et al. 2006).

One has to keep in mind that recombinant GPCRs are not produced in a purely "active" or "inactive form" but instead as an ensemble of functional and nonfunctional proteins, in an equilibrium that can be changed by varying the expression conditions or perhaps others factors (Andre et al. 2006). The production of active protein is a more important indicator for successful expression than the production of total protein that can also include nonnatively folded molecules. The amount of total protein produced does not necessarily correlate with the yields of active, ligand-binding competent protein, as exemplified by the study of Andre et al. and others (Andre et al. 2006; Hassaine et al. 2006; Lundstrom et al. 2006).

2.3 Engineered Expression Systems for GPCR Production

As described in Section 2.1, recombinant GPCR production is typically carried out on a trial and error basis by varying simple expression parameters, such as host strains, expression vectors, media composition. These approaches, however, have met only limited success, especially with the harder-to-express receptors. Furthermore, the optimized expression parameters differ depending on the target receptor and, thus, have to be optimized every time a different GPCR needs to be studied. Due to these reasons, there is an emerging trend to attempt to enhance GPCR production by performing strain engineering and optimizing the cell host itself (Bill et al. 2011). The great advantage of this approach is that it can potentially provide specialized "membrane protein-producing strains" able to provide increased yields of properly folded protein for a variety of different receptors, without having to re-optimize the host each time. Furthermore, characterization of the modifications that lead to enhanced GPCR accumulation could provide invaluable clues about the bottlenecks associated with high-level protein production.

2.3.1 Bacteria

Bacterial hosts, such as *E. coli*, offer the great advantages of simplicity, fast growth, low cost, scalability, and genetic tractability. A large variety of different strains, expression vectors, and inducible promoters are available for these systems, which have allowed the production of a plethora of recombinant proteins, including membrane proteins. In *E. coli*, there are two main strategies for GPCR production: membrane-integrated expression and expression in the form of insoluble protein aggregates or inclusion bodies. Inclusion body formation typically results in low toxicities and higher yields, but the accumulated aggregates are difficult to denature and refold. GPCR expression in the form of inclusion bodies has been reviewed elsewhere (Baneres et al. 2011) and will not described here. Instead, we will focus on membrane-integrated GPCR expression. Apart from *E. coli*, the toolbox of bacterial expression hosts contains also other prokaryotic organisms, which have shown great promise as recombinant membrane protein producers. *L. lactis*, for example, does not typically accumulate recombinant membrane proteins in inclusion bodies but in membrane-embedded form instead, and exhibits relatively low proteolytic activity toward these targets (Pinto et al. 2011).

Despite their simplicity, bacterial systems pose a number of limitations as hosts for the production of eukaryotic integral membrane proteins like GPCRs. First, eukaryotic membrane protein biogenesis is quite different than in prokaryotes, as proteins are synthesized first in the ER and then transported to the plasma membrane rather than the direct incorporation into the cytoplasmic membrane that occurs in prokaryotes. Second, although most integral membrane proteins are synthesized co-translationally via the signal recognition particle (SRP) pathway both in prokaryotic and eukaryotic cells following largely conserved processes, there are significant differences in several aspects of these procedures in these two domains of life. For example, the rates of mRNA translation are much higher in prokaryotes, a factor that typically leads to higher protein production rates and potential protein mistargeting and/or misfolding. Furthermore, the eukaryotic translocon Sec61 contains conserved charged amino acids important for membrane protein integration, which are not necessarily present in bacteria (Goder et al. 2004). Third, membrane-protein-interacting factors, e.g. molecular chaperones, foldases, proteases, vary significantly in the two domains, and these differences can have a profound effect on the synthesis and quality control of GPCRs. For instance, the eukaryotic molecular chaperone calnexin has been found to interact with and assist the expression and stability of a number of GPCRs in humans, such as the thyrotropin receptor and the dopamine receptors D1 and D2 (Siffroi-Fernandez et al. 2002; Free et al. 2007). There are no calnexin homologs in *E. coli*. Fourth, a number of human GPCRs contain posttranslational modifications, such as *N*-glycosylation, which may be important for their folding and function (Zhang et al. 2001), and which do not normally occur in bacterial hosts. Finally, the lipid content of the plasma membrane of higher eukaryotes is significantly different from that of prokaryotes. Lipids specifically found in the plasma membranes of higher animals, such as cholesterol, are thought to play an important role for the structure and function of a variety of GPCRs (Paila and Chattopadhyay 2010; Zocher et al. 2012).

Despite all these limitations, *E. coli* has historically been the most successful recombinant expression host for structural biology studies of membrane proteins that are not GPCRs (Wagner et al. 2006). Solved structures of bacterially produced membrane proteins include sequences of both prokaryotic and eukaryotic origin (Aaij et al. 2014). Most importantly, the structure of a human GPCR, the CXCR1 chemokine receptor, embedded in a phospholipid bilayer has been determined by NMR using protein that has been recombinantly produced in its wild-type form in *E. coli* (Park et al. 2012). Furthermore, the structure of a signaling-competent variant of the neurotensin receptor 1 was determined by X-ray crystallography after overexpression in *E. coli* (Egloff et al. 2014). These reports provide strong indications that bacterial cells can indeed serve as useful protein production hosts for GPCR structural biology studies. These successes, in combination with important developments in bacterial genetic engineering, such as the generation of *E. coli* strains with the ability to form disulfide bonds (Bessette et al. 1999; Faulkner et al. 2008; Skretas et al. 2009) and to carry out protein *N*-glycosylation (Valderrama-Rincon et al. 2012) point to the potential of these hosts for use in the structural analysis of complex GPCRs carrying posttranslational modifications. A number of recent reports have demonstrated that bacterial cells can be appropriately engineered to overcome different limitations associated with recombinant integral membrane production, including that of human GPCRs.

Careful consideration of the available knowledge about the pathways, cellular processes, and factors involved in the biogenesis of integral membrane proteins and high-level recombinant protein production, in general, may allow one to rationally design targeted cellular modifications that result in enhanced membrane protein yields. For example, Chen et al. reported that co-expression of the chaperone/cochaperone DnaK/DnaJ conferred a significant increase in the bacterial production of the magnesium transporter CorA (Chen et al. 2003), while Link et al. demonstrated that coexpression of the membrane-bound protease FtsH can lead to a dramatic enhancement in the accumulation of membrane-incorporated human GPCRs in *E. coli* (Link et al. 2008). In a similar way, Baneyx and coworkers demonstrated that simultaneous deletion of *tig*, the gene encoding for the molecular chaperone trigger factor, and overexpression of *yidC*, the gene encoding for the membrane protein integrase YidC can result in a significant increase in the accumulation of different integral membrane proteins in *E. coli* (Nannenga and Baneyx 2011).

As the impact of membrane protein overexpression on cellular physiology is investigated in greater depth (Wagner et al. 2007; Gubellini et al. 2011; Klepsch et al. 2011; Marreddy et al. 2011), relevant expression bottlenecks are expected to be identified. With this knowledge at hand, targeted genetic engineering approaches will likely become increasingly effective. Currently, however, the bottlenecks associated with the production of recombinant membrane proteins even in the simplest hosts, such as bacteria, remain largely unidentified. Under these circumstances, combinatorial approaches become a very attractive alternative. These library-type approaches offer the great advantage that no *a priori* hypotheses need to be posed regarding which genes/pathways should be modified and how.

Bacterial strains that confer improved GPCR production can be engineered by screening libraries of genomic mutants or plasmid-encoded libraries of heterologous or native genes. In an example of classical strain mutagenesis for enhanced recombinant GPCR production, Skretas and Georgiou isolated *E. coli* gene knockout variants that accumulate increased amounts of the human GPCR central cannabinoid receptor 1 (CB1) (Skretas and Georgiou 2009). First, the authors used insertional mutagenesis of the Tn5 transposon to create a library of single-gene knockouts in *E. coli* MC4100A cells. Then, they transformed this library with an expression vector encoding a fusion protein comprising CB1 at the N terminus and the green fluorescent protein (GFP) as the C-terminal domain. Prior work had shown that the fluorescence of *E. coli* cells expressing membrane protein-GFP fusions correlates with the amount of membrane-incorporated protein (Drew et al. 2001), although it does not provide information about the functionality of the protein. Based on this, the authors used fluorescence-activated cell sorting (FACS) to screen the generated library and isolate the high-expression clones. After multiple rounds of cell culturing, CB1-GFP overexpression, and FACS, they isolated a clone carrying a transposon insertion in *dnaJ*, the gene encoding the molecular cochaperone DnaJ, which could accumulate approximately sixfold more membrane-bound CB1 than the parental strain (Skretas and Georgiou 2009). Furthermore, the genetic lesion contained in the isolated strain acted as a suppressor of the toxicity of CB1 overexpression, enabling increased biomass production. The combined effect of the increased CB1 production on a per cell basis and of the enhanced biomass upon harvesting resulted in dramatically enhanced volumetric yields. Interestingly, the enhancing effects of *dnaJ* inactivation were found to be CB1-specific, as none of the other human GPCRs tested could be produced at higher amounts in the isolated strain. This approach, however, could be applied as described for the identification of inhibitory genes for other GPCRs as well.

Genes, gene fragments, or operon fragments that favorably affect protein expression can be isolated from plasmid libraries coexpressing genomic fragments. Skretas et al. screened an *E. coli* genomic library in order to identify multicopy genes that lead to enhanced accumulation of membrane-embedded and properly folded GPCRs in the same host (Skretas et al. 2012). To isolate the desired genes/gene clusters, they used a high-throughput FACS-based assay that directly monitors the accumulation levels of membrane-embedded and properly folded (ligand-binding competent) receptor. This assay was a modification of a technique termed periplasmic expression followed by cytometric sorting (PECS), first developed by Chen et al. (2001), and then adapted for GPCR expression screening by Sarkar et al. (2008), Dodevski and Pluckthun (2011), and Skretas and Georgiou (2008). In this screen, cells overexpressing the high-expression variant D03 of the rat neurotensin receptor 1 (Sarkar et al. 2008) were incubated in a high-osmolarity buffer that renders the outer membrane of *E. coli* cells permeable to fluorescently labeled neurotensin ligand (Figure 2.3). Cells carrying genes/gene clusters capable of acting as enhancers of the accumulation of membrane-incorporated and ligand-binding-competent receptor exhibited enhanced fluorescence and could be readily isolated after repeated rounds of flow cytometric sorting. In this manner, three enhancer genes/gene clusters were isolated: *nagD*, encoding the ribonucleotide phosphatase NagD; a fragment of *nlpD*, encoding a truncation of the predicted lipoprotein NlpD, and the three-gene cluster

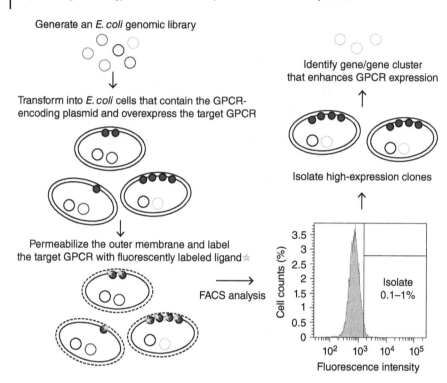

Figure 2.3 Identification of genes/gene clusters whose overexpression results in enhanced GPCR accumulation. Bacterial cells are first transformed with a plasmid-encoded genomic library and a plasmid encoding the target receptor. After GPCR overexpression, cells are incubated in a high-osmolarity buffer that renders the outer membrane permeable to a fluorescently labeled ligand. *E. coli* cells displaying improved expression levels exhibit higher fluorescence and they can be readily isolated by FACS. Finally, the genomic fragments responsible for increased GPCR expression can be easily identified by sequencing the corresponding plasmid.

ptsN–yhbJ–npr, encoding three proteins of the nitrogen phosphotransferase system (Skretas et al. 2012). Coexpression of these genes resulted in a 3–10-fold increase in the yields of different GPCRs (the human bradykinin receptor 2, the human CB1, and the human neurokinin receptor 1) and other unrelated integral membrane proteins. Thus, *E. coli* cells coexpressing these genes could become generalized hosts for recombinant production of membrane-integrated and well-folded GPCRs.

Although their applicability for recombinant GPCR production has not been demonstrated yet, another set of bacterial strains that could serve as general expression hosts for the production of complex integral membrane proteins are the *E. coli* Top10 variants EXP-Rv1337-1–EXP-Rv1337-5, generated by Massey-Gendel et al. (2009). These strains were constructed by applying classical genome mutagenesis using a combination of the mutagenic base analog 2-aminopurine and the mutator gene *mutD5* (a mutated *dnaQ* gene causing a DNA proofreading defect), and a genetic selection system comprising two separate fusions of the integral membrane protein Rv1337 from *Mycobacterium tuberculosis* with two different antibiotic resistance markers at its C terminus, murine dihydrofolate reductase (conferring resistance to trimethoprim) and aminoglycoside 3′-phosphotransferase (conferring resistance to kanamycin). High-expression chromosomal variants were identified by selecting for clones with increased resistance to both antibiotics simultaneously (Figure 2.4). The double antibiotic selection aimed at counterselecting the mutations that lead to host strain antibiotic tolerance and that are unrelated to levels of protein accumulation. The evolved strains were found capable of accumulating markedly enhanced amounts of a variety of different *M. tuberculosis* rhomboid family proteins and other prokaryotic and

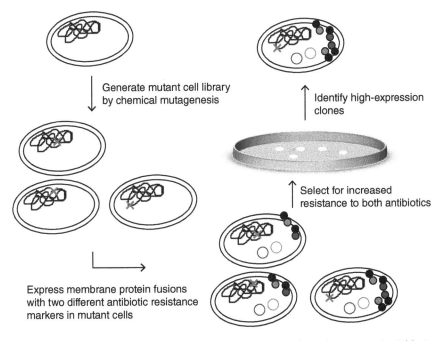

Figure 2.4 Isolation of bacterial mutants that exhibit increased membrane protein yields. A library of mutant cells is first generated by exposing a bacterial culture to a sublethal dose of a chemical mutagen, which introduces randomly distributed genetic lesions throughout the chromosome. The resulting library is then transformed with two expression vectors encoding fusions of a model membrane protein with two different antibiotic resistance markers. The cell library is then challenged for elevated resistance to both antibiotics under membrane protein overexpression conditions, and high-expression chromosomal variants can be readily selected and identified. Each "X" indicates a chromosomal genetic lesion.

eukaryotic integral membrane proteins, in some cases up to 90-fold higher amounts of protein compared to the parental strain TOP10 (Massey-Gendel et al. 2009). The chromosomal lesions leading to enhanced membrane protein production in EXP-Rv1337-1–EXP-Rv1337-5 have not been characterized yet.

Similar classical genetic approaches can also be carried out in other bacterial hosts suitable for integral membrane protein production, which are of potential interest for GPCR production, such as *L. lactis*. In an example of such an attempt, Linares et al. used C-terminal fusions of model membrane proteins with the expression marker ErmC (a protein-imparting resistance to the antibiotic erythromycin) and subjected a *L. lactis* strain to adaptive evolution to select for mutations and identify factors that contribute to enhanced membrane protein productivity in this host (Linares et al. 2010). By using this approach, evolved variants were identified that accumulated two to eightfold more membrane protein than the parental strain. Whole-genome sequencing revealed that the effect of the selected mutations was to enhance transcription for the utilized nisin A promoter (Linares et al. 2010). The generated strains, thus, are expected to be useful for protein production only when using nisin A-regulated expression vectors. Further attempts to engineer *L. lactis* toward high-level membrane protein production and identify relevant expression bottlenecks led to a more detailed, system-level analysis of the organism's response to membrane protein overexpression (Marreddy et al. 2011; Pinto et al. 2011). In these studies, transcriptomics and proteomics analyses were used to identify the two-component system CesSR, which senses cell envelope stresses and assists in maintaining the integrity of this cellular compartment, as an important factor affecting membrane protein yields. Guided by this analysis, Pinto et al. demonstrated that coexpression of *cesSR* could lead to a significant increase in membrane protein yields, primarily by restoring the growth defect caused by the overexpression of these proteins, which led to enhanced biomass production (Pinto et al. 2011).

2.3.2 Yeasts

Yeasts are advantageous expression hosts for mammalian GPCRs as they combine the simplicity, low cost, and potential for large-scale culturing offered by a microbial organism with the biological complexity provided by a eukaryotic cell. Mainly used yeasts are the budding *Saccharomyces cerevisiae*, the methylotrophic *Pichia pastoris* and the fission yeast *Schizosaccharomyces pombe* (Sarramegna et al. 2003). Among the great advantages of yeasts is the ability to perform complex posttranslational modifications, such as protein glycosylation. Although yeasts natively tend to hypermannosylate their glycoproteins and may yield GPCRs with compromised structure or function, glycoengineered yeast strains have been generated in recent years, which allow for fully human-like glycosylation of recombinant proteins (Wildt and Gerngross 2005).

These engineered yeasts can perform remarkably homogeneous glycosylation (Hamilton et al. 2003), a factor that may prove critical for acquiring properly folded protein for crystallization, at least for certain receptors. Additionally, yeasts offer a more native-like membrane lipid environment than bacterial hosts that contains sterols, despite the fact that they also do not contain cholesterol, but ergosterol instead. Recently, metabolically engineered *S. cerevisiae* strains have been generated that preferentially synthesize cholesterol or cholesterol-like lipids over ergosterol (Kitson et al. 2011; Souza et al. 2011), and these could be used for the production of receptors with high sensitivity to the content of their membrane lipid environment. One such lipid-sensitive GPCR is the μ-opioid receptor, which when expressed in *S. cerevisiae* exhibits binding activity to its agonists and antagonists only after the ergosterol of the yeast membranes is removed and replaced with cholesterol *in vitro* (Lagane et al. 2000). Indeed, expression studies of the human μ-opioid and β_3 adrenergic receptors in the yeast strains producing cholesterol-like lipids revealed that active receptor accumulation was enhanced by two- to threefold compared to the parental ergosterol-producing strain (Kitson et al. 2011).

S. cerevisiae is the most extensively studied yeast, but *P. pastoris* can reach very high cell densities and is sometimes preferred (Shiroishi et al. 2012; Routledge et al. 2016). One way of enhancing cell densities in *S. cerevisiae* cultures and, thus, increasing the total volumetric yields for overexpressed membrane proteins and GPCRs is the use of the respiratory strain TM6* (Otterstedt et al. 2004). This *S. cerevisiae* strain carries genetic deletions of the endogenous hexose transporters (Hxts), which control glucose import, and instead expresses an engineered Hxt that is responsible for reduced sugar uptake. The engineered transporter is a chimeric protein consisting of one half of the low-affinity Hxt1, and the high-affinity one, Hxt7 (Otterstedt et al. 2004). In this way, the strain remains purely respiratory irrespective of the levels of glucose, and this results in higher biomass yields at the expense of ethanol production. TM6* has been tested for improved GPCR overexpression, and it was found that its use could enhance the volumetric yields of ligand-binding-competent protein for the human adenosine A_{2a} receptor and the human peripheral cannabinoid receptor 2 compared to the parental respiro-fermentative *S. cerevisiae* by at least fourfold (Ferndahl et al. 2010).

Yeasts have been shown to be efficient GPCR producers in a number of cases. In a recent study, expression of a diverse set of full-length, rhodopsin class GPCRs in *S. cerevisiae* revealed that the vast majority (24 out of 25 tested) could be produced in functional form under optimized conditions (Shiroishi et al. 2012). Results in *P. pastoris* were similar: 20 out of 20 GPCRs tested (not restricted to the rhodopsin class) were shown to accumulate as ligand-binding-competent receptors (Andre et al. 2006). In an even more extended study, the accumulation of 100 GPCRs was tested and showed that for 94 of them, production of full-length protein could be detected by immunoblotting (Lundstrom et al. 2006). Among these, two-thirds were found to accumulate at high or intermediate levels and the vast majority (approximately 85%) exhibited ligand-binding activity (Lundstrom et al. 2006). The microbial nature of yeasts allows for relatively facile scale-up of culturing, which can lead to significantly enhanced yields, as demonstrated when the human adenosine A_{2a} receptor was expressed in a ligand-binding-competent form in *P. pastoris* (Singh et al. 2008). As the ultimate proof of the capabilities of yeasts as an expression platform for GPCR structural biology, high-resolution structures of the human histamine and the adenosine A_{2a} receptor have been solved using protein prepared from *P. pastoris* cultures (Shimamura et al. 2011; Hino et al. 2012).

Despite this very encouraging data, yeast hosts very often fail to accumulate recombinant GPCRs at levels sufficient for crystallization (Sarramegna et al. 2003). For example, in the study of Shiroishi et al. mentioned above, none of the 24 receptors expressed in ligand-binding-competent form accumulated at levels satisfactory for structural biology studies, even after optimization of a number of expression parameters (Shiroishi et al. 2012). Yeasts frequently produce large amounts of total receptor protein but accumulate plasma membrane-embedded, functional GPCR only poorly (O'Malley et al. 2009). Failure of the produced receptor to reach the plasma membrane often results in its accumulation in other cellular compartments, such as the ER, secretory vesicles, or others (Butz et al. 2003; O'Malley et al. 2009). This, in turn, induces the unfolded protein response (UPR), the stress signaling mechanism responsible for sensing the accumulation of unfolded proteins in the ER (Griffith et al. 2003; O'Malley et al. 2009). Griffith et al. have reported that membrane protein expression conditions that lead to high levels of UPR induction could be detrimental for the yields of plasma membrane-incorporated and ligand-binding-competent protein (Griffith et al. 2003). They observed that high-copy expression of the *Trypanosoma equiperdum* P2 H^+/adenosine co-transporter in *S. cerevisiae* resulted in very high levels of UPR and very poor accumulation of membrane-bound protein. On the other hand, reduced copy numbers of the expression vector correlated with reduced UPR and increased levels of membrane-bound and ligand-binding-competent protein. The authors proposed that high levels of UPR and low membrane-bound protein productivity coincide because, under these conditions, the protein folding machinery of the cell is overwhelmed. Based on these results, they suggested that monitoring the levels of the UPR during integral membrane protein production could serve as an effective variable during the process of expression optimization (Griffith et al. 2003).

Once activated, UPR leads to upregulation of the UPR-specific transcription factor Hac1p, which controls the expression of hundreds of genes encoding for ER-residing molecular chaperones, foldases, proteases, factors that assist protein translocation and secretion, enzymes involved in lipid metabolism and other proteins (Mori 2009). Some UPR-related factors could, thus, be beneficial for the folding and/or targeting of recombinantly produced GPCRs, as it has already been reported for the production of secreted proteins in yeast and mammalian cells (Valkonen et al. 2003; Ohya et al. 2008). Indeed, overexpression of *HAC1*, the gene encoding for Hac1p, has been shown to almost double the yields of ligand-binding-competent adenosine A_{2a} receptor in *P. pastoris* (Guerfal et al. 2010). Based on the data described in the previous paragraphs, the UPR pathway could be an important target for GPCR expression optimization. It is not clear yet, however, in what way to interfere with this pathway and how.

In order to identify better defined bottlenecks involved in membrane protein production in yeast, Bonander et al. analyzed the transcriptome profiles of *S. cerevisiae* cells under culturing conditions that had opposing effects on the production of the eukaryotic glycerol facilitator protein Fps1p: one set of conditions that facilitated the accumulation of plasma membrane-embedded protein versus another set that resulted in enhanced yields of total protein but dramatically decreased production of plasma membrane-incorporated protein (Bonander et al. 2005). Several genes were found to be differentially expressed under high-accumulation conditions. One of these, *BMS1*, was found to be significantly upregulated under different conditions that favored Fps1p membrane integration. *BMS1* encodes for the protein Bms1, an essential nucleolar factor previously shown to be involved in ribosome biogenesis (Wegierski et al. 2001). Regulated expression of *BMS1* from a doxycycline-repressible promoter in an engineered yeast strain enabled the identification of *BMS1* transcript levels, which were optimal for the production of membrane-bound Fps1 and the human adenosine A_{2a} receptor (Bonander et al. 2009). Under these conditions, Fps1 production was increased by 78-fold in shake flasks and by 138-fold in a bioreactor, while accumulation of ligand-binding-competent human adenosine A_{2a} receptor was enhanced by 2-fold (Bonander et al. 2009). Increased membrane protein incorporation was shown to be associated with an increase in the ratio of the ribosomal subunits 60S:40S (Figure 2.5) (Bonander et al. 2009). These results revealed ribosome biogenesis as a critical factor affecting recombinant membrane production.

2.3.3 Insect Cells

Plasma membrane-embedded and functional receptors have been produced in insect cell cultures in a number of cases (Massotte 2003; Sarramegna et al. 2003). Expression tests for 16 rhodopsin class GPCRs in insect cells resulted in the production of ligand-binding-competent receptors in all cases, while nine of them accumulated at sufficient levels for crystallization (Akermoun et al. 2005). Insect cells have been immensely successful expression hosts for GPCR structural biology studies as the majority of the unique determined

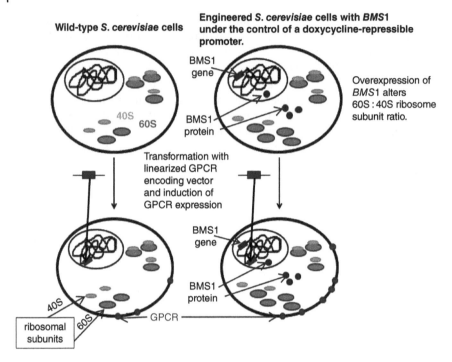

Figure 2.5 Effect of optimal co-expression of BMS1 on the functional expression of a GPCR in *S. cerevisiae* cells. Increased membrane protein accumulation was shown to be associated with an increase in the ratio of the ribosomal subunits 60S:40S. The BMS1 gene and protein are indicated by arrows, as are rectangles and circles identifying the GPCR-encoding gene and ligand-binding-competent GPCRs, respectively. 60S and 40S subunits of ribosome are indicated by arrows. For simplicity other organelles of the *S. cerevisiae* cell apart from the nucleus are not shown.

structures correspond to protein isolated from this type of recombinant source (Table 2.1). The most frequently utilized insect cell lines for recombinant protein production are derived from the ovaries of the fall armyworm *Spodoptera frugiperda* (*Sf*9 and *Sf*21 cells) and from embryonic tissue of the cabbage looper *Trichoplusia ni* (High Five™ cells). GPCR expression in insect cells is mostly carried out by transient transfection using a recombinant *Autographa californica* baculovirus or recombinant baculoviral DNA (e.g. the Bac-to-Bac® system) that express the GPCR transgene usually under the control of the strong, late-stage polyhedrin promoter. The advantage of this host is that it offers an even more mammalian-like environment for recombinant protein production compared to yeasts. For example, insect cells do not hypermannosylate their glycoproteins to the extent that yeasts do, and they can be modified genetically to add human-like, terminally galactosylated or sialylated *N*-glycans (Hollister et al. 2002; Aumiller et al. 2003; Chang et al. 2003). Furthermore, they have the protein folding/quality control machinery of a higher eukaryote, and their plasma membranes contain cholesterol, albeit to small amounts. At the same time, insect cells are easier to culture than mammalian cells as they require simpler growth media and can grow easily in suspension cultures, features that make their culturing much more amenable to scale-up. To this end, the turkey β_1 adrenergic receptor has been expressed in *Sf* 9 and High Five cells cultured in bioreactors at scales between 4 and 32 l and provided yields between 0.5 and 2.5 mg l^{-1} of isolated protein (Warne et al. 2003, 2009). Similarly, the human histamine H1 and M2 muscarinic acetylcholine receptors were produced in *Sf* 9 cells in 10 and 6 l cultures in bioreactors, respectively, with a final yield of 5–7 mg l^{-1} of ligand-binding-competent receptor (Ratnala et al. 2004; Haga et al. 2012). In both of the latter examples, it was demonstrated that cell culturing in wave bioreactors could be used for achieving high-level production of ligand-binding-competent GPCRs, a technology with the potential to lower the complexity and cost of large-scale GPCR production. Importantly, GPCR production in a wave bioreactor has already been shown to yield sufficient amounts of active protein for crystallization studies (Haga et al. 2012).

Despite the enormous success of insect cells as GPCR expression hosts for structural studies, there are some drawbacks. Apart from the high cost of the culture medium (He et al. 2014), a number of receptors for which production in these hosts does not yield sufficient amounts for structural analysis (Akermoun et al. 2005; Bernaudat et al. 2011). Genetically modified insect cell lines that specialize in the high-level accumulation of these GPCRs that are problematic for wild-type cells could, in principle at least, be generated, in a manner analogous to what has been attempted for microbial hosts, such as *E. coli* and yeasts. This potential has already been demonstrated for other membrane proteins, such as the *Drosophila melanogaster* Shaker H4 potassium channel, where *Sf* 9 cells co-expressing the molecular chaperone calnexin were found to accumulate twice as much properly assembled transporter than wild-type cells (Higgins et al. 2003). Similarly, coexpression of calnexin was found to increase the yield of the serotonin transporter by threefold in the same cell line, while coexpression of calreticulin and BiP had a smaller enhancing effect (Tate et al. 1999). To our knowledge, genetic modifications that assist the production of recombinant GPCRs in insect cell lines have not been reported yet. Targeted genetic modifications in these systems are straightforward once the target-for-modification gene/genes has/have been determined, and it seems likely that they could contribute to the creation of better GPCR-producing insect cell lines. On the contrary and unlike microbial cell systems, large libraries of genetically modified insect cells cannot be easily generated and, thus, isolation of high-production variants has not been pursued yet for GPCRs. However, recent advances in genome engineering methodologies, such as the ones based on the bacterial clustered regularly interspaced short palindromic repeats (CRISPRs)/CRISPR-associated (Cas) systems (Mali et al. 2013) could make this a tractable task.

2.3.4 Mammalian Cells

One would expect that mammalian cells would be the preferred type of expression host for structural analysis of recombinant GPCRs of human and mammalian origin as they provide the cellular environment closest to that of the native protein host in terms of posttranslational modifications, membrane lipid composition, and protein folding/quality control machinery. Indeed, mammalian expression studies for more than 100 different GPCRs showed that the vast majority could be produced in ligand-binding-competent form, and that approximately one-third of them accumulated at high-enough yields for potential initiation of crystallization trials (Hassaine et al. 2006; Lundstrom et al. 2006). To date, however, with the exception of the human cytomegalovirus US28 GPCR with bound human cytokine CX3CL1 (Burg et al. 2015), only three structures have been solved using material acquired from mammalian cultures, while all of them correspond to the same protein, rhodopsin (Standfuss et al. 2007, 2011; Deupi et al. 2012). Limitations associated with mammalian GPCR expression are the time and effort required to generate stable cell lines, the need for more complex and expensive growth media, the difficulties associated with their large-scale culturing, and the quality of the produced protein (Andrell and Tate 2013).

For the majority of the GPCRs for which glycosylation is important for folding and function, mammalian expression can result in the attachment of extremely heterogeneous glycans (Reeves et al. 2002), a factor that can prevent the formation of well-ordered crystals (Smyth et al. 2003). This issue can be addressed by using mutants defective in specific parts of the mammalian protein glycosylation pathways. Two such cell lines are Chinese hamster ovary (CHO) and HEK293 cells that lack N-acetylglucosaminyltransferase I (GnTI) activity (Stanley and Chaney 1985; Reeves et al. 2002). These GnTI$^-$ cell lines can accumulate homogeneously glycosylated GPCRs with a GlcNAc$_2$Man$_5$ glycan structure attached to them (Reeves et al. 2002). The use of GnTI-HEK293 cells has already facilitated the determination the structure of bovine rhodopsin in an active state (Reeves et al. 2002).

Thus far, large-scale mammalian production has been reported only for a handful of GPCRs, such as the human olfactory receptor 17-4, which was expressed at a 2.5 l scale with a yield of 3 mg l^{-1} of culture medium (Cook et al. 2009), for mammalian rhodopsins at 1–10 l scales with yields of >1 mg l^{-1} (Reeves et al. 2002; Min et al. 2011; Deupi et al. 2012) and for a variant of rat neurotensin receptor at a 5 l scale with yield 1 mg l^{-1} culture (Xiao et al. 2013).

2.3.5 Transgenic Animals

Transgenic animals have been used for the production of proteins since the early 1980s (Houdebine 2009) and, in the last decade, such systems have been developed for the production and purification of mammalian GPCRs and other membrane proteins. Animals that have been used for such purposes are mice, flies, frogs, and worms (Eroglu et al. 2002, 2003; Kodama et al. 2005; Zhang et al. 2005; Panneels et al. 2011; Salom et al. 2012).

The first attempt to produce GPCRs in transgenic animals targeted the photoreceptor cells in the eyes of the fruitfly *D. melanogaster* (Eroglu et al. 2002). These cells contain extended stacks of membranes, termed rhabdomeres, and are specialized in producing high levels of rhodopsin properly incorporated within these structures. By using an eye-specific driver for expression, different fruitfly, rat, and human GPCRs have been produced in membrane-incorporated, ligand-binding competent, and homogeneous form, at levels which are comparable to those for homologous GPCRs and higher than the levels achieved in yeast or insect cell cultures (Eroglu et al. 2002; Panneels et al. 2011). The high membrane content of the photoreceptor cells, which are formed at the expense of intracellular membranous compartments, contribute to the enhanced productivity, while minimizing the amount of mistargeted, intracellularly retained membrane protein (Eroglu et al. 2002). Once stable transgenic flies have been generated, the receptor production process can be easily scaled up and, contrary to what is frequently observed with cell-culture-based systems, protein yields do not appear to decrease (Eroglu et al. 2002). For the human serotonin transporter, a non-GPCR but hard-to-express integral membrane protein, 0.5 mg of isolated transporter could be produced from 2 g of transgenic fly heads (Panneels et al. 2011).

GPCR production in the eyes of a transgenic animal has also been reported for the frog *Xenopus laevis* and the mouse *Mus musculus* (Kodama et al. 2005; Zhang et al. 2005). Expression of 12 human serotonin and 8 human endothelial differentiation gene receptors under the control of an opsin promoter in retina rod cells of the eyes of transgenic *X. laevis* tadpoles resulted in accumulation of membrane-inserted protein for all receptors tested, although for some of the members of the two families intracellular protein retention and lower yields were observed (Zhang et al. 2005). For the better expressed receptors, each tadpole could produce approximately 1–5 ng of recombinant protein capable of binding ligand and of coupling to G proteins upon agonist binding. Based on these numbers, production of a sufficient amount of protein for structural studies would require a large number of animals. To address this issue, Zhang et al. constructed a robotic system that can generate hundreds of transgenic tadpoles per day with a potential yield about 6 μg of purified GPCR per week (Zhang et al. 2005). One of the great advantages of this expression system is that it produces highly homogeneously glycosylated protein (Zhang et al. 2005).

The nematode worm *C. elegans* has also been tested for GPCR production in its neurons and muscles (Salom et al. 2012). These two tissues were targeted for expression in a pilot study as they are the most abundant type of cells and comprise the largest part of the body mass in the organism, respectively. Expression tests for a variety of mammalian GPCRs showed that they could be produced in a properly localized and ligand-binding-competent form (Salom et al. 2012). By scaling-up the process, where GPCR-expressing transgenic animals were grown in 10 l liquid cultures, close to mg quantities of properly folded isolated receptors could be produced (Salom et al. 2012). Worm-targeted expression offers many advantages as *C. elegans* is an organism easy to manipulate genetically, it has a short life cycle and is scalable at a low cost.

2.3.6 Cell-Free Systems

Cell-free protein synthesis involves the production of proteins in open systems using cell extracts or components from different organisms, such as insects, rabbits, wheat germ, bacteria, instead of intact cells. In these systems, there is no toxicity associated with the protein production process, which in living cell systems often leads to losses of the expression vector or poor protein yields due to lower biomass production (Endo and Sawasaki 2006).

GPCR production in a cell-free reaction can be carried out in the absence or presence of detergents (Klammt et al. 2005, 2007). In the former case, the receptors are produced as precipitates, which are, however, easier to refold than bacterial GPCR inclusion bodies (Klammt et al. 2007). In the presence of detergents, GPCRs can accumulate in the membrane-resembling environment of a micelle. When the appropriate detergent is selected, the receptor can be produced in a well folded, stable and ligand-binding-competent form. Very importantly, the volumetric yields for the target receptor can be extremely high, reaching milligram per milliliter levels (Corin et al. 2011a). At the same time, reaction scale-up above the milliliter scale is currently impossible due to the rapid decreases in protein yields at higher reaction volumes and the high cost, especially when using cell extracts from higher organisms (McCusker et al. 2007). Other limitations involve chaperone-associated folding as chaperones that are normally found in the cytosolic milieu are not necessarily included in the reaction. Also, in the cell-free synthesis reaction, the environment may be significantly different compared to what is encountered *in vivo*. For example, conditions, concentrations of reactants, folding rates, or the oligomerization propensity of the produced protein in this environment may be different. Cell-free protein expression is, in general, complex, and standardization can be difficult due to the large number of reaction components.

By using extracts from *E. coli* cultures and after screening libraries of detergents in order to identify the most effective one, a number of different GPCRs, such as the human and porcine vasopressin type 2, the human β2 adrenergic, the human muscarinic acetylcholine, the human melatonin 1B, and other nonolfactory and olfactory mammalian receptors were produced in ligand-binding-competent form in mg quantities with cell-free expression technology (Ishihara et al. 2005; Klammt et al. 2005, 2007; Kaiser et al. 2008; Corin et al. 2011a). In most cases, the effective detergent was found to be protein-specific, although certain types of detergents, such as the polyoxyethylene-alkylether ones (e.g. brij-35), have been shown to promote cell-free GPCR production for a number of different receptors (Corin et al. 2011a).

It has been demonstrated that detergent micelles are not necessary for successful cell-free GPCR production and can be replaced with other amphipathic environments. For example, the β_2 adrenergic receptor has been successfully expressed in a cell-free system in the presence of nanolipoprotein particles (Yang et al. 2011), which are planar phospholipid membrane bilayers encircled by apolipoproteins (Schlegel et al. 2010). Furthermore, the use of the fructose-based polymer NV10 has allowed the expression of the corticotrophin-releasing factor receptors 1 and 2β in a ligand-binding-competent form (Klammt et al. 2011). Another interesting class of detergent alternatives is short peptide-based surfactants. These have already been shown to be very effective for the production of a variety of GPCRs, such as the human formyl peptide and the human trace amine-associated receptors and a variety of human olfactory receptors (Wang et al. 2011; Corin et al. 2011b). Despite their great potential, cell-free expression systems have not been used yet to determine the structure of a GPCR.

2.4 Conclusion

GPCR structural biology has experienced a tremendous revolution in the last decade, with many unique, nonrhodopsin, and pharmacologically relevant structures having been published and more being reported every few months. The knowledge, experience, and tools necessary for acquiring high-resolution structures by X-ray crystallography and NMR seem now to be in place. Furthermore, rapidly developing techniques, such as cryogenic transmission electron microscopy (cryo-TEM), are anticipated to make important contributions to GPCR structural biology as the acquired resolutions (Bartesaghi et al. 2015) and the size of membrane proteins (Liao et al. 2013; Bai et al. 2015), which can be probed structurally with this technique, approach or even surpass the capabilities of X-ray crystallography and NMR. In parallel, the availability of structural information will accelerate the rational, structure-guided development of GPCR modulators with potentially therapeutic properties. Before the full pharmacological potential of the GPCR superfamily can be realized, however, the repertoire of structurally characterized receptors needs to be expanded further to include a more diverse set of proteins belonging to different classes. Furthermore, more structures belonging to the native, wild-type sequence of a GPCR need to be determined as, in many cases, the protein engineering interventions imposed on these targets in order to enhance their "crystallizability," interfere with their physiologically relevant conformations. Before these goals can be achieved, the expression bottleneck, i.e. the acquiring of sufficient amounts of properly folded protein to initiate structural determination efforts, needs to be addressed. Years of studies have shown that traditional trial-and-error approaches where simple expression parameters are attempted to be optimized, consume both time and effort, and in the end, they typically have limited and receptor-specific success. Motivated by these observations, a number of investigators have claimed that before the problem of recombinant GPCR production is solved, we need to understand where it stems from. Furthermore, they have proposed that the most efficient way to address the issue and generate new and efficient expression systems for GPCRs may be to engineer the cellular machinery of the expression host itself. In the last few years, a number of studies consistent with this rationale have been reported and have already yielded promising results. The majority of these efforts have focused on microbial organisms, such as bacteria and yeasts, primarily because these types of cells are easier to manipulate genetically. A number of new engineered strains with great potential as GPCR expression hosts have been generated, and their characterization has started to reveal specific bottlenecks that interfere with membrane protein production and possible ways about how they could be alleviated. These initial successes in microbial hosts can pave the way for similar efforts in higher organisms, such as insect, mammalian cells, or even transgenic animals. Once current difficulties both in GPCR production and structural determination have been surpassed, this untapped source of targets for potential development of effective therapeutics can be exploited to its full extent.

References

Aaij, R., Adeva, B., Adinolfi, M. et al. (2014). Observation of B(s)(0) → J/ψ f1(1285) decays and measurement of the f1(1285) mixing angle. *Phys. Rev. Lett.* 112 (9): 091802.

Akermoun, M., Koglin, M., Zvalova-Iooss, D. et al. (2005). Characterization of 16 human G protein-coupled receptors expressed in baculovirus–infected insect cells. *Protein Expr. Purif.* 44 (1): 65–74.

Andre, N., Cherouati, N., Prual, C. et al. (2006). Enhancing functional production of G protein-coupled receptors in *Pichia pastoris* to levels required for structural studies via a single expression screen. *Protein Sci.* 15 (5): 1115–1126.

Andrell, J. and Tate, C.G. (2013). Overexpression of membrane proteins in mammalian cells for structural studies. *Mol. Membr. Biol.* 30 (1): 52–63.

Araki, T., Hirayama, M., Hiroi, S., and Kaku, K. (2012). GPR40-induced insulin secretion by the novel agonist TAK-875: first clinical findings in patients with type 2 diabetes. *Diabetes Obes. Metab.* 14 (3): 271–278.

Aumiller, J.J., Hollister, J.R., and Jarvis, D.L. (2003). A transgenic insect cell line engineered to produce CMP-sialic acid and sialylated glycoproteins. *Glycobiology* 13 (6): 497–507.

Bai, X.C. et al. (2015). An atomic structure of human gamma-secretase. *Nature* 525 (7568): 212–217.

Baneres, J.L., Yan, C., Yang, G. et al. (2003). Structure-based analysis of GPCR function: conformational adaptation of both agonist and receptor upon leukotriene B_4 binding to recombinant BLT1. *J. Mol. Biol.* 329 (4): 801–814.

Baneres, J.L., Popot, J.L., and Mouillac, B. (2011). New advances in production and functional folding of G-protein-coupled receptors. *Trends Biotechnol.* 29 (7): 314–322.

Bartesaghi, A., Merk, A., Banerjee, S. et al. (2015). 2.2 A resolution cryo-EM structure of beta-galactosidase in complex with a cell-permeant inhibitor. *Science* 348 (6239): 1147–1151.

Bernaudat, F., Frelet-Barrand, A., Pochon, N. et al. (2011). Heterologous expression of membrane proteins: choosing the appropriate host. *PLoS ONE* 6 (12): e29191.

Bernier, V., Bichet, D.G., and Bouvier, M. (2004). Pharmacological chaperone action on G-protein-coupled receptors. *Curr. Opin. Pharmacol.* 4 (5): 528–533.

Bessette, P.H., Aslund, F., Beckwith, J., and Georgiou, G. (1999). Efficient folding of proteins with multiple disulfide bonds in the *Escherichia coli* cytoplasm. *Proc. Natl. Acad. Sci. U.S.A.* 96 (24): 13703–13708.

Bill, R.M., Henderson, P.J., Iwata, S. et al. (2011). Overcoming barriers to membrane protein structure determination. *Nat. Biotechnol.* 29 (4): 335–340.

Bonander, N., Hedfalk, K., Larsson, C. et al. (2005). Design of improved membrane protein production experiments: quantitation of the host response. *Protein Sci.* 14 (7): 1729–1740.

Bonander, N., Darby, R.A.J., Grgic, L. et al. (2009). Altering the ribosomal subunit ratio in yeast maximizes recombinant protein yield. *Microb. Cell Fact.* 8: 10.

Burg, J.S., Ingram, J.R., Venkatakrishnan, A.J. et al. (2015). Structural biology. Structural basis for chemokine recognition and activation of a viral G protein-coupled receptor. *Science* 347 (6226): 1113–1117.

Busuttil, B.E., Turney, K.L., and Frauman, A.G. (2001). The expression of soluble, full-length, recombinant human TSH receptor in a prokaryotic system. *Protein Expr. Purif.* 23 (3): 369–373.

Butz, J.A., Niebauer, R.T., and Robinson, A.S. (2003). Co-expression of molecular chaperones does not improve the heterologous expression of mammalian G-protein coupled receptor expression in yeast. *Biotechnol. Bioeng.* 84 (3): 292–304.

Cabrera-Vera, T.M., Vanhauwe, J., Thomas, T.O. et al. (2003). Insights into G protein structure, function, and regulation. *Endocr. Rev.* 24 (6): 765–781.

Chang, G.D., Chen, C.J., Lin, C.Y. et al. (2003). Improvement of glycosylation in insect cells with mammalian glycosyltransferases. *J. Biotechnol.* 102 (1): 61–71.

Chelikani, P., Reeves, P.J., Rajbhandarry, U.L., and Khorana, H.G. (2006). The synthesis and high-level expression of a beta2-adrenergic receptor gene in a tetracycline-inducible stable mammalian cell line. *Protein Sci.* 15 (6): 1433–1440.

Chen, G., Hayhurst, A., Thomas, J.G. et al. (2001). Isolation of high-affinity ligand-binding proteins by periplasmic expression with cytometric screening (PECS). *Nat. Biotechnol.* 19 (6): 537–542.

Chen, Y., Song, J., Sui, S.F., and Wang, D.N. (2003). DnaK and DnaJ facilitated the folding process and reduced inclusion body formation of magnesium transporter CorA overexpressed in *Escherichia coli. Protein Expr. Purif.* 32 (2): 221–231.

Cherezov, V., Rosenbaum, D.M., Hanson, M.A. et al. (2007). High-resolution crystal structure of an engineered human beta2-adrenergic G protein-coupled receptor. *Science* 318 (5854): 1258–1265.

Chien, E.Y., Liu, W., Zhao, Q. et al. (2010). Structure of the human dopamine D_3 receptor in complex with a D_2/D_3 selective antagonist. *Science* 330 (6007): 1091–1095.

Congreve, M. and Marshall, F. (2010). The impact of GPCR structures on pharmacology and structure-based drug design. *Br. J. Pharmacol.* 159 (5): 986–996.

Congreeve, M., Langmead, C.J., Mason, J.S., and Marshall, F.H. (2011). Progress in structure based drug design for G protein-coupled receptors. *J. Med. Chem.* 54 (13): 4283–4311.

Conn, P.M., Ulloa-Aguirre, A., Ito, J., and Janovick, J.A. (2007). G protein-coupled receptor trafficking in health and disease: lessons learned to prepare for therapeutic mutant rescue in vivo. *Pharmacol. Rev.* 59 (3): 225–250.

Cook, B.L., Steuerwald, D., Kaiser, L. et al. (2009). Large-scale production and study of a synthetic G protein-coupled receptor: human olfactory receptor 17-4. *Proc. Natl. Acad. Sci. U.S.A.* 106 (29): 11925–11930.

Corin, K., Baaske, P., Ravel, D.B. et al. (2011a). A robust and rapid method of producing soluble, stable, and functional G-protein coupled receptors. *PLoS ONE* 6 (10): e23036.

Corin, K., Baaske, P., Ravel, D.B. et al. (2011b). Designer lipid-like peptides: a class of detergents for studying functional olfactory receptors using commercial cell-free systems. *PLoS ONE* 6 (11): e25067.

Deupi, X., Edwards, P., Singhal, A. et al. (2012). Stabilized G protein binding site in the structure of constitutively active metarhodopsin-II. *Proc. Natl. Acad. Sci. U.S.A.* 109 (1): 119–124.

Dodevski, I. and Pluckthun, A. (2011). Evolution of three human GPCRs for higher expression and stability. *J. Mol. Biol.* 408 (4): 599–615.

Dore, A.S., Robertson, N., Errey, J.C. et al. (2011). Structure of the adenosine A_{2A} receptor in complex with ZM241385 and the xanthines XAC and caffeine. *Structure* 19 (9): 1283–1293.

Drew, D.E., Von Heijne, G., Nordlund, P., and de Gier, J.W. (2001). Green fluorescent protein as an indicator to monitor membrane protein overexpression in *Escherichia coli. FEBS Lett.* 507 (2): 220–224.

Egloff, P., Hillenbrand, M., Klenk, C. et al. (2014). Structure of signaling-competent neurotensin receptor 1 obtained by directed evolution in *Escherichia coli. Proc. Natl. Acad. Sci. U.S.A.* 111 (6): E655–E662.

Endo, Y. and Sawasaki, T. (2006). Cell-free expression systems for eukaryotic protein production. *Curr. Opin. Biotechnol.* 17 (4): 373–380.

Eroglu, C., Cronet, P., Panneels, V. et al. (2002). Functional reconstitution of purified metabotropic glutamate receptor expressed in the fly eye. *EMBO Rep.* 3 (5): 491–496.

Eroglu, C., Bruggerr, B., Wieland, F., and Sinning, I. (2003). Glutamate-binding affinity of *Drosophila* metabotropic glutamate receptor is modulated by association with lipid rafts. *Proc. Natl. Acad. Sci. U.S.A.* 100 (18): 10219–10224.

Faulkner, M.J., Veeravalli, K., Gon, S. et al. (2008). Functional plasticity of a peroxidase allows evolution of diverse disulfide-reducing pathways. *Proc. Natl. Acad. Sci. U.S.A.* 105 (18): 6735–6740.

Ferndahl, C., Bonander, N., Logez, C. et al. (2010). Increasing cell biomass in *Saccharomyces cerevisiae* increases recombinant protein yield: the use of a respiratory strain as a microbial cell factory. *Microb. Cell Fact.* 9: 47.

Fredriksson, R., Lagerstrom, M.C., Lundin, L.G., and Schioth, H.B. (2003). The G-protein-coupled receptors in the human genome form five main families. Phylogenetic analysis, paralogon groups, and fingerprints. *Mol. Pharmacol.* 63 (6): 1256–1272.

Free, R.B., Hazelwood, L.A., Cabrera, D.M. et al. (2007). D_1 and D_2 dopamine receptor expression is regulated by direct interaction with the chaperone protein calnexin. *J. Biol. Chem.* 282 (29): 21285–21300.

George, S.R., O'Dowd, B.F., and Lee, S.P. (2002). G-protein-coupled receptor oligomerization and its potential for drug discovery. *Nat. Rev. Drug Discov.* 1 (10): 808–820.

Goder, V., Junne, T., and Spiess, M. (2004). Sec61p contributes to signal sequence orientation according to the positive-inside rule. *Mol. Biol. Cell* 15 (3): 1470–1478.

Granier, S., Manglik, A., Kruse, A.C. et al. (2012). Structure of the delta-opioid receptor bound to naltrindole. *Nature* 485 (7398): 400–404.

Griffith, D.A., Delipala, C., Leadsham, J. et al. (2003). A novel yeast expression system for the overproduction of quality-controlled membrane proteins. *FEBS Lett.* 553 (1–2): 45–50.

Grisshammer, R. and Tate, C.G. (1995). Overexpression of integral membrane proteins for structural studies. *Q. Rev. Biophys.* 28 (3): 315–422.

Grisshammer, R., Duckworth, R., and Henderson, R. (1993). Expression of a rat neurotensin receptor in *Escherichia coli. Biochem. J.* 295 (Pt 2): 571–576.

Gubellini, F., Verdon, G., Karpowich, N.K. et al. (2011). Physiological response to membrane protein overexpression in *E. coli. Mol. Cell. Proteomics* 10 (10): M111.007930.

Guerfal, M., Ryckaert, S., Jacobs, P.P. et al. (2010). The HAC1 gene from *Pichia pastoris*: characterization and effect of its overexpression on the production of secreted, surface displayed and membrane proteins. *Microb. Cell Fact.* 9: 49.

Haga, K., Kruse, A.C., Asada, H. et al. (2012). Structure of the human M_2 muscarinic acetylcholine receptor bound to an antagonist. *Nature* 482 (7386): 547–551.

Hamilton, S.R., Bobrowicz, P., Bobrowicz, B. et al. (2003). Production of complex human glycoproteins in yeast. *Science* 301 (5637): 1244–1246.

Hassaine, G., Wagner, R., Kempf, J. et al. (2006). Semliki Forest virus vectors for overexpression of 101 G protein-coupled receptors in mammalian host cells. *Protein Expr. Purif.* 45 (2): 343–351.

He, Y., Wang, K., and Yan, N. (2014). The recombinant expression systems for structure determination of eukaryotic membrane proteins. *Protein Cell* 5 (9): 658–672.

Higgins, M.K., Demir, M., and Tate, C.G. (2003). Calnexin co-expression and the use of weaker promoters increase the expression of correctly assembled shaker potassium channel in insect cells. *Biochim. Biophys. Acta* 1610 (1): 124–132.

Hino, T., Arakawa, T., Iwanari, H. et al. (2012). G-protein-coupled receptor inactivation by an allosteric inverse-agonist antibody. *Nature* 482 (7384): 237–240.

Hollenstein, K., Kean, J., Bortolato, A. et al. (2013). Structure of class B GPCR corticotropin-releasing factor receptor 1. *Nature* 499 (7459): 438–443.

Hollister, J., Grabenhorst, E., Nimtz, M. et al. (2002). Engineering the protein *N*-glycosylation pathway in insect cells for production of biantennary, complex *N*-glycans. *Biochemistry* 41 (50): 15093–15104.

Houdebine, L.M. (2009). Production of pharmaceutical proteins by transgenic animals. *Comp. Immunol. Microbiol. Infect. Dis.* 32 (2): 107–121.

Ishihara, G., Goto, M., Saeki, M. et al. (2005). Expression of G protein coupled receptors in a cell-free translational system using detergents and thioredoxin-fusion vectors. *Protein Expr. Purif.* 41 (1): 27–37.

Jaakola, V.P., Griffith, M.T., Hanson, M.A. et al. (2008). The 2.6 angstrom crystal structure of a human A_{2A} adenosine receptor bound to an antagonist. *Science* 322 (5905): 1211–1217.

Jacoby, E., Bouhelal, R., Gerspacher, M., and Seuwen, K. (2006). The 7 TM G-protein-coupled receptor target family. *ChemMedChem* 1 (8): 761–782.

Kaiser, L., Graveland-Bikker, J., Steuerwald, D. et al. (2008). Efficient cell-free production of olfactory receptors: detergent optimization, structure, and ligand binding analyses. *Proc. Natl. Acad. Sci. U.S.A.* 105 (41): 15726–15731.

Kajiya, K., Inaki, K., Tanaka, M. et al. (2001). Molecular bases of odor discrimination: reconstitution of olfactory receptors that recognize overlapping sets of odorants. *J. Neurosci.* 21 (16): 6018–6025.

Kaku, K., Enya, K., Nakaya, R. et al. (2015). Efficacy and safety of fasiglifam (TAK-875), a G protein-coupled receptor 40 agonist, in Japanese patients with type 2 diabetes inadequately controlled by diet and exercise: a randomized, double-blind, placebo-controlled, phase III trial. *Diabetes Obes. Metab.* 17 (7): 675–681.

Kitson, S.M., Mullen, W., Cogdell, R.J., and Fraser, N.J. (2011). GPCR production in a novel yeast strain that makes cholesterol-like sterols. *Methods* 55 (4): 287–292.

Klammt, C., Schwarz, D., Fendler, K. et al. (2005). Evaluation of detergents for the soluble expression of alpha-helical and beta-barrel-type integral membrane proteins by a preparative scale individual cell-free expression system. *FEBS J.* 272 (23): 6024–6038.

Klammt, C., Schwarz, D., Eifler, N. et al. (2007). Cell-free production of G protein-coupled receptors for functional and structural studies. *J. Struct. Biol.* 158 (3): 482–493.

Klammt, C., Perrin, M.H., Maslennikov, I. et al. (2011). Polymer-based cell-free expression of ligand-binding family B G-protein coupled receptors without detergents. *Protein Sci.* 20 (6): 1030–1041.

Klepsch, M.M., Persson, J.O., and Gier, J.W.D. (2011). Consequences of the overexpression of a eukaryotic membrane protein, the human KDEL receptor, in *Escherichia coli*. *J. Mol. Biol.* 407 (4): 532–542.

Kodama, T., Imai, H., Doi, T. et al. (2005). Expression and localization of an exogenous G protein-coupled receptor fused with the rhodopsin C-terminal sequence in the retinal rod cells of knockin mice. *Exp. Eye Res.* 80 (6): 859–869.

Kristiansen, K. (2004). Molecular mechanisms of ligand binding, signaling, and regulation within the superfamily of G-protein-coupled receptors: molecular modeling and mutagenesis approaches to receptor structure and function. *Pharmacol. Ther.* 103 (1): 21–80.

Kruse, A.C., Hu, J., Pan, A.C. et al. (2012). Structure and dynamics of the M_3 muscarinic acetylcholine receptor. *Nature* 482 (7386): 552–556.

Lagane, B., Gaibelet, G., Meihoc, E. et al. (2000). Role of sterols in modulating the human mu-opioid receptor function in *Saccharomyces cerevisiae*. *J. Biol. Chem.* 275 (43): 33197–33200.

Lappano, R. and Maggiolini, M. (2011). G protein-coupled receptors: novel targets for drug discovery in cancer. *Nat. Rev. Drug Discov.* 10 (1): 47–60.

Lebon, G., Warne, T., Edwards, P.C. et al. (2011). Agonist-bound adenosine A_{2A} receptor structures reveal common features of GPCR activation. *Nature* 474 (7352): 521–525.

Liao, M., Cao, E., Julius, D., and Cheng, Y. (2013). Structure of the TRPV1 ion channel determined by electron cryo-microscopy. *Nature* 504 (7478): 107–112.

Linares, D.M., Geertsma, E.R., and Poolman, B. (2010). Evolved *Lactococcus lactis* strains for enhanced expression of recombinant membrane proteins. *J. Mol. Biol.* 401 (1): 45–55.

Link, A.J., Skretas, G., Strauch, E.M. et al. (2008). Efficient production of membrane-integrated and detergent-soluble G protein-coupled receptors in *Escherichia coli*. *Protein Sci.* 17 (10): 1857–1863.

Liu, W., Chun, E., Thompson, A.A. et al. (2012). Structural basis for allosteric regulation of GPCRs by sodium ions. *Science* 337 (6091): 232–236.

Lundstrom, K., Wagner, R., Reinhart, C. et al. (2006). Structural genomics on membrane proteins: comparison of more than 100 GPCRs in 3 expression systems. *J. Struct. Funct. Genom.* 7 (2): 77–91.

McCusker, E.C., Bane, S.E., O'Malley, M.A., and Robinson, A.S. (2007). Heterologous GPCR expression: a bottleneck to obtaining crystal structures. *Biotechnol. Progr.* 23 (3): 540–547.

Magalhaes, A.C., Dunn, H., and Ferguson, S.S. (2012). Regulation of GPCR activity, trafficking and localization by GPCR-interacting proteins. *Br. J. Pharmacol.* 165 (6): 1717–1736.

Mali, P., Yang, L., Esvelt, K.M. et al. (2013). RNA-guided human genome engineering via Cas9. *Science* 339 (6121): 823–826.

Manglik, A., Kruse, A.C., Kobika, T.S. et al. (2012). Crystal structure of the micro-opioid receptor bound to a morphinan antagonist. *Nature* 485 (7398): 321–326.

Marreddy, R.K.R., Geertsma, E.R., Permentier, H.P. et al. (2010). Amino acid accumulation limits the overexpression of proteins in *Lactococcus lactis*. *PLoS ONE* 5 (4): e10317.

Marreddy, R.K.R., Pinto, J.P.C., Wolters, J.C. et al. (2011). The response of *Lactococcus lactis* to membrane protein production. *PLoS ONE* 6 (8): e24060.

Massey-Gendel, E., Zhao, A., Boulting, G. et al. (2009). Genetic selection system for improving recombinant membrane protein expression in *E. coli*. *Protein Sci.* 18 (2): 372–383.

Massotte, D. (2003). G protein-coupled receptor overexpression with the baculovirus-insect cell system: a tool for structural and functional studies. *Biochim. Biophys. Acta* 1610 (1): 77–89.

Min, K.C., Jin, Y., and Hendrickson, W.A. (2011). Large-scale production of a disulfide-stabilized constitutively active mutant opsin. *Protein Expr. Purif.* 75 (2): 236–241.

Mori, K. (2009). Signalling pathways in the unfolded protein response: development from yeast to mammals. *J. Biochem.* 146 (6): 743–750.

Nannenga, B.L. and Baneyx, F. (2011). Reprogramming chaperone pathways to improve membrane protein expression in *Escherichia coli*. *Protein Sci.* 20 (8): 1411–1420.

Norholm, M.H., Light, S., Virkki, M.T. et al. (2012). Manipulating the genetic code for membrane protein production: what have we learnt so far? *Biochim. Biophys. Acta* 1818 (4): 1091–1096.

Ohya, T., Hayashi, T., Kiyama, E. et al. (2008). Improved production of recombinant human antithrombin III in Chinese hamster ovary cells by ATF4 overexpression. *Biotechnol. Bioeng.* 100 (2): 317–324.

O'Malley, M.A., Mancini, J.D., Young, C.L. et al. (2009). Progress toward heterologous expression of active G-protein-coupled receptors in *Saccharomyces cerevisiae*: linking cellular stress response with translocation and trafficking. *Protein Sci.* 18 (11): 2356–2370.

Otterstedt, K., Larsson, C., Bill, R.M. et al. (2004). Switching the mode of metabolism in the yeast *Saccharomyces cerevisiae*. *EMBO Rep.* 5 (5): 532–537.

Paila, Y.D. and Chattopadhyay, A. (2010). Membrane cholesterol in the function and organization of G-protein coupled receptors. *Subcell. Biochem.* 51: 439–466.

Palczewski, K., Kumasaka, T., Hori, T. et al. (2000). Crystal structure of rhodopsin: a G protein-coupled receptor. *Science* 289 (5480): 739–745.

Panneels, V., Kock, I., Krijnse-Locker, J. et al. (2011). *Drosophila* photoreceptor cells exploited for the production of eukaryotic membrane proteins: receptors, transporters and channels. *PLoS ONE* 6 (4): e18478.

Park, S.H., Das, B.B., Casagrande, F. et al. (2012). Structure of the chemokine receptor CXCR1 in phospholipid bilayers. *Nature* 491 (7426): 779–783.

Piali, L., Froidevaux, S., Hess, P. et al. (2011). The selective sphingosine 1-phosphate receptor 1 agonist ponesimod protects against lymphocyte-mediated tissue inflammation. *J. Pharmacol. Exp. Ther.* 337 (2): 547–556.

Pinto, J.P.C., Kuipers, O.P., Marreddy, R.K.R. et al. (2011). Efficient overproduction of membrane proteins in *Lactococcus lactis* requires the cell envelope stress sensor/regulator couple CesSR. *PLoS ONE* 6 (7): e21873.

Rajagopal, S., Rajagopal, K., and Lefkowitz, R.J. (2010). Teaching old receptors new tricks: biasing seven-transmembrane receptors. *Nat. Rev. Drug Discov.* 9 (5): 373–386.

Rasmussen, S.G., Choi, H.J., Rosenbaum, D.M. et al. (2007). Crystal structure of the human beta2 adrenergic G-protein-coupled receptor. *Nature* 450 (7168): 383–387.

Rasmussen, S.G., DeVree, B.T., Zou, Y. et al. (2011a). Crystal structure of the beta2 adrenergic receptor-Gs protein complex. *Nature* 477 (7366): 549–555.

Rasmussen, S.G., Choi, H.J., Fung, J.J. et al. (2011b). Structure of a nanobody-stabilized active state of the beta(2) adrenoceptor. *Nature* 469 (7329): 175–180.

Ratnala, V.R., Swarts, H.G., VanOostrum, J. et al. (2004). Large-scale overproduction, functional purification and ligand affinities of the His-tagged human histamine H1 receptor. *Eur. J. Biochem.* 271 (13): 2636–2646.

Reeves, P.J., Callewaert, N., Contreras, R., and Khorana, H.G. (2002). Structure and function in rhodopsin: high-level expression of rhodopsin with restricted and homogeneous N-glycosylation by a tetracycline-inducible N-acetylglucosaminyltransferase I-negative HEK293S stable mammalian cell line. *Proc. Natl. Acad. Sci. U.S.A.* 99 (21): 13419–13424.

Ren, H., Yu, D., Ge, B. et al. (2009). High-level production, solubilization and purification of synthetic human GPCR chemokine receptors CCR5, CCR3, CXCR4 and CX3CR1. *PLoS ONE* 4 (2): e4509.

Rosenbaum, D.M., Cherezov, V., Hanson, M.A. et al. (2007). GPCR engineering yields high-resolution structural insights into beta2-adrenergic receptor function. *Science* 318 (5854): 1266–1273.

Routledge, S.J., Mikaliunaite, L., Patel, A. et al. (2016). The synthesis of recombinant membrane proteins in yeast for structural studies. *Methods* 95: 26–37.

Russ, A.P. and Lampel, S. (2005). The druggable genome: an update. *Drug Discov. Today* 10 (23–24): 1607–1610.

Salom, D., Cao, P., Sun, W. et al. (2012). Heterologous expression of functional G-protein-coupled receptors in *Caenorhabditis elegans*. *FASEB J.* 26 (2): 492–502.

Salon, J.A., Lodowski, D.T., and Palczewski, K. (2011). The significance of G protein-coupled receptor crystallography for drug discovery. *Pharmacol. Rev.* 63 (4): 901–937.

Sarkar, C.A., Dodevski, I., Kenig, M. et al. (2008). Directed evolution of a G protein-coupled receptor for expression, stability, and binding selectivity. *Proc. Natl. Acad. Sci. U.S.A.* 105 (39): 14808–14813.

Sarramegna, V., Talmont, F., Demange, P., and Milon, A. (2003). Heterologous expression of G-protein-coupled receptors: comparison of expression systems from the standpoint of large-scale production and purification. *Cell. Mol. Life Sci.* 60 (8): 1529–1546.

Schlegel, S., Klepsch, M., Gialama, D. et al. (2010). Revolutionizing membrane protein overexpression in bacteria. *Microb. Biotechnol.* 3 (4): 403–411.

Schlinkmann, K.M., Honegger, A., Tureci, E. et al. (2012a). Critical features for biosynthesis, stability, and functionality of a G protein-coupled receptor uncovered by all-versus-all mutations. *Proc. Natl. Acad. Sci. U.S.A.* 109 (25): 9810–9815.

Schlinkmann, K.M., Hillenbrand, M., Rittner, A. et al. (2012b). Maximizing detergent stability and functional expression of a GPCR by exhaustive recombination and evolution. *J. Mol. Biol.* 422 (3): 414–428.

Schlyer, S. and Horuk, R. (2006). I want a new drug: G-protein-coupled receptors in drug development. *Drug Discov. Today* 11 (11–12): 481–493.

Schrodinger, LLC (2010). The PyMOL Molecular Graphics System, Version 1.3r1.

Shimamura, T., Shiroishi, M., Weyand, S. et al. (2011). Structure of the human histamine H1 receptor complex with doxepin. *Nature* 475 (7354): 65–70.

Shiroishi, M., Tsujimoto, H., Makyio, H. et al. (2012). Platform for the rapid construction and evaluation of GPCRs for crystallography in *Saccharomyces cerevisiae*. *Microb. Cell Fact.* 11: 78.

Siffroi-Fernandez, S., Giraud, A., Lanet, J., and Franc, J.L. (2002). Association of the thyrotropin receptor with calnexin, calreticulin and BiP. Efects on the maturation of the receptor. *Eur. J. Biochem.* 269 (20): 4930–4937.

Singh, S., Gras, A., Fiez-Vandal, C. et al. (2008). Large-scale functional expression of WT and truncated human adenosine A_{2A} receptor in *Pichia pastoris* bioreactor cultures. *Microb. Cell Fact.* 7: 28.

Siu, F.Y., He, M., de Graaf, C. et al. (2013). Structure of the human glucagon class B G-protein-coupled receptor. *Nature* 499 (7459): 444–449.

Skretas, G. and Georgiou, G. (2008). Engineering G protein-coupled receptor expression in bacteria. *Proc. Natl. Acad. Sci. U.S.A.* 105 (39): 14747–14748.

Skretas, G. and Georgiou, G. (2009). Genetic analysis of G protein-coupled receptor expression in *Escherichia coli*: inhibitory role of DnaJ on the membrane integration of the human central cannabinoid receptor. *Biotechnol. Bioeng.* 102 (2): 357–367.

Skretas, G., Carroll, S., DeFrees, S. et al. (2009). Expression of active human sialyltransferase ST6GalNAcI in *Escherichia coli*. *Microb. Cell Fact.* 8: 50.

Skretas, G., Makino, T., Varadarajan, N. et al. (2012). Multi-copy genes that enhance the yield of mammalian G protein-coupled receptors in *Escherichia coli*. *Metab. Eng.* 14 (5): 591–602.

Smyth, D.R., Mrozkiewicz, M.K., McGrath, M.J. et al. (2003). Crystal structures of fusion proteins with large-affinity tags. *Protein Sci.* 12 (7): 1313–1322.

Souza, C.M., Schwabe, T.M., Pichler, H. et al. (2011). A stable yeast strain efficiently producing cholesterol instead of ergosterol is functional for tryptophan uptake, but not weak organic acid resistance. *Metab. Eng.* 13 (5): 555–569.

Spyropoulos, I.C., Liakopoulos, T.D., Bagos, P.G., and Hamodrakas, S.J. (2004). TMRPres2D: high quality visual representation of transmembrane protein models. *Bioinformatics* 20 (17): 3258–3260.

Standfuss, J., Xie, G., Edwards, P.C. et al. (2007). Crystal structure of a thermally stable rhodopsin mutant. *J. Mol. Biol.* 372 (5): 1179–1188.

Standfuss, J., Edwards, P.C., D'Antona, A. et al. (2011). The structural basis of agonist-induced activation in constitutively active rhodopsin. *Nature* 471 (7340): 656–660.

Stanley, P. and Chaney, W. (1985). Control of carbohydrate processing: the lec1A CHO mutation results in partial loss of *N*-acetylglucosaminyltransferase I activity. *Mol. Cell Biol.* 5 (6): 1204–1211.

Subei, A.M. and Cohen, J.A. (2015). Sphingosine 1-phosphate receptor modulators in multiple sclerosis. *CNS Drugs* 29 (7): 565–575.

Tate, C.G., Whiteley, E., and Betenbaugh, M.J. (1999). Molecular chaperones stimulate the functional expression of the cocaine-sensitive serotonin transporter. *J. Biol. Chem.* 274 (25): 17551–17558.

Thompson, A.A., Liu, W., Chun, E. et al. (2012). Structure of the nociceptin/orphanin FQ receptor in complex with a peptide mimetic. *Nature* 485 (7398): 395–399.

Tucker, J. and Grisshammer, R. (1996). Purification of a rat neurotensin receptor expressed in *Escherichia coli*. *Biochem. J.* 317 (Pt 3): 891–899.

Valderrama-Rincon, J.D., Fisher, A.C., Merritt, J.H. et al. (2012). An engineered eukaryotic protein glycosylation pathway in *Escherichia coli*. *Nat. Chem. Biol.* 8 (5): 434–436.

Valkonen, M., Penttila, M., and Saloheimo, M. (2003). Effects of inactivation and constitutive expression of the unfolded-protein response pathway on protein production in the yeast *Saccharomyces cerevisiae*. *Appl. Environ. Microbiol.* 69 (4): 2065–2072.

Venkatakrishnan, A.J., Deupi, X., Lebon, G. et al. (2013). Molecular signatures of G-protein-coupled receptors. *Nature* 494 (7436): 185–194.

Wagner, S., Bader, M.L., Drew, D., and de Gier, J.W. (2006). Rationalizing membrane protein overexpression. *Trends Biotechnol.* 24 (8): 364–371.

Wagner, S., Baars, L., Ytterberg, A.J. et al. (2007). Consequences of membrane protein overexpression in *Escherichia coli*. *Mol. Cell. Proteomics* 6 (9): 1527–1550.

Wang, X., Corin, K., Baaske, P. et al. (2011). Peptide surfactants for cell-free production of functional G protein-coupled receptors. *Proc. Natl. Acad. Sci. U.S.A.* 108 (22): 9049–9054.

Warne, T., Chirnside, J., and Schertler, G.F. (2003). Expression and purification of truncated, non-glycosylated turkey beta-adrenergic receptors for crystallization. *Biochim. Biophys. Acta* 1610 (1): 133–140.

Warne, T., Serrano-Vega, M.J., Baker, J.G. et al. (2008). Structure of a beta1-adrenergic G-protein-coupled receptor. *Nature* 454 (7203): 486–491.

Warne, T., Serrano-Vega, M.J., Tate, C.G., and Schertler, G.F. (2009). Development and crystallization of a minimal thermostabilised G protein-coupled receptor. *Protein Expr. Purif.* 65 (2): 204–213.

Warne, T., Moukhametzianov, R., Baker, J.G. et al. (2011). The structural basis for agonist and partial agonist action on a beta(1)-adrenergic receptor. *Nature* 469 (7329): 241–244.

Wegierski, T., Billy, E., Nasr, F., and Filipowicz, W. (2001). Bms1p, a G-domain-containing protein, associates with Rcl1p and is required for 18S rRNA biogenesis in yeast. *RNA* 7 (9): 1254–1267.

Weiss, H.M. and Grisshammer, R. (2002). Purification and characterization of the human adenosine A_{2A} receptor functionally expressed in *Escherichia coli*. *Eur. J. Biochem.* 269 (1): 82–92.

White, J.F., Noinaj, N., Shibata, Y. et al. (2012). Structure of the agonist-bound neurotensin receptor. *Nature* 490 (7421): 508–513.

Wildt, S. and Gerngross, T.U. (2005). The humanization of N-glycosylation pathways in yeast. *Nat. Rev. Microbiol.* 3 (2): 119–128.

Wu, B., Chien, E.Y., Mol, C.D. et al. (2010). Structures of the CXCR4 chemokine GPCR with small-molecule and cyclic peptide antagonists. *Science* 330 (6007): 1066–1071.

Xiao, S., White, J.F., Betenbaugh, M.J. et al. (2013). Transient and stable expression of the neurotensin receptor NTS1: a comparison of the baculovirus-insect cell and the T-REx-293 expression systems. *PLoS ONE* 8 (5): e63679.

Xu, F., Wu, H., Katritch, V. et al. (2011). Structure of an agonist-bound human A_{2A} adenosine receptor. *Science* 332 (6027): 322–327.

Yang, J.P., Cirico, T., Federico, K. et al. (2011). Cell-free synthesis of a functional G protein-coupled receptor complexed with nanometer scale bilayer discs. *BMC Biotechnol.* 11: 57.

Young, C.L., Britton, Z.T., and Robinson, A.S. (2012). Recombinant protein expression and purification: a comprehensive review of affinity tags and microbial applications. *Biotechnol. J.* 7 (5): 620–634.

Zhang, Z., Austin, S.C., and Smyth, E.M. (2001). Glycosylation of the human prostacyclin receptor: role in ligand binding and signal transduction. *Mol. Pharmacol.* 60 (3): 480–487.

Zhang, L., Salom, D., He, J. et al. (2005). Expression of functional G protein-coupled receptors in photoreceptors of transgenic *Xenopus laevis*. *Biochemistry* 44 (44): 14509–14518.

Zhao, Q. and Wu, B.L. (2012). Ice breaking in GPCR structural biology. *Acta Pharmacol. Sin.* 33 (3): 324–334.

Zocher, M., Zhang, C., Rasmussen, S.G. et al. (2012). Cholesterol increases kinetic, energetic, and mechanical stability of the human beta2-adrenergic receptor. *Proc. Natl. Acad. Sci. U.S.A.* 109 (50): E3463–E3472.

3

Glycosylation

Maureen Spearman[1], Erika Lattová[2], Hélène Perreault[2], and Michael Butler[1]

[1] Department of Microbiology, University of Manitoba, Winnipeg, Manitoba, Canada
[2] Department of Chemistry, University of Manitoba, Winnipeg, Manitoba, Canada

3.1   Introduction

Glycosylation is the most prevalent form of posttranslational modification of recombinant glycoproteins. Defining glycosylation as a critical quality attribute by regulatory agencies is due to glycan's substantial influence on the function and properties of many therapeutic proteins. Solubility, thermal stability (Zheng et al. 2011), serum half-life (Wright and Morrison 1997; Egrie et al. 2003), protease resistance (Sareneva et al. 1995), immunogenicity (Noguchi et al. 1995; Chung et al. 2008), aggregation (Rodriguez et al. 2005), and efficacy (Sola and Griebenow 2010) are all affected by the glycosylation profile. A highly relevant example of the structure–function relationship of glycosylation is how small modification of glycosylation of monoclonal antibodies (Mabs) can dramatically affect their function. Glycosylation, in general, is typically heterogeneous, with variation in amount of glycosylation at specific sites (macroheterogeneity) and also variation in structure of the glycans at each site (microheterogeneity). The goal in the production of biotherapeutics is to limit this variation to provide batch-to-batch consistency, but also to ensure the efficacy of the drug and limit immunogenic reactions.

The introduction of quality by design (QbD) standards in cell culture manufacturing process development (Marasco et al. 2014) and identification of glycosylation as a critical quality attribute emphasizes the need to understand the relationship of culture conditions and media components with glycosylation outcomes. Numerous factors affect the glycosylation profile, including the glycoprotein itself, the culture conditions, clonal variation, and the recombinant host species. While many of these factors can be controlled to reduce glycosylation heterogeneity, other means such as genetically modified host cell systems, glycoprotein processing inhibitors, and chemoenzymatic

Bioprocessing Technology for Production of Biopharmaceuticals and Bioproducts, First Edition.
Edited by Claire Komives and Weichang Zhou.

and *in vitro* modification, result in highly defined and restricted sets of glycans. N-linked glycosylation is the predominant type of glycosylation and therefore has been the focus of much of the work, but many glycoproteins also have smaller O-linked glycans.

Improved methods of glycosylation analysis have aided in defining the structure–function relationship of glycans in biotherapeutics. Advancements in a variety of techniques and introduction of high throughput technologies will allow for better quality control and design of newer therapies.

3.2   Types of Glycosylation

3.2.1   N-linked Glycans

N-linked glycans are added to protein as a co-translational modification in the endoplasmic reticulum (ER). Oligosaccharide chains are transferred "enbloc" to the nascent polypeptide chain from a lipid-oligosaccharide precursor, dolichol (Dol) phosphate (Kornfeld and Kornfeld 1985; Lennarz 1987; Stanley et al. 2009). Localized on the cytosolic face of the ER membrane, dolichol phosphate acts as a carrier molecule during the initial synthesis of the oligosaccharide. A high mannose precursor molecule consisting of two *N*-acetylglucosamine (GlcNAc) molecules with a branched nine mannose (Man) structure and three additional glucose (Glc) residues is produced through the sequential enzymatic addition of sugars via specific glycosyltransferases that use nucleotide sugars (UDP-GlcNAC, GDP-Man) as substrates. Following the addition of the fifth mannose (Dol-P-P-GlcNac2-Man5), the entire Dol-glycan structure is enzymatically flipped from the cytosol into the endoplasmic reticulum where an additional four Man residues are added through a Dol-P-Man carrier and three Glc molecules are then added using Dol-P-Glc. This glycan structure is then transferred from the dolichol carrier to the protein by the oligosaccharyltransferase complex.

N-linked oligosaccharides are attached through the R-group nitrogen of asparagine at specific sites or "sequons" on the protein. Sequons are comprised of a three amino acid sequence of asparagine-X-serine/threonine (Asn-X-Ser/Thr), where X can be any amino acid with the exception of proline, and the third amino acid can be either Ser or Thr. Not all sequons are glycosylated, but typically a specific glycoprotein will have a set of sequons that are glycosylated. Variation in occupied glycan sequons from one molecule to the next leads to macroheterogeneity within a specific glycoprotein. Glucose residues on the high mannose structure are necessary to influence proper folding of the glycoprotein and interact with the calnexin/calreticulin system (Bedard et al. 2005). Improperly folded Man9 glycans interact with UGGT (UDP-glucose: glycoprotein glucosyltransferase) that re-glucosylates the glycan, allowing it to re-bind to calreticulin to aid in folding (Hirano et al. 2015). Once properly folded, the glucose residue is removed in the ER along with one mannose residue producing a GlcNAc2Man8 glycan (Bischoff and Kornfeld 1983).

Microheterogeneity, the variation in the structure of individual N-linked glycans, occurs during the processing reactions, as the glycoprotein passes through the Golgi (Figure 3.1). Some oligosaccharides are maintained as high mannose glycans following the removal of the glucose residues in the ER, but most glycans are trimmed by mannosidases to reduce the GlcNAc2Man8 structure down to the core molecule GlcNAc2Man5. Additional sugar residues are added along with removal of two more mannose residues in the Golgi to convert the structures into complex glycans that contain a terminal triplet of sugars consisting of *N*-acetylglucosamine, galactose (Gal) and sialic acid (*N*-acetylneuraminic acid, NeuAc or *N*-acetylglycolylneuraminic acid, Neu5Gc). In mammalian cells, fucose (Fuc) can also be added in an α1,6 linkage to the first core GlcNAc residue or outer arm GlcNAc residue. A diverse array of glycosyltransferases catalyzes the addition of sugar residues using nucleotide-sugars as substrates and thus introduce branching and heterogeneity into the complex glycans. Complex glycans are normally defined by their antennarity with two (biantennary), three (triantennary), or four

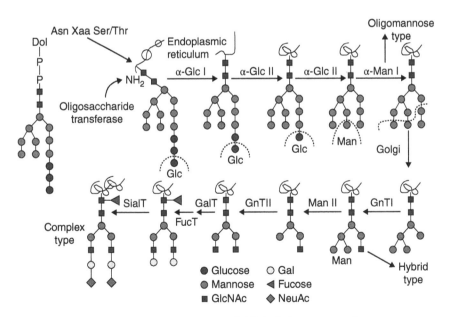

Figure 3.1 Glycoprotein processing reactions of N-linked glycans create the microheterogeneity within the oligosaccharide by removal of the glucose and mannose residues and attachment of additional sugar residues (GlcNAc, Gal, NeuAc, and fucose). Source: Butler 2006. Adapted with permission from Springer Nature.

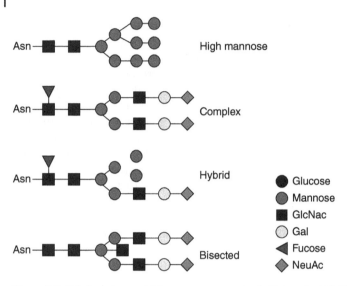

Figure 3.2 N-linked glycans: high mannose, complex (with terminal GlcNAc, Gal, NeuAc), hybrid form with one arm high mannose and one arm with complex monosaccharides, and complex with bisecting GlcNAc.

(tetraantennary) branches. Sugar residues can also have different linkages, such as sialic acid that are generally attached via an α2,3 or α2,6 linkage. Control of the transferase activity resulting in sugar addition and branching patterns and the ultimate microheterogeneity of the glycans in a specific glycoprotein are not well understood, but pools of intracellular nucleotide sugars, Golgi transit time, pH, temperature, and enzyme levels are known to affect the glycosylation (Butler 2006; Stanley 2011). There are three common types of oligosaccharides: high mannose, complex, and hybrid (Figure 3.2). But within each of these groups, large amounts of variation can occur in antennarity, chain length, additional sugars (e.g. fucose, lactosamine), as key enzymatic control points of sugar additions influence the final structure. For example, addition of a bisecting GlcNAc between biantennary arms will prevent the addition of α1,6-fucose to the core GlcNAc adjacent to the sequon (Schachter 1986). Single modifications of sugar residues such as this can have important impact on function. For example, the effector function of monoclonal antibodies is significantly reduced with the addition of a fucose residue to the core GlcNAc (Shields et al. 2002). Recent work has found that competition between glycosyltransferase enzymes affect the antennarity of the glycan (McDonald et al. 2014).

3.2.2 O-linked Glycans

Recombinant proteins may also contain O-linked glycans attached through an *N*-acetylgalactosamine (GalNAc) residue (also known as mucin *O*-glycans)

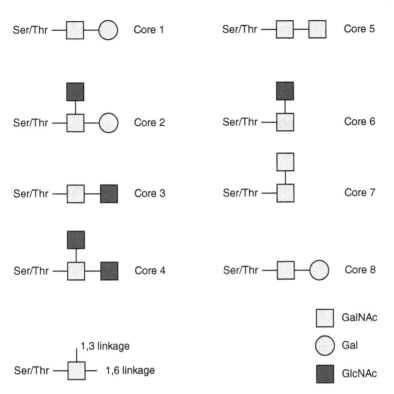

Figure 3.3 O-linked glycans, linked to serine or threonine, are grouped into eight groups based on their structural core. Additional sugars such as lactosamine (Gal-GlcNAc) can be added to elongate the glycans and then terminated with sugars such as fucose and sialic acid.

that are structurally different than N-linked glycans, but include many of the same monosaccharides and linkages. O-linked glycosylation is a posttranslational event, involving only glycosyltransferases. There are eight groups of mucin glycans that are based on their core structures linked through an GalNAc residue via a family of GalNAc transferase enzymes to the oxygen of Ser or Thr (Figure 3.3) (Tian and Ten Hagen 2009). The consensus sequences of O-linked glycans are not as defined as the N-linked glycans, but regions of proline and alanine along with Ser and Thr may affect glycan addition. In the synthesis pathway, glycosyltransferases add individual sugar residues to produce the oligosaccharide on the protein using nucleotide sugar substrates, which differs from the *en bloc* transfer and processing in N-linked glycosylation. Core structures can be extended with additional sugars structures such as polylactosamine residues (GlcNAc-Gal) and fucose, and the terminal sugar residues are commonly sialic acids (e.g. NeuAc). Although control of the biosynthetic pathway is not well understood, patterns of sugar addition may be controlled by localization of specific transferases through the Golgi compartments (Hang and Bertozzi 2005). Other types of O-linked glycosylation can occur, but these are highly specific with alternate sugars attached at other amino acid sites (Moremen et al. 2012).

3.3 Factors Affecting Glycosylation

Glycosylation is generally known to have effects on stability of the protein, such as inhibition of proteolysis, solubility, aggregation, clearance, uptake, pharmacokinetics, and efficacy (Li and D'Anjou 2009). For example, recombinant proteins with atypical glycosylation are cleared from circulation via a mannose-binding receptor (Liu et al. 2011) and asialoglycoprotein receptor (ASGPR) in the liver (Wright et al. 2000). Immunoglobulin G (IgG) with complex glycan (with or without core fucose) has higher serum half-life in mice than IgG with high mannose or hybrid glycan, suggesting that the terminal sugar residues (Gal-GlcNAc) also contributes to maintaining the IgG in circulation (Kanda et al. 2007b). Hyperglycosylated and sialylated erythropoietin (EPO) has increased bioactivity and half-life (Egrie et al. 2003). One of the most notable examples is the enhanced effector function-associated IgG glycans lacking core fucosylation (Ferrara et al. 2006a).

The necessity of function-appropriate glycosylation has driven production and monitoring of biotherapeutics to the use of QbD to ensure the end product in manufacturing results in batch-to-batch consistency in glycan micro- and macroheterogeneity (Horvath et al. 2010; Elliott et al. 2013). Thus, the understanding of how culture conditions affect glycosylation is of great importance.

In a batch culture, some parameters may be controlled in a bioreactor to a predetermined set point. These include pH, oxygen, stirring rate, and temperature, but most components of the media change continuously over time through nutrient depletion or metabolic by-product accumulation. Any of these parameters can affect the intracellular glycosylation process that in turn can alter the glycan profile of a secreted glycoprotein during the course of the culture. Some of the most prominent effects of culture parameters on protein glycosylation are described.

3.3.1 Nutrient Depletion

The gradual depletion of nutrients during the course of culture has been found to cause time-dependent glycosylation effects. A reduction of glycan site occupancy leading to macroheterogeneity of γ-interferon has been shown during the batch culture of Chinese hamster ovary (CHO) cells (Curling et al. 1990). Glucose or glutamine is the key nutrient likely to cause this effect (Hayter et al. 1992; Jenkins and Curling 1994; Nyberg et al. 1999). Several studies have demonstrated the relationship between low glucose and reduced glycosylation. A glucose-limited chemostat CHO with a high proportion of nonglycosylated γ-interferon was rapidly restored to normal levels of glycosylation by pulsed addition of glucose (Hayter et al. 1992). Reduced site-occupancy of N-glycans in IgG was shown in mouse myeloma cells at low glucose concentration (<0.5 mM) (Tachibana et al. 1997). Under-glycosylation and abnormal truncated glycans of viral proteins occurred in glucose-depleted CHO cells (Davidson and Hunt 1985). Glucose starvation may result in an intracellular depleted state or a shortage of glucose-derived precursors of glycans giving rise to a higher proportion of high mannose structures (Rearick et al. 1981). Depletion of glucose has been shown to reduce the size of lipid-linked oligosaccharide from the typical dolichol-GlcNAc2-Man9-Glc3 structure to a shorter dolichol-GlcNAc2-Man5, correlating with a reduction of overall glycosylation of a Mab (Liu et al. 2014a). Glucose depletion increases the amount of nonglycosylated Mabs and also reduces galactosylation, and this was correlated with reduced GDP-sugars and UDP-hexosamines (Villacres et al. 2015). Intracellular nucleotide-sugars are the immediate precursors for protein glycosylation in the endoplasmic reticulum and modifications to glycosylation microhetergeneity can be related to decreased concentrations of these precursors in cultures with low levels or depleted glucose or glutamine (Nyberg et al. 1999; Chee Furng Wong et al. 2005; Kochanowski et al. 2008; Gonzalez-Leal et al. 2011; Villacres et al. 2015). Also, studies have suggested that there is a relationship between macroheterogeneity and adenylate energy charge, with a reduction in site-specific glycosylation (Kochanowski et al. 2008; Villacres et al. 2015). Glutamine depletion causes decreased intracellular concentration of UDP-GlcNAc due to reduced formation of glucosamine phosphate via the glutamine: fructose 6-phosphate amidotransferase (GFAT) reaction or through glucose depletion by reduced synthesis of UTP (Nyberg et al. 1999). Both UTP and glucosamine-phosphate are the key precursors of UDP-GlcNAc, which in turn is required for glycosylation of proteins. Cells adapted to glutamine-free growth have elevated UDP-sugars and increased antennarity (Taschwer et al. 2012). Supplementation of cultures with nucleotide precursors such as glucosamine and uridine for UDP-GlcNAc synthesis (Baker et al. 2001; Hills et al. 2001) and uridine and galactose for UDP-Gal synthesis (Gramer et al. 2011; Grainger and James 2013), galactose, glucosamine, and ManNAc (Wong et al. 2010b) have been successful in increasing the nucleotide sugar availability and affecting glycosylation. However, other factors such as the nucleotide sugar biosynthetic enzymes and transporters may be limiting in some cell lines (Fan et al. 2015). Some recent work in HEK293 cells indicates that the hexosamine biosynthetic pathway and the GlcNAc transferases that control glycan branching may influence the uptake of glutamine and essential amino acids under low nutrient conditions, allowing increased cell growth (Abdel Rahman et al. 2015).

Specific transferase reactions can be related to the availability of substrates or cofactors. For example, terminal galactosylation of an antibody can be modulated by the supply of uridine, manganese chloride, or galactose in the media (Gramer et al. 2011; Grainger and James 2013). Galactosyltransferase requires manganese for activity and manganese addition alone can increase galactosylation in later-day cultures (Crowell et al. 2007). Clearly, the control of microheterogeneity by nutrient feeding is critical in producing consistent biopharmaceuticals, avoiding significant batch-to-batch product variation, and diminished therapeutic efficacy. However, each cell line and clone may have specific metabolic characteristics that can affect protein glycosylation (van Berkel et al. 2009). Metabolic analyses of culture parameters, along with high throughput glycan analyses, are necessary to monitor factors that affect glycosylation to ensure consistent product quality (see Section 3.5).

As metabolic modeling studies are used to improve productivity and cell growth during process development (Gerdtzen 2012; Sellick et al. 2015), detailed analysis is also necessary of to understand how shifts in nutrients can affect key cellular metabolites in glycan synthesis and glycosylation outcomes. Several modeling systems for glycosylation have been developed (Hossler et al. 2006; Krambeck et al. 2009; Jimenez del Val et al. 2011; Moseley et al. 2011; St. Amand et al. 2014; Rathore et al. 2015). However, because glycosylation is such a complex system, affected by many aspects of cellular metabolism, some of which may yet be unknown, this is a difficult task. Also, because glycosylation is highly variable from between cell types and clones (van Berkel et al. 2009), new parameters must be established for each one. A recent study has linked a modeling system to levels of extracellular nutrient. An *in silico* metabolic model of nucleotide sugar donors, linked to a glycosylation model for the Fc of IgG, allows the prediction of glycosylation from levels of extracellular metabolites (Jedrzejewski et al. 2014). The model gave good results compared to experimental data in a murine hybridoma. Metabolic flux analysis in continuous culture has been useful in understanding the affects of altering the key nutrients, glucose, and glutamine. Under low glutamine conditions reduced sialylation and antennarity of human chorionic gonadotropin were correlated with reduced UPD-GlcNAc (Burleigh et al. 2011). Two metabolic models, a dynamic model based on flux analysis and the GLYCOVIS software model (Hossler et al. 2006), were used to study the relationships of glutamine, glucose, pH, ammonia, and glycosylation in batch cultures (Aghamohseni et al. 2014). Reducing glutamine levels can lower glucose consumption along with cell yield, but increases galactosylation and sialylation. Ammonia was correlated with UDP-GlcNAc synthesis and pH with inhibition of sialylation. Also, multivariate analysis of several bioprocess parameters is able to show a correlation between media components and glycosylation outcomes in murine hybridoma cultures (Rathore et al. 2015). Modeling may also be used as a useful tool in defining specific sites for glycoengineering (Spahn et al. 2015).

3.3.2   Fed-batch Cultures and Supplements

Feeding strategies in batch cultures have been highly successful for bioprocess development, allowing product yields in excess of $5\,g\,l^{-1}$ (Wurm 2004). Numerous commercial feed media are now available. The principle is that a slow feed of essential nutrients such as glucose and glutamine can allow an efficient cellular metabolism with minimal accumulation of by-products. This allows cells to reach a high cell density ($>10^7\,ml^{-1}$), which can be maintained over a prolonged period. From this, target proteins can be secreted into the culture medium over an extended time period. Critical to this strategy is the maintenance of nutrients at a low concentration set point. This can be achieved by several methods. For example, stoichiometric feeding can be based upon the projected cell growth over daily time points (12–24 hours). Alternatively, a dynamic nutrient-feeding regime may be based upon regular sample analysis over shorter time periods (1–2 hours). The choice of method will directly affect the variability of nutrient concentration about the predefined set point. However, it is important to recognize that the critical nutrient concentrations needed to maximize product titers may be different from those that affect glycosylation. Also, significant fluctuations around the set point may result in short-term depletion of critical nutrients that could cause variability of product glycosylation. Early fed-batch strategies were designed to ensure that the concentrations of these key nutrients do not decrease below a critical level that could compromise productivity (Xie and Wang 1997). In these experiments, the concentrations of key nutrients were optimized. Glucose in glycolysis and glutamine in glutaminolysis was the focus in order to maximize growth and productivity, but optimization of glycosylation was not a key factor. Later, it was realized that there were critical nutrient concentrations for glycosylation. From a series of fed-batch culture studies, the lower levels of nutrients for the production of γ-interferon from CHO cells were found to be 0.1 mM glutamine and 0.7 mM glucose, and below these concentrations reduced sialylation and increased hybrid and high mannose glycans occurred (Chee Furng Wong et al. 2005). Further studies coupling gene expression and glycosylation analysis found 0.5 mM glutamine were optimal for maintaining consistent product quality (Wong et al. 2010a). A recent fed-batch study included amino acid analysis and nucleotide sugar analysis has found that glucose and glutamine levels correlate with intracellular levels of UDP-Gal (Fan et al. 2015). Also, in one cell line, the presence of Man5 structures was attributed to insufficient glutamine leading to a reduction in UDP-GlcNac biosynthesis, but in another cell line with a different medium, increased Man5 was linked to ammonia effects on GnT1 activity and transporter efficiency. These differences in responses of two different cell lines emphasize the need to analyze metabolic responses and customize feed media to ensure optimum glycosylation.

Osmolality is important to control in bioreactors and feed media and correlates with increased Man5 structures in late days of cultures (Pacis et al. 2011), but this could be abrogated with the addition of MnCl2. However, the authors suggest that high osmolality affect the ammonia levels of the culture, which in turn affects intracellular pH, thereby altering the activity of glycosyltransferases, and increasing the amount of Man5, although further work is needed to verify this relationship. Manganese is required by many glycosyltransferases for efficient activity. Osmolality is also inversely related to the de-fucosylated glycan of Mabs in a hybridoma cell line, regardless of the type of compound used to control osmolality (Konno et al. 2012).

Metabolic analysis using stable isotope tracers coupled with biochemical analysis (Dean and Reddy 2013), metabolic flux analysis (Ahn and Antoniewicz 2011, 2012; Templeton et al. 2013), metabolic profiling (Sellick et al. 2011), and gene expression analysis (transciptomics) (Schaub et al. 2010) has given a better understanding of utilization of key media components such as glutamine, glucose, and amino acids during different phases of fed batch cultures. However, fine tuning our understanding of key metabolic pathways and how they interplay and affect glycosylation is key to maximizing growth and productivity, but maintaining appropriate glycosylation.

3.3.3 Specific Culture Supplements

It may be possible to target specific intermediates of the metabolic network to enhance certain critical attributes of glycosylation. For example, glucosamine is a precursor for UDP-GlcNAc, which is an important intracellular nucleotide-sugar substrate for a range of GlcNAc transferases present in the Golgi. Supplementation of cultures with glucosamine can lead to an elevated level of intracellular UDP-GlcNAc, particularly if added in conjunction with uridine (Baker et al. 2001). The elevated level of UDP-GlcNAc has been shown to enhance the antennarity of glycan structures produced in baby hamster kidney (BHK) cells, probably through a stimulation of the specific GlcNAc TIV and TV (Gawlitzek et al. 1998; Grammatikos et al. 1998; Valley et al. 1999). However, the phenomenon is not universal for all cell lines. Baker et al. found enhanced antennarity following glucosamine supplementation in CHO cells but not NS0 cells that produced the same recombinant glycoprotein (Baker et al. 2001). However, in all cases, an elevated UDP-GlcNAc appears to cause a decrease in sialylation, which may be explained metabolically by the inhibition of CMP-sialic acid transport (Pels Rijcken et al. 1995). Also, metabolic analysis has found that supplementation of glucosamine in CHO cultures can cause decreased cell growth due to the diversion of acetyl-CoA to produce GlcNAc, and also shift glycosylation to nongalactosylated structures in Mabs (Blondeel et al. 2015). Supplementation with GlcNAc yielded the same impact on glycosylation, but without the affect on growth. Tachibana et al. (1997) showed that by replacing glucose with GlcNAc in the media of a human hybridoma, they were able to change the glycosylation profile of the hypervariable region of the light chain of the antibody, with a 10× increase in affinity binding to its antigen. This change was associated with a lower level of sialylation of the light chain glycans.

Metal ions influence glycosylation. The addition of manganese and iron increased site occupancy of tissue plasminogen activator, t-PA (Gawlitzek et al. 2009). Higher levels of manganese have been found to increase overall galactosylation and also increase sialylation, as well as increase site occupancy of huEPO (Crowell et al. 2007). The addition of up to 12× the normal concentrations of manganese chloride, uridine, and galactose synergistically increased the levels of galactosylation of a monoclonal antibodies produced in two CHO cell lines without an effect on antibody titer (Gramer et al. 2011). Copper is required as an enzymatic co-factor to ensure an efficient oxidative (Nargund et al. 2015) and may increase the sialic acid content of antibodies (Ryll 2003), although this effect is not universal (McCracken et al. 2014).

One of the strategies that can be used to enhance sialylation specifically is to supply a precursor for CMP-NANA. Because of its location in the metabolic pathway and high cell membrane permeability, *N*-acetyl mannosamine (ManNAc) is the best candidate for this, as sialic acid or CMP-sialic acid has poor membrane permeability. Supplementation of ManNAc to CHO cell culture was shown to increase the intracellular pool of CMP-sialic acid and enhance the sialylation of γ-interferon (Gu and Wang 1998). However, this strategy has not increased sialylation in other culture systems. An increase of intracellular CMP-NeuAc was reported as a result of ManNAc feeding to either CHO or NS0 cells but without an increase in overall sialylation of a recombinant protein (TIMP1) (Baker et al. 2001). However, this did change the ratio of *N*-glycolylneuraminic acid (NeuGc)/*N*-acetylneuraminic acid from 1 : 1 to 1 : 2 in the NS0 cells. Another study found that several culture parameters (sodium butyrate addition, reduced culture temperature, high pCO_2, and use of sodium carbonate instead of sodium hydroxide for pH control) reduced the amount of Neu5Gc incorporated into a recombinant protein produced in CHO cells (Borys et al. 2010).

The addition of unusual sugars, tagatose, and sucrose (up to 70 mM) was found to increase high mannose glycans (15% and 37%, respectively) and reduce fucosylation to the same degree, with an associated increase in antibody-dependent cell-mediated cytotoxicity of IgG (Hossler et al. 2014). Although cell density was reduced at higher concentrations, titer was not affected. Others have found an increase in high mannose glycans when fructose and galactose were substituted for glucose, and this increased by almost 30% with the addition of 16 µM manganese (Surve and Gadgil 2015). This effect could be abrogated with the addition of glucose to the media.

A high asparagine concentration was shown to decrease galactosylation of a recombinant glycoprotein and was correlated with increased amounts of ammonia in the culture (McCracken et al. 2014). The authors proposed that ammonia changes the intracellular pH, thereby modifying galactosyltransferase activity (see below).

3.3.4 Ammonia

Ammonia (NH_3) or the ammonium ion (NH_4^+) accumulates in culture as a by-product from cellular glutamine metabolism (glutaminolysis) and from the nonenzymatic decomposition of glutamine in the medium. It has been known for some time that the accumulated ammonia is inhibitory to cell growth, an effect that is greater at high pH (Doyle and Butler 1990).

The major effect that ammonia exerts on glycosylation is to decrease terminal sialylation, a phenomenon observed in the production of a variety of recombinant proteins (Andersen and Goochee 1994; Gawlitzek et al. 1995; Zanghi et al. 1998; Yang and Butler 2000). For EPO, this results in a shift of all glycoforms to higher pI values in the presence of high levels of ammonia (Yang and Butler 2000). It was reported that even a low level of ammonia (2 mM) could affect the sialylation of O-glycans (Andersen and Goochee 1995).

There are two possible mechanisms to explain this effect of ammonia. The first of these is the observed increase in the UDP-GlcNAc/UTP ratio that is brought about by the enhanced incorporation of ammonia into glucosamine, a precursor for UDP-GlcNAc. This nucleotide sugar competes with the transport of CMP-NeuAc into the Golgi and would therefore decrease the available substrate concentration for sialylation. The second plausible mechanism for the effect of ammonia is that it raises the pH of the Golgi thereby shifting from the optimal pH of galactosyltransferase and sialyltransferase enzymes (Valley et al. 1999; Gawlitzek et al. 2000; Hong et al. 2010). However, another study has found that ammonia decreases the expression of sialyltransferase, galactosyltransferase, and a CMP-sialic acid transporter (Chen and Harcum 2006). Substitution of glutamate for glutamine reduced ammonia concentrations, thereby increasing galactosylation, increased productivity, but inhibited cell growth (Hong et al. 2010). Feed media containing glutamine substitutes (pyruvate, a glutamine dipeptide, glutamate, and wheat gluten hydrolysate) were found to reduce ammonia production by over 45% and increase productivity, while maintaining glycosylation (Kim et al. 2013).

3.3.5 pH

Under adverse external pH conditions, the internal pH of the Golgi is likely to change resulting in a reduction of the activities of key glycosylating enzymes. The pH of the medium was shown to have some effect on the distribution of glycoforms of IgG secreted by a murine hybridoma (Rothman et al. 1989b). Borys et al. (1993) related the extracellular pH to the specific expression rate and glycosylation pattern of recombinant mouse placental lactogen-I (mPL-I) by CHO cells, with the maximum specific mPL-I expression rates occurring between pH 7.6 and 8.0. The level of site occupancy was maximum within this pH range, decreasing at lower (<6.9) and higher (>8.2) pH values. In another study, maximum site occupancy was found at pH 7.05 for *t*-PA, with decreasing site occupancy as the pH decreased to 7.15 (Gawlitzek et al. 2009). The use of sodium hydroxide, instead of sodium carbonate, to control pH can also reduce the amount of the antigenic sialic acid, NeuGc, by 33% (Borys et al. 2010).

3.3.6 Oxygen

The control of the dissolved oxygen (DO) level is an important parameter to maintain optimal metabolism and growth of producer cells in bioprocesses (Jan et al. 1997; Heidemann et al. 1998). The effect of DO on the glycosylation of a recombinant protein from CHO cells was observed by a changing glycoform profile (Chotigeat et al. 1994). In particular, an increase in sialyltransferase was observed at high oxygen levels that translated into increased sialylation of recombinant follicle stimulating hormone (FSH).

By controlling DO set points between 1% and 100% air saturation in a culture, the terminal galactosylation of an immunoglobulin (IgG) was changed significantly with a gradual decrease in the digalactosylated glycans (G2) from 30% at the higher oxygen level to around 12% under low oxygen conditions. The mechanism for the effect of DO is unclear, but it is unlikely to be due to a change in the activity of the transferase enzyme (Kunkel et al. 1998). The explanation could be that reduced DO causes a decline in the availability of UDP-Gal, or the galactosylation might be sterically impeded by the early formation of an inter-heavy chain disulfide. It has been proposed previously that the timing and rate of formation of the disulfide bond in the hinge region of IgG is critical to the extent of galactosylation (Rademacher et al. 1996). The redox potential of the culture has also been shown to affect disulfide bond formation, which is critical for glycosylation at sites close to disulfide bridges (Allen et al. 1995; Gawlitzek et al. 2009). DO levels in CHO cultures at extreme levels (3%, 10%, and 200%) resulting in a reduction in fucsylation of EPO from 80% fucosylation at 50–100% to 75–77% fucosylation (Restelli et al. 2006), but antennarity and sialylation were not significantly different.

3.3.7 Host Cell Systems

CHO cell lines have become the predominant host cell system for the production of recombinant proteins because of similar glycosylation patterns to human cell lines (Dumont et al. 2015). However, they do not express the α2,6 sialyltransferase enzyme, and attach sialic acid in only the 2,3 linkage (Xu et al. 2011), and are also lacking the α1-3/4 fucosyltransferase and bisecting *N*-acetylglucosamine transferase (Grabenhorst et al. 1999).

Cell type	*N*-Glycan	Structure
Yeast	High mannose	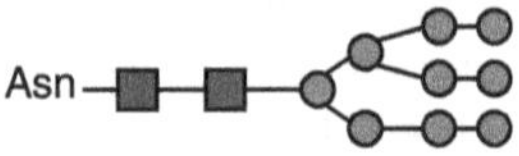
Insect	Fucosylated core	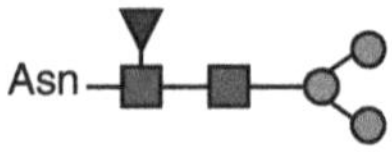
Plant	Xylosylated and α1,3 fucosylated core	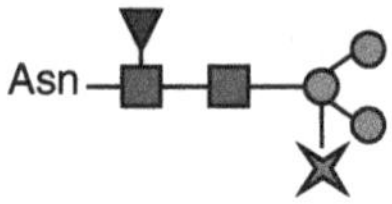
Mammalian	Complex biantennary	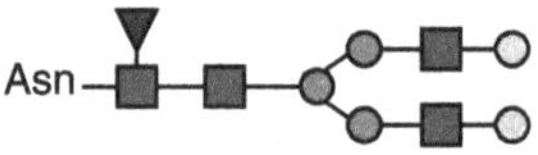

Figure 3.4 Comparison of N-linked glycosylation in different cell expression systems.

Variations in N-linked glycosylation exist between mammalian species and greater variability occurs across eukaryotes (Brooks 2004) (Figure 3.4). Hence, recombinant glycoproteins produced in nonhuman cell lines may contain glycosylated structures that are antigenic in humans (Durocher and Butler 2009). Mouse cells (NSO) (Jenkins et al. 1996) and CHO cells (Bosques et al. 2010) have an α1,3-galactosyltransferase enzyme that produces glycans containing Galα1,3-Gal residues that are immunogenic in humans. Human cells do not produce NeuGc, a terminal neuraminic acid, but it is common in all nonprimate mammalian cells, thus potentially immunogenic (Ghaderi et al. 2012). Plants have additional sugars (xylose and α1,3-linked fucose) (Webster and Thomas 2012); insects have short mannose glycans (paucimannose) (Ahn et al. 2008a). Yeasts produce predominantly high mannose structures (Chiba and Akeboshi 2009). In order to reduce the antigenicity attributed to nonhuman glycosylation, genetic engineering is used to humanize glycosylation in recombinant glycoproteins produced in other cell types (Wildt and Gerngross 2005; Chiba and Akeboshi 2009; Durocher and Butler 2009; Castilho et al. 2011; Loos and Steinkellner 2012).

Human cell lines for expression of recombinant proteins are now available as well and have the potential for wider use (Fliedl et al. 2015). HEK 293 (human embryonic kidney) is approved for the production of human coagulation factor VIII with a similar glycosylation pattern to the natural plasma protein, lacking alpha-gal and NeuGc (Kannicht et al. 2013), but has glycosylation patterns with high sialylation and no terminal GalNAc. Therefore, CHO was suggested as a better expression system for increased half-life (Bohm et al. 2015). HT-1080 (human fibrosarcoma cell line) have received regulatory approval for the production of recombinant proteins for the treatment of lysosomal storage diseases (Dumont et al. 2015). EPO produced in HT-1080 is lacking O-acetylation of sialic acid and the potentially immunogenic NeuGc, which is found at low levels in EPO produced in CHO cells (Shahrokh et al. 2011). However, there are also other significant differences, such as 20–30% higher amounts of tetraantennary glycans and a decrease in polylactosamine extensions. PER.C6 (adenovirus transformed human embryonic retinoblasts) produces IgG with glycosylation patterns very similar to human IgG, and with no immunogenic glycans (Jones et al. 2003). A human recombinant alpha-1-antitrypsin produced in PER.C6 showed higher fucosylation and a higher degree of glycan branching but similar sialylation compared to plasma-derived molecule (Wang et al. 2013). RS, an immortalized human epithelial cell line, produced EPO with much high levels of sialylation than CHO but no NeuGc (Fliedl et al. 2015). HKB-11, a hybrid of human embryonic kidney cells and Burkitts lymphoma cells, and CAP, adenovirus-immortalized primary human amniocytes, are two other cell lines that produce human-like glycosylation patterns of recombinant proteins (Swiech et al. 2015). With many human cell lines in the initial stages of testing, more commercial recombinant proteins with human glycosylation patterns are predictable. However, low production levels are still a problem in some of these cell lines (Fliedl et al. 2015).

3.3.8 Other Factors

Reducing culture temperature is commonly used to boost productivity in cultures. Lowering the culture temperature to 33 or 31 °C increased the site occupancy of tPA by up to 4% (Gawlitzek et al. 2009). Glycosylation microheterogeneity of EPO remained unaffected up to a reduction of 32 °C, but below 32 °C, sialylation and tetraantennary sialylated glycoforms were reduced (Ahn et al. 2008b). Metabolic analysis under mild hypothermic conditions (32 °C) found reduced nucleotide sugar donors and reduced glycosyltransferase expression, correlated with a reduction in galactosylation (Sou et al. 2015). The authors suggest that flux analysis could be used to establishing better fed media under hypothermic conditions, and glycosylation changes may be cell line-specific. However, recombinant β-interferon under hypothermic conditions (32 °C) in a perfusion culture had elevated sialylation compared to batch culture (Rodriguez et al. 2010). The use of microcarriers and low temperature in a bioreactor increases sialylation and decreases fucosylation (Nam et al. 2008). Temperature shift can also be used to reduce the amount of NeuGc, an antigenic sialic acid, but timing of the temperature shift is also a factor. Lowering the culture temperature in early stationary phase compared to exponential phase, reduced NeuGc by 59% (Borys et al. 2010).

3.4 Modification of Glycosylation

3.4.1 siRNA and Gene Knockout/Knockin

Functional roles of glycosylation have now been directly related to structure through refinements in methods of structural analysis. A goal now is to modify the microheterogeneity and macroheterogeneity of biotherapeutic glycoproteins to optimize their efficacy through increased activity and half-life and reduce immunogenicity without compromising productivity (Sinclair and Elliott 2005). Although modifications to culture conditions can be used to achieve this in many cases, particular functional aspects of biotherapeutics are known to require very specific glycosylation structures and modifications and thus require a more targeted approach using glycoengineering (Dicker and Strasser 2015) and other methods.

A very important example of how structure is important in function is the glycosylation of monoclonal antibodies (IgG). N-linked glycans at Asn297 of the Fc region of the heavy chain affect effector functions and hold the Fc region in the required conformation (Deisenhofer 1981; Nose and Wigzell 1983; Arnold et al. 2007). Loss of glycan allows a "closing" between the CH_2 domains causing a reduction in FcγR-binding capacity (Krapp et al. 2003; Ferrara et al. 2006b). Furthermore, α1,6 fucosylation of the core GlcNAc of the glycan limits the interactions of the Fc region with the FcγIIIa receptor and also reduces the effectiveness of the Mab for eliciting an ADCC response with NK and peripheral blood monocytes (Shields et al. 2002). An increase in bisecting GlcNAc with a glycolengineered increase in N-acetylglucosaminyltransferase III (GnTIII) can increase ADCC by reducing α1,6 core fucosylation (Umana et al. 1999; Shinkawa et al. 2003). Galactosylation of the Mab is also important in complement-dependent cytotoxicity, CDC (Hodoniczky et al. 2005). Therefore, much research has focused on modification of the glycosylation pathway in the host cells to produce Mabs that are afucosylated or reduced in fucosylation, with increased bisecting GlcNAc and have consistent galactosylation patterns. Gene knockout of FUT8, coding for the α1,6 fucosyltransferase enzyme, has successfully created a CHO host cell line that produces Mabs with reduced fucose (Yamane-Ohnuki et al. 2004). Recent work using specific genome-editing techniques, zinc fingers, transcription activator-like effector nuclease (TALENs), and the clustered regularly interspaced palindromic repeat (CRISPR)-Cas9, to inactivate the GDP-fucose transporter (SLC35C1) in CHO cells and eliminated fucosylation of IgG1 (Chan et al. 2015). Others have reduced the fucosylation by using a tandem sequence interfering RNA (siRNA) to prevent the expression of the 1,6 core fucosyltransferase (FUT8) and GDP-mannose 4,6 dehydratase (GMD, an enzyme required for the production of GDP-fucose) (Imai-Nishiya et al. 2007). However, GMD knockout alone can also reduce fucosylated Mab (Kanda et al. 2007a). An example of unique glycoengineering introduces a diversion pathway for GDP-fucose precursors into GDP-rhamnose through the bacterial oxidoreductase GDP-6-deoxy-D-lyxo-4-hexulose reductase (RMD) pathway, creating a dead-end product, and thereby reducing core fucosylation (von Horsten et al. 2010; Ogorek et al. 2012).

Sialylation of IgG can attenuate the ADCC response (Scallon et al. 2007) and glycoengineering of CHO and NS0 may be used to express a secretable sialidase A to trim the sialic acid from recombinant proteins (Naso et al. 2010). Sialylated IgG has anti-inflammatory activity (Kaneko et al. 2006), and the ADCC response is undesirable when used in anti-inflammatory therapy. In such cases, enhanced sialylation is desirable. Transfection of a gene for α2,6-sialyltranferase into a recombinant CHO cell line increased the sialylation of a bispecific antibody (Onitsuka et al. 2012), thus allowing an increased anti-inflammatory response. This potentially may be used in other recombinant glycoproteins requiring increased sialylation. Several other studies have increased sialylation of recombinant proteins by over expression of sialyltransferase (Bragonzi et al. 2000; Fukuta et al. 2000; Jassal et al. 2001) or by overexpressing the CMP-sialic acid transporter (Wong et al. 2006) or transfection of genes for key steps in sialic acid synthesis, GNE (uridine diphosphate-N-acetyl glucosamine 2-epimerase) and MNK (N-acetyl mannosamine kinase) (Son et al. 2011). Co-transfection of human α2,6-sialyltransferase 1 (ST6) and β1,4-galactosyltransferase 1 (GT) in CHO cells resulted in higher sialylation of IgG than transfection of ST6 alone, and with 85% as the α2,6-sialic acid, a linkage which is not present in CHO cells (Raymond et al. 2015).

Targeting recombinant proteins to mannose receptors for increased uptake of vaccine and other recombinant proteins requires the production of high mannose glycans (Van Patten et al. 2007; Sealover et al. 2013). Several studies have used CHO cell lines deficient in mannosyl (alpha-1,3-)-glycoprotein beta-1,2-N-acetylglucosaminyltransferase (Mgat1, also called GnTI) to produce high mannose glycans (Van Patten et al. 2007; Zhong et al. 2012), and recent work has used zinc-finger nucleases (ZFNs) to knockout the enzyme resulting in predominantly Man5 glycans (Sealover et al. 2013). Alternate glycoengineering uses 293GnT1(−) cells with a predominant GlcNAc2-Man5 glycan and further genetic modification with expression of an endo T enzyme, which can cleave this structure to one remaining GlcNac (Meuris et al. 2014). Further additions of galactose and sialic acid result in a simplified trisaccharide glycan, which would be useful in creating more homogeneous biopharmaceuticals where functional glycosylation such as ADCC is not necessary.

Another important example of glycoengineering is the hyperglycosylation of recombinant human erythropoietin (huEPO) to increase stability, activity, solubility, and serum half-life (Egrie et al. 2003) reviewed in Sinclair and Elliott (2005). Two additional N-linked glycosylation sites were introduced into the protein backbone, which increased the sialic acid content, allowing for a three-fold increase in half-life and higher potency with a reduced frequency of administration in patients (Elliott et al. 2004). Introduction of a silkworm gene coding for the 30Kc6 protein into CHO, or addition of the protein into culture media significantly increases sialylation and fucosylation of huEPO as well as productivity (Park et al. 2012). Sialylation can also be increased by a coexpression of α2,3 sialyltransferase and β1,4 galatosyltransferase (Jeong et al. 2008).

Additions of novel glycans to biotherapeutics can be used to alter their pharmacokinetics properties. Using protein engineering, the polysialic acid domain of NCAM was added to a single-chain Fv antibody fragment to increase the serum half-life of the protein by 30-fold (Chen et al. 2012).

Glycoengineering in other host systems (Loos and Steinkellner 2012), such as yeast (Wildt and Gerngross 2005; Chiba and Akeboshi 2009; Irani et al. 2015), plants (Ko et al. 2009; Castilho et al. 2011; Webster and Thomas 2012), insects (Ahn et al. 2008a; Aumiller et al. 2012; Geisler et al. 2015; Mabashi-Asazuma et al. 2015), and also bacteria (Chiba and Jigami 2007) have been used to humanize the glycosylation of recombinant proteins. This often results in more homogenous glycosylation, which may be advantageous for specific therapeutic functions. Additionally, established mutant CHO cell lines with modified glycosylation have been studied as useful hosts for recombinant glycoproteins (Zhang et al. 2013).

Newer computational modeling systems may be used to better predict glycoengineering outcomes (Spahn et al. 2015), with the potential for more refined glycosylation host systems. Also, the use of more efficient knockout systems such as ZFNs (Sealover et al. 2013) and CRISPR-Cas9 technology (Chan et al. 2015; Dicker and Strasser 2015; Mabashi-Asazuma et al. 2015; Wang et al. 2015) will allow very specific and stable modifications of glycosylation pathways. A recent elaborate study was performed using ZFNs gene to knockout 19 glycosyltransferase genes of CHO-K1 involved in galactosylation, sialylation, polylactosamine addition, and core α6 fucosylation using EPO as a model glycoprotein (Yang et al. 2015). Knockouts of individual and combinations of genes gave new insights into controlling glycosylation to produce more homogenous glycoproteins in these key groups of enzymes and is an important step in creating "designer" glycosylation.

3.4.2 Glycoprotein Processing Inhibitors and *In Vitro* Modification of Glycans

Modification of glycan structure can also be accomplished through the addition of glycoprotein-processing inhibitors to the culture media during production.

This can result in alternative glycan structures. Glucosidase inhibitors (castanospermine and methydeoxynojirimycin) result in GlcNAc2-Man8-9Glc1-3 glycan, mannosidase I inhibitors (deoxymannojirimycin and kifunensin) gives high mannose structures (GlcNAc2-Man8 or 9) and swainsonine, and a mannosidase II inhibitor results in a hybrid glycans. Studies with altered glycan structures of IgG found the glucosidase and mannosidase I inhibitors could enhance ADCC or FcγIIIa binding (Kanda et al. 2007b; Zhou et al. 2008; van Berkel et al. 2010; Yu et al. 2012). X-ray crystallography identified a more open conformation of the Fc region with kifunensin-treated high mannose glycosylation that is proposed to facilitate receptor binding (Crispin et al. 2009). A decrease in the core fucosylation in the presence of the mannosidase I and glucosidase inhibitors may explain the increased ADCC response. Swainsonine, a mannosidase II inhibitor producing hybrid glycans, does not enhance ADCC, which may be partially due to the presence of fucose on the hybrid glycan (Rothman et al. 1989a). A small molecule of ionophore, monensin, has been shown to increase the high mannose content of IgG glycans to 35% under perfusion conditions without an effect on cell culture and productivity (Pande et al. 2015).

Chemoenzymatic modifications of glycans *in vitro* have been useful in defining the structure–function relationship of recombinant protein glycosylation (Wang and Lomino 2012; Lomino et al. 2013). *N*-glycans can be trimmed to the core GlcNAc, and then glycosylated sequentially, using appropriate nucleotide sugars and glycosyltransferase enzymes. This approach may still produce some heterogeneity. Alternatively, homogeneous glycosylation of rituximab has been accomplished by a combination of chemical and enzymatic synthesis of the glycan with an *en bloc* transfer of activated glycan oxazolines to the protein (Huang et al. 2012). The resulting product is highly homogeneous.

Antibody glycans can be modified enzymically as they are attached to the solid support of a Protein A column. This has the advantage that sequential modifications can be performed without the need for purification at each stage. The resulting single glycoform antibody can be eluted from the column under the conditions that are normally used during purification (Tayi and Butler 2014).

3.5 Glycosylation Analysis

The development and refinement of glycan structural analysis has been a key asset in defining the relationship between glycoprotein structure and function of biopharmaceuticals. The necessity for batch-to-batch consistency of glycosylation in production of biotherapeutics and the optimization of glycosylation also requires detailed structural analysis in a QbD approach. Many methods of glycan analysis are available, each with its own advantage and disadvantage and levels of accuracy and sensitivity. The choice of method may depend on the required expertise, cost of equipment, amount of material required, and number of samples. Structural analysis can vary from whole glycoprotein, glycopeptide, or isolated glycan and each will provide different types of information. Whole glycoprotein analysis will provide information on macroheterogeneity, whereas glycan analysis will define microheterogeneity. Glycopeptide analysis can provide both. If glycans are removed from the peptide backbone, they must be isolated, fluorescently labeled and then identified using various types of chromatography such as high-performance liquid chromatography (HPLC), high-performance anion exchange chromatography with pulsed amperometric detection (HPAEC-PAD), capillary electrophoresis (CE), and FACE (fluorophore-assisted carbohydrate electrophoresis). Liquid chromatography and CE can be coupled inline with mass spectroscopy (MS) analysis and tandem mass spectroscopy (MS/MS) can give detailed information on oligosaccharide sequence and branching (Butler and Perreault 2010; Zaia 2013). Automation and high throughput analysis are now the focus of many techniques to reduce labor and increase analytical output, particularly for industrial purposes where QbD requires larger numbers of analysis, and a quick turn around time (Shubhakar et al. 2015). The combination of ion mobility with mass spectrometry has also become a useful tool in glycosylation analysis (Harvey et al. 2015). Some analytical approaches can provide structural information without cleavage of glycans from the peptides or proteins, allowing site-specific analysis. For other methods of glycan analysis, removal of the glycan from the peptide backbone is necessary.

3.5.1 Release of Glycans from Glycoproteins

Glycan removal typically involves enzymatic (peptide N-glycosidases, endoglycosidases) or chemical release (hydrazinolysis, β-elimination) of the glycans from the protein. PNGaseF (peptide N-glycosidase F) is the most widely used enzyme to hydrolyze the β-aspartylglycosamine linkage between the GlcNAc of the N-glycan and the side chain amine of asparagine (Tarentino et al. 1985). Hydrolysis liberates amino oligosaccharides, which are slowly hydrolyzed by water and converted into forms with a reducing terminal GlcNAc. PNGaseF has the ability to release all types of N-linked oligosaccharides, except for glycans with fucose linked at the alpha 3-position in GlcNAc attached to asparagine, as commonly found in plants and insect glycoproteins. In this case, PNGaseA can be successfully applied (O'Neill 1996). EndoH recombinant glycosidase is a more specific enzyme that cleaves a glycosidic bond within the N-glycan chitobiose (GlcNAc–GlcNAc) core, leaving terminal GlcNAc attached to the protein (Maley et al. 1989). However, EndoH glycosidase cleaves only high-mannose and some hybrid N-oligosaccharides, but not highly processed complex N-oligosaccharides.

Hydrazinolysis is the most used chemical cleavage method and can be applied successfully for releasing both O- and N-linked glycans (Patel and Parekh 1994). This is the only chemical deglycosylation method to release glycans with the intact reducing terminal GlcNAc, which is necessary for glycan labeling. However, hydrazinolysis suffers from disadvantages such as requiring extra caution during the preparation of reagent and sample, low overall reaction yields, toxicity, and volatility of hydrazine and results in destruction of the protein chain. Alkaline β-elimination is another widely used chemical approach for the release of O-glycans (Carlson 1968). To prevent isomerization or degradation of the carbohydrates by "peeling" reactions, hydrolysis is performed in the presence of a reducing agent, which leads to the formation of the reduced (alditol) form of the glycans. This method is more effective than hydrazinolysis; however, the lack of chromophore in these glycans makes their detection more difficult.

3.5.2 Derivatization of Glycans

For many methods of glycan analysis, attachment of a chromophore to the reducing end is necessary to enhance detection (for details see Section 6.5.2). Among the most popular derivatization reagents often used are 2-aminoacridine (AMAC) (Camilleri et al. 1998), 3-acetamido-6-aminoacridine (Charlwood et al. 2000), 2-aminopyridine (Okamoto et al. 1997), 4-aminobenzoic acid ethyl ester (Suzuki et al. 1996), 2-aminobenzoic acid (2-AA) (Anumula and Dhume 1998), and 2-aminobenzamide (2-AB) (Bigge et al. 1995). Reductive amination requires a clean-up step after derivatization to remove reducing reagent (e.g. $NaBH_3CN$) and the free fluorescent label prior to analysis. Methods based on hydrazide chemistry have also been described where clean up can be avoided (Lattova and Perreault 2013).

3.6 Methods of Analysis

3.6.1 Lectin Arrays

Lectin microarrays are a relatively new analytical technique modeled after DNA microarrays and use the unique carbohydrate binding patterns of plant and animal-derived lectins to distinguish differences in carbohydrate structure (Kuno et al. 2005; Hirabayashi et al. 2015). Arrays of lectins are immobilized on solid surfaces, such as epoxy-coated glass slides. Each lectin recognizes and binds a specific carbohydrate structure or group of carbohydrate structures with varying intensities (K_d of 10^{-7}–10^{-3} M) (Hirabayashi et al. 2011) (Figure 3.5). The microarray detects differences in overall oligosaccharide structure and antennarity by the binding intensity of the fluorescently labeled glycoprotein(s) to the array. Comparison of microarrays to LC–MS found they are unable to distinguish between some similar oligosaccharides, such

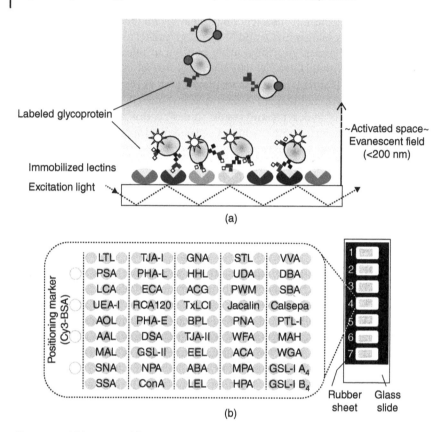

Figure 3.5 (a) Diagram of fluorescently labeled glycoproteins attaching to immobilized lectins on a lectin microarray. (b) An example of a layout of 45 lectins in a LecChip from GP Biosciences. Source: Hirabayashi et al. 2011. Adapted with permission from John Wiley & Sons.

as different sialic acids and some linkages, and thus this method is not yet able to give good structural data, and the reproducibility is low (18% error at 95% confidence) (Hayes et al. 2012). Recent work comparing GlycoScope lectin-affinity microarray with hydrophobic interaction liquid chromatography (HILIC) and MS analysis found very comparable analysis among the different methods (<6% coefficient of variation) with several lots of IgG (Cook et al. 2015). However, the microarray analysis can only give semiquantitative results as the glycans are grouped into subclasses and not identified individually. IgGs generally have simpler glycan patterns than other recombinant glycoproteins, so this may be ideal for rapid IgG glycan screening. The advantage of this analysis is that it does not require costly equipment, it has high throughput capability, and there is little preparation of the sample required (intact proteins can be analyzed) (Hirabayashi et al. 2011). Also, both O-linked and N-linked glycosylation can be analyzed simultaneously.

3.6.2 Liquid Chromatography

3.6.2.1 HILIC Analysis

A standard, widely used quantitative method for glycoprotein glycan analysis uses glycan release, fluorescent labeling, and identification using normal phase HPLC (NP-HPLC), also known as HILIC separation (Guile et al. 1996; Rudd et al. 2001; Domann et al. 2007; Royle et al. 2007). The first step of analysis is removal of the glycan from the protein either chemically (hydrazinolysis) or by using the broad specificity glycosidase such as PNGaseF. PNGaseF digestion, the most preferred method, can be done in-solution for purified glycoproteins, but an in-gel PNGaseF digestion method can be used to analyze glycoprotein bands separated on SDS-PAGE.

Glycans are labeled with a fluorescent tag at the reducing end most commonly with 2-AB. HILIC analysis separates the glycans using a gradient with increasing hydrophilicity on an amide column. The elution is compared to the retention time of a dextran ladder (increasing linear glucose units (GU)) using a fifth-order polynomial standard curve of GU values versus retention time (Figure 3.6). The GU values of the glycans are then compared to a standard database of glycans (such as https://www.hsls.pitt.edu/obrc/index.php?page= URL1263237902) (Guile et al. 1996; Royle et al. 2007; Campbell et al. 2008, 2015) for preliminary identification based on their GU value. Because many glycans may have similar GU values, further analysis is required for structural confirmation. Digestion arrays of exoglycosidases can confirm structure and linkages by their removal of specific sugar residues monitored by shifts in retention times or GU values. MS analysis may also be combined with HILIC for verification of structure (Tharmalingam et al. 2013). HILIC has several advantages over other current techniques for glycan analysis including quantitative values at the femtomole range with stoichiometric 2-AB labeling; the ability to analyze for neutral and sialic acid-containing glycans at the same time; provides information on sequence and linkage specificity (Royle et al. 2008). Early methods used a 5 μm particle size amide column (Guile et al. 1996; Rudd et al. 2001). New methods using amide columns with smaller particle size of 3 μm for HPLC (Melmer et al. 2010) or 1.7 μm with a UPLC system (Clarke et al. 2009; Ahn et al. 2010; Zauner et al. 2011) can reduce run times from 180 minutes to 60 or 30 minutes, respectively, which is a major advantage in higher throughput analysis. The resolution with the smaller particle-sized columns is much improved over the 5 μm particle-sized columns and allows separation of the G1F and the Man6 isomers that are often hard to resolve (Clarke et al. 2009).

Newer high throughput variations of HILIC have recently been developed, which allows for increased numbers of samples and a reduced time

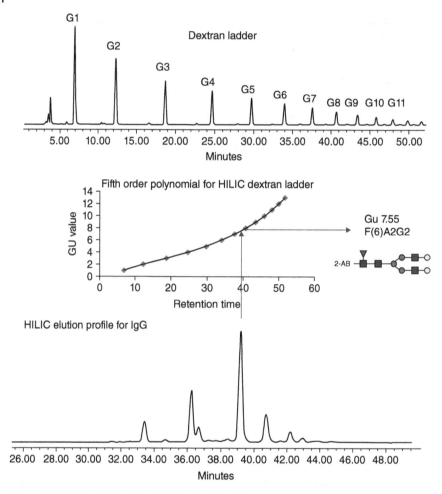

Figure 3.6 HILIC analysis of 2-AB labeled N-linked glycans. A dextran ladder is used to create a fifth-order polynomial standard curve, which is used to determine the GU (glucose unit value) for each sample peak. A structure can then be assigned from a database based on the GU value.

of analysis, which is essential in industry for culture monitoring. A detailed comparison of some of these methods is available (Royle et al. 2008; Doherty et al. 2012). A method using in-gel blocks captures the glycoprotein in a 96-well gel plate format, followed by a transfer to a washing plate. The gels are treated and washed and PNGase F added in two successive steps to ensure adequate uptake, followed by overnight incubation. The 96-well format aids in the multiple alternate washing steps with acetonitrile and water to release the glycan from the gel. Clean-up cartridges in a 96-well format also speed the removal of the free 2-AB following the labeling step. However, this method releases all glycans in the sample; therefore, samples must be purified prior to loading on the gel (Royle et al. 2008).

Another high throughput method, specific for IgG glycan analysis, uses a robotic liquid handling system with protein A capture 96-well plate for the analysis of IgG glycans and can be completed in five hours (Doherty et al. 2012, 2013). Following direct capture of the IgG from the cultured media onto the plate, plates are washed and samples digested with PNGase F for only an hour. The glycan is then washed from the plates and concentrated for 2-AB labeling. Also, a high throughput method for IgG glycan analysis with robotic automation uses flow-through ultrafiltration plates to aid in rapid buffer exchange between steps and removal of small molecules, and a capture of the glycan on solid-supported hydrazide to allow better cleanup prior to 2-AB labeling (Stockmann et al. 2013). Recent work allowing real-time glycan analysis (within 83 minutes) uses a micro sequential injection system (µSI) for sample preparation, coupled to UPLC analysis (Tharmalingam et al. 2015). Each µSI can monitor up to four bioreactors in parallel with sampling every 12 hours, or one bioreactor more frequently.

Several methods using HILIC coupled with CE or electrospray ionization mass spectrometry (ESI-MS) analysis have been developed, allowing better separation and identification, and the potential for high throughput capabilities (Zauner et al. 2011). A semiautomated method uses HILIC interaction chromatography combined with nano-reverse-phase LC-MS/MS to analyze glycopeptides, and couples the analysis to a glycan database search (Pompach et al. 2012).

3.6.2.2 Reversed Phase (RP) and Porous Graphitic Carbon (PGC) Chromatography

RP-HPLC analysis using a C18 column can be used to separate and identify fluorescently labeled (e.g. 2-AB) glycans and can also be combined with ESI-MS (Prater et al. 2009). RP columns coupled to ion-mobility MS have been used for the analysis of nonderivatized glycans (Lareau et al. 2015). RP-nanoLC-MS with 2-AA labeling of glycans in a 96-well formal allows a high throughput analysis with attomolar sensitivity of glycans, making it useful in clone selection and routine analysis (Higel et al. 2013).

Porous graphitic carbon has been used for fluorescent-labeled glycan separation, but the inability to predict glycan structure based on monosaccharide composition makes it limiting on its own (Melmer et al. 2011). However, graphitized carbon HPLC coupled to ESI-MS of native glycans has been useful in resolving differences between $\alpha2,3$ and $\alpha2,6$-linked NeuAc residues and $\beta1,4$ and $\beta1,3$-linked Gal residues (Stadlmann et al. 2008). In tandem with RP-HPLC, porous graphitic carbon is useful in separating more hydrophilic species of glycopeptides and glycans (Lam et al. 2011), and retaining both small hydrophilic and large hydrophobic glycopeptides (Liu et al. 2014b). Also, it can be fully automated for high throughput glycoproteomics (Zhao et al. 2014) and can be used to analyze O- and N-linked glycopeptides in the same sample analysis (Stavenhagen et al. 2015).

3.6.2.3 Weak Anion Exchange (WAX) HPLC Analysis

WAX HPLC analysis can be used to provide detailed information not only on negatively charged glycan species, typically the sialic containing structures, but also sulfates or phosphates. Glycans that are fluorescently labeled with 2-AB, are separated on an anion exchange column (Prozyme Glycosep C polymeric DEAE anion exchange column 75×7.5 mm) using 0.1 M ammonium acetate buffer in 20% acetronitrile in a gradient with 20% acetonitrile (Royle et al. 2007; Doherty et al. 2012). Glycans elute as nonsialylated, mono-, di-, tri, and tetra-sialylated species, and are compared to standard, 2-AB-labeled peaks isolated from the glycoprotein, fetuin. Care must be taken to ensure there is efficient removal of the 2-AB label, or the free 2-AB may mask the peaks in the elution profile.

3.6.2.4 High pH Anion Exchange Chromatography with Pulsed Amperometric Detection (HPAEC-PAD)

HPAEC-PAD utilizes anion exchange chromatography with a specialized detector for a highly sensitive method of separation of mono- and oligosaccharides (Cataldi et al. 2000). Glycan separation is based on negative charges on the oligosaccharide created with high pH buffers (pH 12–14) that result in differential deprotonation of the hydroxyl groups. The anion exchange CarboPac columns (Dionex) contain nonporous polymer beads that are stable under the extreme pH conditions. The pulsed amperometric detector (PAD) uses a gold electrode that catalyzes the oxidation of the oligosaccharides, allowing for electrochemical detection. Deposits of oxidation products on the electrode can lead to erroneous results and consistency in analysis requires cleaning steps involving full oxidation followed by full reduction to strip contaminants (Cataldi et al. 2000). Other disadvantages are interference by other oxidizable molecules, noisy baseline, and it is not compatible with methods where high salt interferes (Dhume et al. 2008). One advantage is that it allows for analysis of glycans in the native state (nonlabeled) down to the pmol range, although fluorescent labeled glycans such as 2-AB, 2-AA, or APTS (8-aminopyrene-1,3,6-trisulfonate) can also be analyzed with increased sensitivity of detection and eliminates the need for the specialized detector (Dhume et al. 2008). For biopharmaceutical quality control purposes, HPAEC-PAD is a useful analytical system, as it is highly sensitive with high accuracy, precision, and specificity required in routine analysis of glycotherapeutics. However, peak areas are not directly proportional to absolute quantity, as the peak response is dependent upon the glycan (Cook et al. 2015). HPAEC-PAD can be used to determine lot-to-lot variability and the complex sialylation pattern of EPO from different manufacturing processes is well separated and distinguishable using this method (Kandzia et al. 2010) (Figure 3.7). HPAEC-PAD is also able to monitor both substrate and products in exoglycosidase reactions for structural determination (Weitzhandler et al. 1998). Modifications to the mobile phase allowed separation of isomers of neutral glycans and provide good resolution of sialic acid-containing glycans from monoclonal antibodies with results consistent with MS analysis (Grey et al. 2009). O-linked glycans can also be analyzed with HPAEC-PAD, following removal from the protein using beta-elimination (Stadheim et al. 2008).

3.6.3   Capillary Electrophoresis (CE)

CE is a highly specialized analytical tool that can be used to separate molecules based on different characteristics, dependent upon the type of capillary used for analysis. CE analysis of glycans in the biopharmaceutical industry has several advantages: high resolution, high sensitivity (up to 10^{-15} to 10^{-18} mol) (Kamoda and Kakehi 2006), small sample size and a relatively fast analysis, but the disadvantage is that it requires a highly specialized instrument. CE can be used for analysis of whole glycoproteins (Alahmad et al. 2011) or released oligosaccharides (reviewed in Nakano et al. 2011). Capillary zone electrophoresis (CZE) of whole glycoproteins, the most common method, separates on a charge to mass ratio. Interfacing of CE technology with ESI-MS links rapid separation of glycoproteins with the powerful identification tool of mass spectrometry, and thus is very advantageous for the biotechnology industry (Neususs and Pelzing 2009). In recent work, CZE coupled (off-line) with ESI-MS allows detections of different glycoforms within the Fc/2 and detection of NeuGc on the glycan of F(ab')2 subunits of cetuximab (Biacchi et al. 2015). Glycopeptide analysis using CZE-MS/MS allows semiquantitative analysis of glycoforms at different glycosylation sites of Mabs, and detection of NeuGc and α1,3-Gal (Gahoual et al. 2014). Capillary (SDS) gel electrophoresis (CGE) separates glycoproteins similar to SDS-PAGE, and is useful in detecting *N*- and *O*-glycan occupancy (Rustandi et al. 2013). Capillary isoelectro focusing (CIEF) separates glycoproteins on the basis of isoelectric points (pI) using carrier ampholytes that establish a pH gradient in an electric field and is useful in detection of differences in sialylation, sulfonation, and phosphorylation (Rustandi et al. 2013). Glycoproteins migrate to a position where pI is equivalent to pH, followed by elution toward a detector. A newer technique, imaged capillary isoelectric focusing (ICIEF), uses whole capillary imaging technology that eliminates the elution phase, retains high resolution and is much faster (Anderson et al. 2012). For example, separation of glycoforms of EPO can easily distinguish different sources based on the resulting profile in only five minutes (Dou et al. 2008). CE of intact glycoproteins is very useful

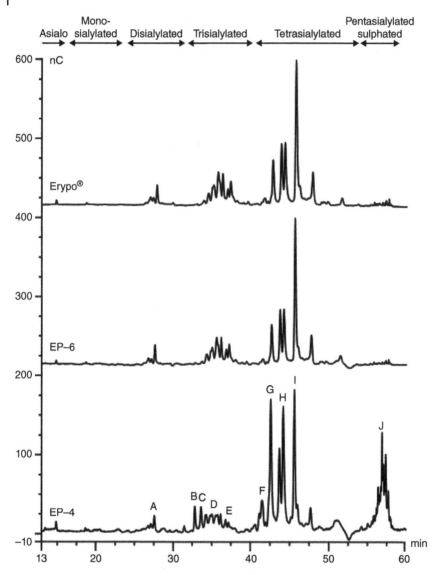

Figure 3.7 HPAEC-PAD analysis of three types of erythropoietin using a CarboPac PA200 column (Dionex) shows differential sialylation and antennarity of the products, without the need for fluorescent tagging. Source: Kandzia et al. 2010. Adapted with permission from Springer Nature.

for quality control monitoring throughout the production process and can provide information on both macroheterogeneity and microheterogeneity of the glycan with qualitative and quantitative results.

With removal of the glycan from the protein, nonlabeled glycans can be detected at low wavelength UV (<200 nm) (Ruhaak et al. 2010), but their limits of detection are relatively low. For more sensitive structural analysis, the oligosaccharide can be fluorescently labeled and detected by laser-induced fluorescence (LIF) known as CE-LIF. Elution profiles can be standardized using a dextran ladder, allowing for conversion of peak elution time to GU values, as in HILIC analysis. Fluorescent derivatization with APTS that imparts negative charge through the sulphonyl groups is commonly used, but other derivitization tags are 2- or 3-AA or 8-aminonapthaene-1,3,6-trisulfonic acid (ANTS). Newer labeling methods reduce the amount of label required and improve analysis (Szabo et al. 2010), allow increased sensitivity (10^{-18} range) (Ijiri et al. 2011), allow a one-pot reaction for releasing, labeling, and extracting of the oligosaccharide (Oyama et al. 2011), and may push the boundaries for sensitivity and high throughput analysis. Modified elution buffers can separate G0, A1F, and Man5 that can sometimes co-elute (Hamm et al. 2013). CGE analysis of fluorescence-labeled glycans coupled with a LIF detector has the advantage of very fast analysis (Domann et al. 2007), and 96-well formats have been developed that will significantly increase throughput of samples for routine analysis (Rodig et al. 2009; Szabo et al. 2011). A recent multi-lab analysis demonstrated high reliability and low variability of CE-LIF analysis (Szekrenyes et al. 2016). Direct coupling of CE to electrospray time-of-flight mass spectrometry (ESI-TOF-MS) enhances the ability to obtain detailed structural information of the glycan (Gennaro and Salas-Solano 2008; Haselberg et al. 2011, 2013; Nakano et al. 2011; Zhong et al. 2015).

Capillary affinity electrophoresis (CAE) is a highly specialized method that utilizes lectin affinity and can also be used to determine oligosaccharide structure. A lectin coating within the capillary recognizes specific oligosaccharide structures and slows the progress of that oligosaccharide, thus providing structural information. CE can also be coupled with exoglycosidase digestion of samples to look for shifts in elution profiles to determine structure and sugar linkages. A partial filling technique, which injects either lectin or exoglycosidase into the capillary prior to running the sample, can provide quick structural information of the glycan, eliminating the need for a separate exoglycosidase digestion prior to analysis (Yagi et al. 2011).

3.6.4 Fluorophore-assisted Carbohydrate Electrophoresis (FACE) and CGE-LIF

FACE is an older method that separates and quantifies glycans based on their size (Starr et al. 1996) using high-density gels. Equipment is relatively inexpensive, and it does not require a high level of expertise. The glycan's reducing end is typically labeled with a fluorescent tag, ANTS, using reductive amination. This provides a stoichiometric (1 : 1) labeling and imparts a strong, uniform negative charge utilized to separate the glycans through a dense polyacrylamide gel (20–40%) using a high electrical current. The resolution between glycan bands is related to their hydrodynamic volume and charge to mass ratio and migration is compared to a dextran ladder and glycans are then assigned GU values, similar to HILIC analysis. Removal of the sialic acid decreases the mobility due to decreased charge and results in increased resolution. The fluorescent intensity of the band is proportional to the relative abundance of the specific glycan in the sample and can be quantified by densitometry relative to a standard with a typical sensitivity in the range of 10–500 pmol/band.

A more recent novel modification of FACE, called DSA-FACE, uses a multicapillary DNA-sequencer for separation and quantitation of the glycans (Callewaert et al. 2001; Defrancq et al. 2004; Laroy et al. 2006; Szabo et al. 2011; Rodig et al. 2012). The technique is also called capillary gel electrophoresis coupled to LIF detection (CGE–LIF) and uses ANTS or APTS labeling of glycans. Exoglycosidase digests of samples allow detection of isomers and linkages such as the immunogenic α-Gal epitope (Defrancq et al. 2004; Rodig et al. 2012). An automated high throughput system uses a 96-sample labeling format combined with multiplexing with capillaries that have a capacity of 96 samples at one time (Reusch et al. 2013). As little as 0.25 µg of IgG is required for detection. Automated software analysis is linked to a database can be used for identification of glycans. Figure 3.8 shows a DSA-FACE separation of APTS-labeled glycans released from a monoclonal antibody. In order to determine elution order, the sample is spiked in turn with standard glycans and results of different electropherograms are compared.

A recent comparison of several of the above methods (HILIC, DSA-FACE, CE-LIF, CGCE, and HPAEC-PAD) found that all had high precision and accuracy, but varied in their ability to detect some minor glycan species, and some sialic acid could be lost under acidic labeling conditions (Reusch et al. 2015). The advantages and disadvantages of each technique are discussed.

3.6.5 Mass Spectrometry (MS)

Over the past decade, MS has become a fundamental analytical tool for analysis of biomolecules. It is based on ionization to generate charged molecules and measuring their mass-to-charge ratio (m/z). This technique provides many advantages over traditional analytical methods, such as low sample consumption and high sensitivity with very precise results. MS instruments with capabilities of MS/MS allow sequencing of molecules and give information for fine structural assignments.

3.6.5.1 Ionization

Two basic instrumentation techniques, described often as "soft ionization methods": electrospray (ESI) and matrix-assisted laser desorption/ionization

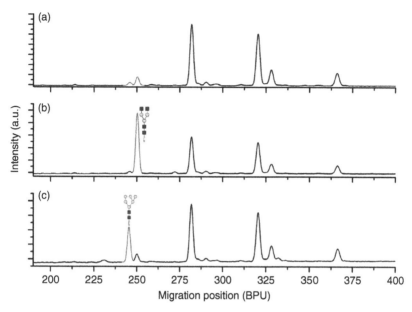

Figure 3.8 Spiking of HILIC-UPLC separated APTS-labeled glycans into capillary gel electrophoresis–laser-induced fluorescence glycan profile. (a) CGE-LIF electropherogram of Mab glycan pool without spiking. (b) CGE-LIF electropherogram of MAb glycan pool with spiking of G0-F (H3N4). (c) CGE-LIF electropherogram of Mab glycan pool with spiking of Man5-F (H5N2). Source: Reusch et al. 2013. Adapted with permission from Taylor & Francis.

(MALDI) are generally used to promote gentle desorption/ionization with detection in the femtomole range. ESI sources for use in combination with mass spectrometers allow dispersion of analyzed sample in the liquid form into a fine aerosol that results in the formation of charged particles (Fenn et al. 1989). Because the ion formation involves extensive solvent evaporation, the typical solvents for electrospray ionization are usually prepared by mixing water with volatile organic compounds such as methanol or acetonitrile.

In the MALDI technique, desorption is first triggered by a UV laser beam. Matrix, spotted on the target together with the sample to be analyzed, absorbs UV laser light and then ionization of the sample occurs (Karas and Hillenkamp 1988). Matrices are low molecular weight compounds that allow facile vaporization, but are not large enough to evaporate during sample preparation or during measurement. Most matrices are acidic and act as a proton source to promote ionization of the sample. They have polar groups, allowing their use in aqueous solutions and typically contain a chromophore that efficiently absorbs the laser irradiation. The choice of appropriate matrix depends on the type of sample analyzed. Various organic compounds have been used for MS analyses, but still three most commonly used are 3,5-dimethoxy-4-hydroxycinnamic acid known as sinapinic acid, α-cyano-4-hydroxycinnamic acid (alpha-cyano), and 2,5-dihydroxybenzoic acid (DHB). Sinapic acid is preferred for proteins and peptides of high masses, whereas DHB matrix is very often used for all classes of saccharides and lower weight peptides. The matrix solution is typically mixed with the sample in an organic solvent (e.g. acetonitrile, methanol) and water, allowing both hydrophobic and hydrophilic molecules to dissolve. The solution is spotted onto a MALDI target (usually a metal plate) from which evaporation of solvent, co-crystallization of the sample and matrix occurs. MALDI is generally less sensitive to contaminants such as salts from biological buffers than ESI.

3.6.5.2 Derivatization Techniques Used for MS Analysis of Glycans

Glycans released enzymatically or chemically from the protein (discussed in Section 5.1) are often derivatized to enhance their mass spectrometric detection. Tagging can also influence MS/MS fragmentation patterns and makes structural analysis of oligosaccharides easier. Each derivatization procedure has its advantages and the final result depends on the sample and type of targeted oligosaccharide. In this respect, a large number of derivatization procedures for oligosaccharides have been described in the literature and among them permethylation, reductive amination, and labeling based on hydrazone linkage have been often applied for analysis of glycans obtained from biological material (Figure 3.9). Useful labeling and derivatization methods were listed in a recent review article (Harvey 2015).

Permethylation is one of the oldest derivatization technique used for carbohydrate analysis and is based on the use of iodomethane catalyzed by sodium hydroxide (Figure 3.9a) (Ciucanu and Kerek 1984). This derivatization technique was originally used for MS analysis to enhance carbohydrate ionization efficiency on older MS instruments equipped with fast atom bombardment (FAB) or liquid secondary ion (LSI) sources, but permethylated oligosaccharides are still advantageous to analyze over native ones by ESI-MS and MALDI-MS (Dell 1990; Jang et al. 2015; Park et al. 2015). Permethylation also provides additional clean up of the sample from salts while producing increased ion signals over their nonderivatized counterparts, allowing the sensitive analysis of minor components. Another advantage of permethylation is that sialic acid residues become stabilized, which otherwise have a tendency to be cleaved during MS analysis of native glycans. Permethylated derivatives are thus suitable for the quantification of glycans by MS (Hu et al. 2013).

Reductive amination is based on the on the reaction of carbonyl group of a reducing sugar with aromatic amine yielding a Schiff base in the first step. These so-called nonreduced derivatives are unstable and, therefore, they are stabilized by the reduction (e.g. $NaBH_3CN$) under acidic conditions at increased temperature, providing open ring structures at the former reducing ends (Figure 3.9b). Different types of aromatic amines have been used for MS analysis of carbohydrates and some of them are discussed above (Section 5.2).

Hydrazone type labeling uses substituted hydrazine reagents that allow coupling to the reducing terminus of saccharides under nonreductive conditions

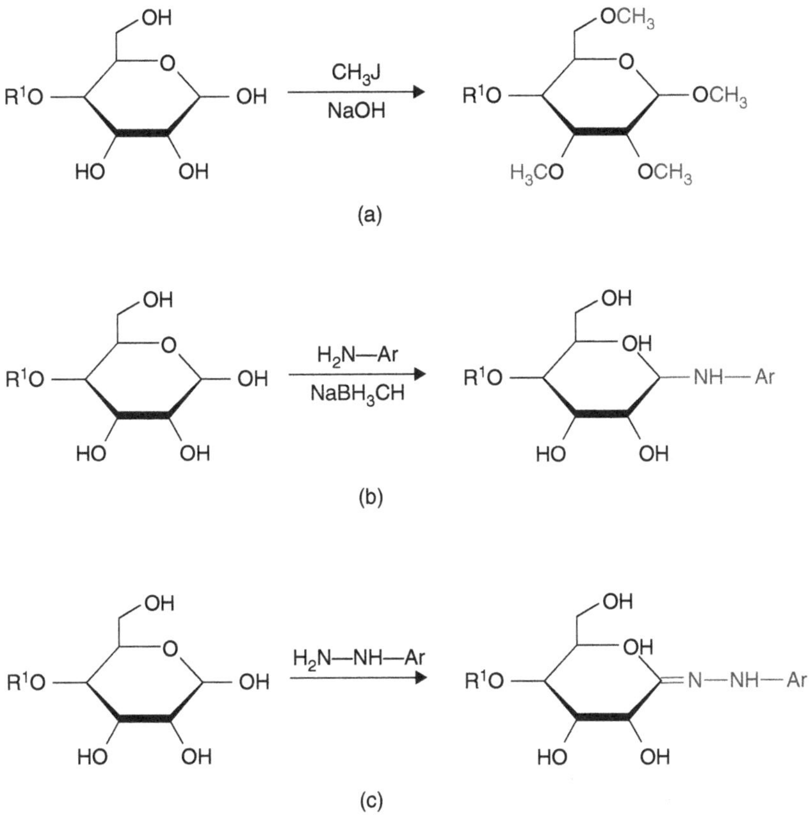

(a)

(b)

(c)

Figure 3.9 Derivatization reactions of glycans: (a) permethylation; (b) reductive amination; (c) labeling based on hydrazone-linkage.

(Lattova and Perreault 2013) (Figure 3.9c). This labeling method offers an advantage over other procedures; the hydrazone products can be injected into an HPLC or analyzed by MS immediately after derivatization. The general characteristic that hydrazone formation does not require a purification step makes this labeling suitable for direct on-target derivatization for MALDI-MS (Figure 3.10).

3.6.5.3 Fragmentation of Carbohydrates

The mass spectrometric fragmentation behavior of carbohydrates has been studied for many years, and it has been revealed that oligosaccharides produce characteristic fragmentation patterns under collision-induced dissociation (CID) MS/MS conditions (Rodrigues et al. 2007; Michael and Rizzi 2015). In general, oligosaccharide fragmentation pathways can be classified into two groups: glycosidic cleavages resulting from the break of a bond between

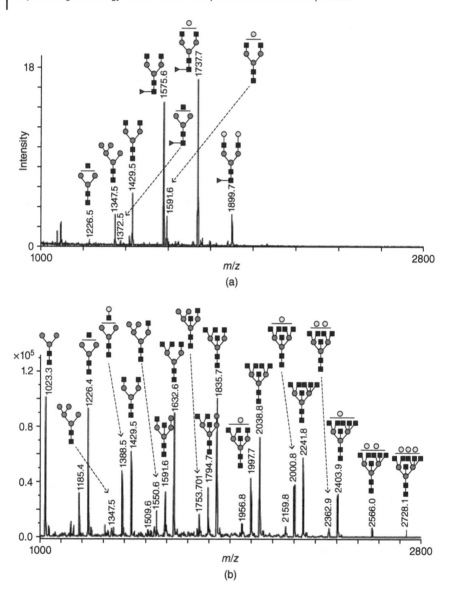

Figure 3.10 MALDI-MS spectra of neutral oligosaccharides obtained by PNGase F digestion from (a) IgG1 (Herceptin), recorded on a QTOF instrument and (b) ovalbumin, recorded on UltrafleXtreme (Bruker). Oligosaccharides were labeled on-target with PHN immediately after PNGaseF digestion without purification. All ions are [M+Na]⁺. Not all glycan structures are shown.

two sugar rings (B, C, Y, or Z ions) and cross-ring fragment ions (A, X ions) that involve the cleavage of two bonds within a monosaccharide constituent (Domon and Costello 1988; Reinhold et al. 1995). It was demonstrated that CID spectra of [M+H]$^+$ ions provided information about the sequence of the monosaccharide units, while the dissociation of [M+Na]$^+$ ions by two-bond ring cleavage processes gives rise to the fragment ions that allowed differentiation of $1 \rightarrow 2$ and $1 \rightarrow 4$ linkages (Kovacik et al. 1995). In the spectra of larger glycans, the preferential losses of groups attached to the 3-position of hexose have frequently been observed in the fragmentation of different oligosaccharides. Some specific fragment ions can be used to distinguish between 3- and 6-antenna of N-linked glycans (Richter et al. 1990; Laine et al. 1991). These ions are usually formed through the loss of the 3-antenna together with the chitobiose core (B/Y3) and are accompanied with characteristic loss of water (Harvey 2000; Lattova et al. 2004). Cross-ring fragments are important for determining linkage positions and among them, the most useful ions in the spectra of N-linked glycans are the ^{3,5}A and ^{0,4}A ions produced from cleavages of the core-branching mannose residue (Figure 3.11).

Recent models of mass spectrometers often include fragmentation modes other than CID, and use electron transfer dissociation (Zhu et al. 2013) or electron capture dissociation (Bourgoin-Voillard et al. 2014) methods that yield information complementary to CID experiments. Fragmentation patterns using ECD display cross-ring cleavages (Han and Costello 2013) that are useful to study the different linkages that make up a glycan molecule.

MS analysis of oligosaccharides usually does not provide complete information about the linkage positions between the individual monosaccharides. However, the linkages can be confirmed by sequential treatment with specific exoglycosidases to cleave monosaccharide residues from the nonreducing end (Prime et al. 1996). Exoglycosidase-treated oligosaccharides can then be re-analyzed by MS to observe any corresponding mass losses.

Glycopeptide analysis is gaining popularity for the determination of glycosylation contents and profiles. Especially with the advent of ECD as a fragmentation technique, it has become possible to use ESI-MS/MS and obtain insightful spectra. Figure 3.12 compares fragmentation patterns obtained for the same tryptic *O*-glycopeptide from cerebrospinal fluid (CSF) (Nilsson and Larson 2013). In Figure 3.12a, a CID experiment was conducted on precursor ions [M+3H]$^{3+}$ at *m/z* 707.35. In Figure 3.12b, ECD was used to fragment [M+4H]$^{4+}$ ions at *m/z* 530.76. For the same molecule, two different and complementary patterns are obtained. Typically, as seen in Figure 3.11a, ESI-CID-MS/MS brings a limited amount of structural information. On the other hand, MALDI-CID-MS/MS spectra of glycopeptide precursor ions have been shown to contain valuable information on both peptide and

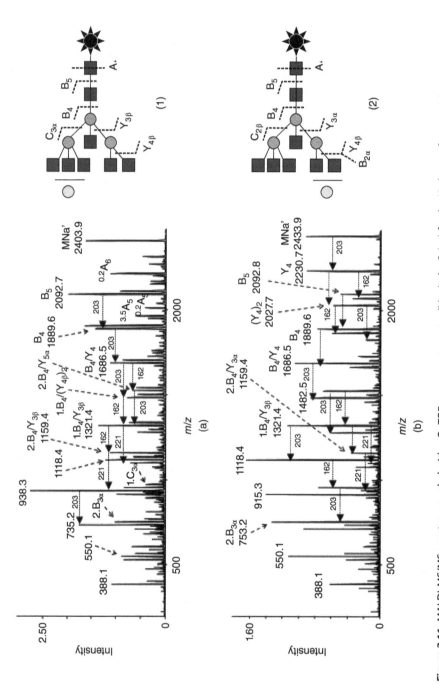

Figure 3.11 MALDI-MS/MS spectra recorded with a QqTOF mass spectrometer (Manitoba Sciex) for the *N*-glycan of composition Gal1GlcNAc6Man3GlcNAc2 after (a) derivatization with PHN; (b) AB derivatization. Fragment ions are sodiated. Star symbol: derivatization label. Source: Lattova et al. 2005. Adapted with permission from Elsevier.

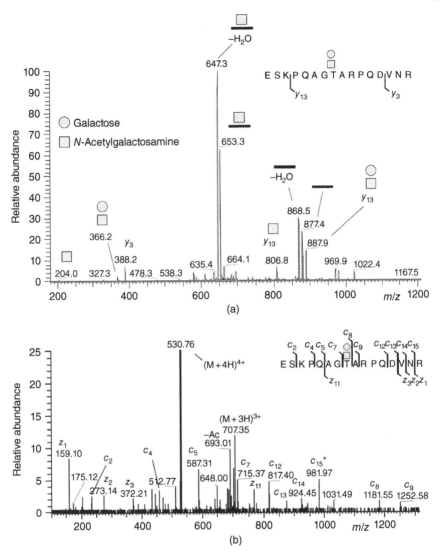

Figure 3.12 Tandem mass spectra an *O*-glycopeptide from human cerebrospinal fluid (CSF). (a) Collision-induced dissociation CID MS^2 spectrum of the $[M+3H]^{3+}$ precursor ions at m/z 707; (b) electron capture dissociation (ECD) fragmentation of $[M+4H]^{4+}$ precursors at m/z 530. −Ac, loss of an acetyl group from the charge-reduced precursor. Source: Nilsson and Larson 2013. Adapted with permission from Springer Nature.

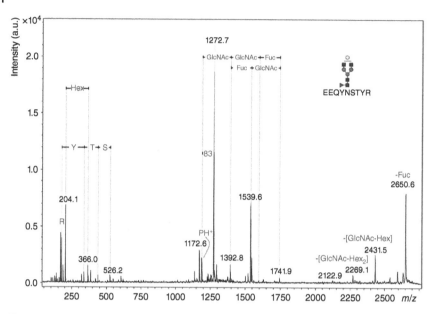

Figure 3.13 MALDI-MS/MS spectrum of the [M+H]⁺ ions of a tryptic glycopeptide from human IgG1. The experiment was conducted on a Bruker UltraFleXtreme instrument using LIFT™ technology. The sample was in dihydroxy benzoic acid matrix. PH⁺, protonated peptide backbone fragment; Hex, hexose; Fuc, fucose; GlcNAc, *N*-acetyl galactosamine.

glycan moieties (Krokhin et al. 2004; Bodnar et al. 2015). Figure 3.13 shows the MALDI-MS/MS spectrum of [M+H]⁺ precursors of tryptic glycopeptide EEQYNSTYR from human IgG1. The glycoform is biantennary, mono-galactosylated. Fragment ions at *m/z* 2431 correspond to losses of either Gal-GlcNAc or GlcNAc-Man branch, while *m/z* 2269 ions show the loss of Gal-GlcNAc-Man. Ions at *m/z* 1272 result from a cross ring cleavage of the GlcNAc residue attached to asparagine and can be described as PH+83 ions, where PH is the protonated bare peptide backbone (*m/z* 1189). In the low mass range, *m/z* 204 fragments indicate the presence of GlcNAc residues.

Quantitative analysis of glycosylation by mass spectrometry has also made considerable advances in recent years. The introduction of stable isotope labeling of tandem mass tags (TMT) as currently used in proteomics analysis (Cheng et al. 2016) has allowed to perform relative quantitation of glycopeptides (Bodnar and Perreault 2013, 2015) based on MS/MS production of specific marker ions. As for quantitation of released glycans, methods have included the use of specifically adapted TMT reagents (Zhong et al. 2015) and derivatization agents labeled with stable isotopes (Zhang et al. 2011). A recent review article gives a global description of most quantitative methods available in the field of glycomics (Mechref et al. 2013).

3.7   Conclusion

Our understanding of the functionality of glycosylation has grown rapidly with advances in the analytical techniques for glycoproteins and glycans over the past few decades. We now understand that what appears to be very minor changes in glycan structures can have a large impact on functionality. The application of this knowledge in the biopharmaceutical industry has lead to products with increased serum half-life, higher efficacy, lower immunogenicity, and greater homogeneity, and ultimately higher quality. The challenge now is to further refine our understanding of structure–function relationship in new products, perhaps improve current products by refinement of glycosylation, and increase our understanding of glycosylation machinery within the cell to better manipulate these pathways in production.

References

Abdel Rahman, A.M., Ryczko, M., Nakano, M. et al. (2015). Golgi N-glycan branching N-acetylglucosaminyltransferases I, V and VI promote nutrient uptake and metabolism. *Glycobiology* 25: 225–240.

Aghamohseni, H., Ohadi, K., Spearman, M. et al. (2014). Effects of nutrient levels and average culture pH on the glycosylation pattern of camelid-humanized monoclonal antibody. *J. Biotechnol.* 186: 98–109.

Ahn, W.S. and Antoniewicz, M.R. (2011). Metabolic flux analysis of CHO cells at growth and non-growth phases using isotopic tracers and mass spectrometry. *Metab. Eng.* 13: 598–609.

Ahn, W.S. and Antoniewicz, M.R. (2012). Towards dynamic metabolic flux analysis in CHO cell cultures. *Biotechnol. J.* 7: 61–74.

Ahn, M.-.H., Song, M., Oh, E.-.Y. et al. (2008a). Production of therapeutic proteins with baculovirus expression system in insect cell. *Entomol. Res.* 38: S71–S78.

Ahn, W.S., Jeon, J.J., Jeong, Y.R. et al. (2008b). Effect of culture temperature on erythropoietin production and glycosylation in a perfusion culture of recombinant CHO cells. *Biotechnol. Bioeng.* 101: 1234–1244.

Ahn, J., Bones, J., Yu, Y.Q. et al. (2010). Separation of 2-aminobenzamide labeled glycans using hydrophilic interaction chromatography columns packed with 1.7 microm sorbent. *J. Chromatogr. B Anal. Technol. Biomed. Life Sci.* 878: 403–408.

Alahmad, Y., Tran, N.T., and Taverna, M. (2011). Analysis of intact glycoprotein biopharmaceuticals by capillary electrophoresis. In: *Capillary Electrophoresis of Carbohydrates: From Monosaccharides to Complex Polysaccharides* (ed. N. Volpi). Springer Science+Business Media.

Allen, S., Naim, H.Y., and Bulleid, N.J. (1995). Intracellular folding of tissue-type plasminogen activator. Effects of disulfide bond formation on N-linked glycosylation and secretion. *J. Biol. Chem.* 270: 4797–4804.

Andersen, D.C. and Goochee, C.F. (1994). The effect of cell-culture conditions on the oligosaccharide structures of secreted glycoproteins. *Curr. Opin. Biotechnol.* 5: 546–549.

Andersen, D.C. and Goochee, C.F. (1995). The effect of ammonia on the O-linked glycosylation of granulocyte colony-stimulating factor produced by Chinese hamster ovary cells. *Biotechnol. Bioeng.* 47: 96–105.

Anderson, C.L., Wang, Y., and Rustandi, R.R. (2012). Applications of imaged capillary isoelectric focussing technique in development of biopharmaceutical glycoprotein-based products. *Electrophoresis* 33: 1538–1544.

Anumula, K.R. and Dhume, S.T. (1998). High resolution and high sensitivity methods for oligosaccharide mapping and characterization by normal phase high performance liquid chromatography following derivatization with highly fluorescent anthranilic acid. *Glycobiology* 8: 685–694.

Arnold, J.N., Wormald, M.R., Sim, R.B. et al. (2007). The impact of glycosylation on the biological function and structure of human immunoglobulins. *Annu. Rev. Immunol.* 25: 21–50.

Aumiller, J.J., Mabashi-Asazuma, H., Hillar, A. et al. (2012). A new glycoengineered insect cell line with an inducibly mammalianized protein N-glycosylation pathway. *Glycobiology* 22: 417–428.

Baker, K.N., Rendall, M.H., Hills, A.E. et al. (2001). Metabolic control of recombinant protein N-glycan processing in NS0 and CHO cells. *Biotechnol. Bioeng.* 73: 188–202.

Bedard, K., Szabo, E., Michalak, M., and Opas, M. (2005). Cellular functions of endoplasmic reticulum chaperones calreticulin, calnexin, and ERp57. *Int. Rev. Cytol.* 245: 91–121.

van Berkel, P.H.C., Gerritsen, J., Perdok, G. et al. (2009). N-linked glycosylation is an important parameter for optimal selection of cell lines producing biopharmaceutical human IgG. *Biotechnol. Progr.* 25: 244–251.

van Berkel, P.H., Gerritsen, J., van Voskuilen, E. et al. (2010). Rapid production of recombinant human IgG With improved ADCC effector function in a transient expression system. *Biotechnol. Bioeng.* 105: 350–357.

Biacchi, M., Gahoual, R., Said, N. et al. (2015). Glycoform separation and characterization of cetuximab variants by middle-up off-line capillary zone electrophoresis-UV/electrospray ionization-MS. *Anal. Chem.* 87: 6240–6250.

Bigge, J.C., Patel, T.P., Bruce, J.A. et al. (1995). Nonselective and efficient fluorescent labeling of glycans using 2-amino benzamide and anthranilic acid. *Anal. Biochem.* 230: 229–238.

Bischoff, J. and Kornfeld, R. (1983). Evidence for an alpha-mannosidase in endoplasmic reticulum of rat liver. *J. Biol. Chem.* 258: 7907–7910.

Blondeel, E.J., Braasch, K., McGill, T. et al. (2015). Tuning a MAb glycan profile in cell culture: supplementing N-acetylglucosamine to favour G0 glycans without compromising productivity and cell growth. *J. Biotechnol.* 214: 105–112.

Bodnar, E.D. and Perreault, H. (2013). Qualitative and quantitative assessment on the use of magnetic nanoparticles for glycopeptide enrichment. *Anal. Chem.* 85: 10895–10903.

Bodnar, E.D. and Perreault, H. (2015). Synthesis and evaluation of carboxymethyl chitosan for glycopeptide enrichment. *Anal. Chim. Acta* 891: 179–189.

Bodnar, E., Nascimento, T.F., Girard, L. et al. (2015). An integrated approach to analyze EG2-hFc monoclonal antibody *N*-glycosylation by MALDI-MS. *Can. J. Chem.* 93: 754–763.

Bohm, E., Seyfried, B.K., Dockal, M. et al. (2015). Differences in *N*-glycosylation of recombinant human coagulation factor VII derived from BHK, CHO, and HEK293 cells. *BMC Biotechnol.* 15: 87.

Borys, M.C., Linzer, D.I., and Papoutsakis, E.T. (1993). Culture pH affects expression rates and glycosylation of recombinant mouse placental lactogen proteins by Chinese hamster ovary (CHO) cells. *Biotechnology (N. Y.)* 11: 720–724.

Borys, M.C., Dalal, N.G., Abu-Absi, N.R. et al. (2010). Effects of culture conditions on N-glycolylneuraminic acid (Neu5Gc) content of a recombinant fusion protein produced in CHO cells. *Biotechnol. Bioeng.* 105: 1048–1057.

Bosques, C.J., Collins, B.E., Meador, J.W. III et al. (2010). Chinese hamster ovary cells can produce galactose-alpha-1,3-galactose antigens on proteins. *Nat. Biotechnol.* 28: 1153–1156.

Bourgoin-Voillard, S., Leymarie, N., and Costello, C.E. (2014). Top-down tandem mass spectrometry on RNase A and B using a Qh/FT-ICR hybrid mass spectrometer. *Proteomics* 14: 1174–1184.

Bragonzi, A., Distefano, G., Buckberry, L.D. et al. (2000). A new Chinese hamster ovary cell line expressing alpha2,6-sialyltransferase used as universal host for the production of human-like sialylated recombinant glycoproteins. *Biochim. Biophys. Acta* 1474: 273–282.

Brooks, S.A. (2004). Appropriate glycosylation of recombinant proteins for human use: implications of choice of expression system. *Mol. Biotechnol.* 28: 241–255.

Burleigh, S.C., van de Laar, T., Stroop, C.J. et al. (2011). Synergizing metabolic flux analysis and nucleotide sugar metabolism to understand the control of glycosylation of recombinant protein in CHO cells. *BMC Biotechnol.* 11: 95.

Butler, M. (2006). Optimisation of the cellular metabolism of glycosylation for recombinant proteins produced by Mammalian cell systems. *Cytotechnology* 50: 57–76.

Butler, M. and Perreault, H. (2010). *Approaches and Methods for Determining Glycosylation*. Wiley.

Callewaert, N., Geysens, S., Molemans, F., and Contreras, R. (2001). Ultrasensitive profiling and sequencing of N-linked oligosaccharides using standard DNA-sequencing equipment. *Glycobiology* 11: 275–281.

Camilleri, P., Tolson, D., and Birrell, H. (1998). Direct structural analysis of 2-aminoacridone derivatized oligosaccharides by high-performance liquid chromatography/mass spectrometric detection. *Rapid Commun. Mass Spectrom.* 12: 144–148.

Campbell, M.P., Royle, L., Radcliffe, C.M. et al. (2008). GlycoBase and autoGU: tools for HPLC-based glycan analysis. *Bioinformatics* 24: 1214–1216.

Campbell, M.P., Royle, L., and Rudd, P.M. (2015). GlycoBase and autoGU: resources for interpreting HPLC-glycan data. *Methods Mol. Biol.* 1273: 17–28.

Carlson, D.M. (1968). Structures and immunochemical properties of oligosaccharides isolated from pig submaxillary mucins. *J. Biol. Chem.* 243: 616–626.

Castilho, A., Gattinger, P., Grass, J. et al. (2011). *N*-glycosylation engineering of plants for the biosynthesis of glycoproteins with bisected and branched complex *N*-glycans. *Glycobiology* 21: 813–823.

Cataldi, T.R., Campa, C., and de Benedetto, G.E. (2000). Carbohydrate analysis by high-performance anion-exchange chromatography with pulsed amperometric detection: the potential is still growing. *Fresenius J. Anal. Chem.* 368: 739–758.

Chan, K.F., Shahreel, W., Wan, C. et al. (2015). Inactivation of GDP-fucose transporter gene (Slc35c1) in CHO cells by ZFNs, TALENs and CRISPR-Cas9 for production of fucose-free antibodies. *Biotechnol. J.* 11: 399–414.

Charlwood, J., Birrell, H., Gribble, A. et al. (2000). A probe for the versatile analysis and characterization of N-linked oligosaccharides. *Anal. Chem.* 72: 1453–1461.

Chee Furng Wong, D., Tin Kam Wong, K., Tang Goh, L. et al. (2005). Impact of dynamic online fed-batch strategies on metabolism, productivity and *N*-glycosylation quality in CHO cell cultures. *Biotechnol. Bioeng.* 89: 164–177.

Chen, P. and Harcum, S.W. (2006). Effects of elevated ammonium on glycosylation gene expression in CHO cells. *Metab. Eng.* 8: 123–132.

Chen, C., Constantinou, A., Chester, K.A. et al. (2012). Glycoengineering approach to half-life extension of recombinant biotherapeutics. *Bioconjug. Chem.* 23: 1524–1533.

Cheng, L., Pisitkun, T., Knepper, M.A., and Hoffert, J.D. (2016). Peptide labeling using isobaric tagging reagents for quantitative phosphoproteomics. *Methods Mol. Biol.* 1355: 53–70.

Chiba, Y. and Akeboshi, H. (2009). Glycan engineering and production of 'humanized' glycoprotein in yeast cells. *Biol. Pharm. Bull.* 32: 786–795.

Chiba, Y. and Jigami, Y. (2007). Production of humanized glycoproteins in bacteria and yeasts. *Curr. Opin. Chem. Biol.* 11: 670–676.

Chotigeat, W., Watanapokasin, Y., Mahler, S., and Gray, P.P. (1994). Role of environmental conditions on the expression levels, glycoform pattern and levels of sialyltransferase for hFSH produced by recombinant CHO cells. *Cytotechnology* 15: 217–221.

Chung, C.H., Mirakhur, B., Chan, E. et al. (2008). Cetuximab-induced anaphylaxis and IgE specific for galactose-alpha-1,3-galactose. *N. Engl. J. Med.* 358: 1109–1117.

Ciucanu, I. and Kerek, F. (1984). A simple and rapid method for the permethylation of carbohydrates. *Carbohydr. Research* 131: 209–217.

Clarke, A., Harmon, B., and Defelippis, M.R. (2009). Analysis of 3-(acetylamino)-6-aminoacridine-derivatized oligosaccharides from recombinant monoclonal antibodies by liquid chromatography–mass spectrometry. *Anal. Biochem.* 390: 209–211.

Cook, M.C., Kaldas, S.J., Muradia, G. et al. (2015). Comparison of orthogonal chromatographic and lectin-affinity microarray methods for glycan profiling of a therapeutic monoclonal antibody. *J. Chromatogr. B Anal. Technol. Biomed. Life Sci.* 997: 162–178.

Crispin, M., Bowden, T.A., Coles, C.H. et al. (2009). Carbohydrate and domain architecture of an immature antibody glycoform exhibiting enhanced effector functions. *J. Mol. Biol.* 387: 1061–1066.

Crowell, C.K., Grampp, G.E., Rogers, G.N. et al. (2007). Amino acid and manganese supplementation modulates the glycosylation state of erythropoietin in a CHO culture system. *Biotechnol. Bioeng.* 96: 538–549.

Curling, E.M., Hayter, P.M., Baines, A.J. et al. (1990). Recombinant human interferon-gamma. Differences in glycosylation and proteolytic processing lead to heterogeneity in batch culture. *Biochem. J.* 272: 333–337.

Davidson, S.K. and Hunt, L.A. (1985). Sindbis virus glycoproteins are abnormally glycosylated in Chinese hamster ovary cells deprived of glucose. *J. Gen. Virol.* 66 (Pt 7): 1457–1468.

Dean, J. and Reddy, P. (2013). Metabolic analysis of antibody producing CHO cells in fed-batch production. *Biotechnol. Bioeng.* 110: 1735–1747.

Defrancq, L., Callewart, N., Zhu, J. et al. (2004). High-throughput analysis of the N-glycans of NSO cell-secreted antibodies. *BioProcess Int.* 2: 60–68.

Deisenhofer, J. (1981). Crystallographic refinement and atomic models of a human Fc fragment and its complex with fragment B of protein A from *Staphylococcus aureus* at 2.9- and 2.8-A resolution. *Biochemistry* 20: 2361–2370.

Dell, A. (1990). Preparation and desorption mass-spectrometry of permethyl and peracetyl derivatives of oligosaccharides. *Methods Enzymol.* 193: 647–660.

Dhume, S.T., Saddic, G.N., and Anumula, K.R. (2008). Monitoring glycosylation of therapeutic glycoproteins for consistency by HPLC using highly fluorescent anthranilic acid (AA) tag. *Methods Mol. Biol.* 446: 317–331.

Dicker, M. and Strasser, R. (2015). Using glyco-engineering to produce therapeutic proteins. *Expert Opin. Biol. Ther.* 15: 1501–1516.

Doherty, M., McManus, C.A., Duke, R., and Rudd, P.M. (2012). High-throughput quantitative *N*-glycan analysis of glycoproteins. *Methods Mol. Biol.* 899: 293–313.

Doherty, M., Bones, J., McLoughlin, N. et al. (2013). An automated robotic platform for rapid profiling oligosaccharide analysis of monoclonal antibodies directly from cell culture. *Anal. Biochem.* 442: 10–18.

Domann, P.J., Pardos-Pardos, A.C., Fernandes, D.L. et al. (2007). Separation-based glycoprofiling approaches using fluorescent labels. *Proteomics* 7 (Suppl. 1): 70–76.

Domon, B. and Costello, C.E. (1988). Structure elucidation of glycosphingolipids and gangliosides using high-performance tandem mass spectrometry. *Biochemistry* 27: 1534–1543.

Dou, P., Liu, Z., He, J. et al. (2008). Rapid and high-resolution glycoform profiling of recombinant human erythropoietin by capillary isoelectric focusing with whole column imaging detection. *J. Chromatogr. A* 1190: 372–376.

Doyle, C. and Butler, M. (1990). The effect of pH on the toxicity of ammonia to a murine hybridoma. *J. Biotechnol.* 15: 91–100.

Dumont, J., Euwart, D., Mei, B. et al. (2015). Human cell lines for biopharmaceutical manufacturing: history, status, and future perspectives. *Crit. Rev. Biotechnol.* 36 (6): 1110–1122.

Durocher, Y. and Butler, M. (2009). Expression systems for therapeutic glycoprotein production. *Curr. Opin. Biotechnol.* 20: 700–707.

Egrie, J.C., Dwyer, E., Browne, J.K. et al. (2003). Darbepoetin alfa has a longer circulating half-life and greater in vivo potency than recombinant human erythropoietin. *Exp. Hematol.* 31: 290–299.

Elliott, S., Egrie, J., Browne, J. et al. (2004). Control of rHuEPO biological activity: the role of carbohydrate. *Exp. Hematol.* 32: 1146–1155.

Elliott, P., Billingham, S., Bi, J., and Zhang, H. (2013). Quality by design for biopharmaceuticals: a historical review and guide for implementation. *Pharm. Bioprocess.* 1: 105–122.

Fan, Y., Jimenez Del Val, I., Muller, C. et al. (2015). Amino acid and glucose metabolism in fed-batch CHO cell culture affects antibody production and glycosylation. *Biotechnol. Bioeng.* 112: 521–535.

Fenn, J.B., Mann, M., Meng, C.K. et al. (1989). Electrospray ionization for mass spectrometry of large biomolecules. *Science* 246: 64–71.

Ferrara, C., Brunker, P., Suter, T. et al. (2006a). Modulation of therapeutic antibody effector functions by glycosylation engineering: influence of Golgi enzyme localization domain and co-expression of heterologous beta1, 4-*N*-acetylglucosaminyltransferase III and Golgi alpha-mannosidase II. *Biotechnol. Bioeng.* 93: 851–861.

Ferrara, C., Stuart, F., Sondermann, P. et al. (2006b). The carbohydrate at FcgammaRIIIa Asn-162. An element required for high affinity binding to non-fucosylated IgG glycoforms. *J. Biol. Chem.* 281: 5032–5036.

Fliedl, L., Grillari, J., and Grillari-Voglauer, R. (2015). Human cell lines for the production of recombinant proteins: on the horizon. *New Biotechnol.* 32: 673–679.

Fukuta, K., Yokomatsu, T., Abe, R. et al. (2000). Genetic engineering of CHO cells producing human interferon-gamma by transfection of sialyltransferases. *Glycoconj. J.* 17: 895–904.

Gahoual, R., Biacchi, M., Chicher, J. et al. (2014). Monoclonal antibodies biosimilarity assessment using transient isotachophoresis capillary zone electrophoresis-tandem mass spectrometry. *MAbs* 6: 1464–1473.

Gawlitzek, M., Conradt, H.S., and Wagner, R. (1995). Effect of different cell culture conditions on the polypeptide integrity and *N*-glycosylation of a recombinant model glycoprotein. *Biotechnol. Bioeng.* 46: 536–544.

Gawlitzek, M., Valley, U., and Wagner, R. (1998). Ammonium ion and glucosamine dependent increases of oligosaccharide complexity in recombinant glycoproteins secreted from cultivated BHK-21 cells. *Biotechnol. Bioeng.* 57: 518–528.

Gawlitzek, M., Ryll, T., Lofgren, J., and Sliwkowski, M.B. (2000). Ammonium alters *N*-glycan structures of recombinant TNFR-IgG: degradative versus biosynthetic mechanisms. *Biotechnol. Bioeng.* 68: 637–646.

Gawlitzek, M., Estacio, M., Furch, T., and Kiss, R. (2009). Identification of cell culture conditions to control *N*-glycosylation site-occupancy of recombinant glycoproteins expressed in CHO cells. *Biotechnol. Bioeng.* 103: 1164–1175.

Geisler, C., Mabashi-Asazuma, H., Kuo, C.W. et al. (2015). Engineering beta1,4-galactosyltransferase I to reduce secretion and enhance *N*-glycan elongation in insect cells. *J. Biotechnol.* 193: 52–65.

Gennaro, L.A. and Salas-Solano, O. (2008). On-line CE-LIF-MS technology for the direct characterization of N-linked glycans from therapeutic antibodies. *Anal. Chem.* 80: 3838–3845.

Gerdtzen, Z.P. (2012). Modeling metabolic networks for mammalian cell systems: general considerations, modeling strategies, and available tools. *Adv. Biochem. Eng. Biotechnol.* 127: 71–108.

Ghaderi, D., Zhang, M., Hurtado-Ziola, N., and Varki, A. (2012). Production platforms for biotherapeutic glycoproteins. Occurrence, impact, and challenges of non-human sialylation. *Biotechnol. Genet. Eng. Rev.* 28: 147–175.

Gonzalez-Leal, I.J., Carrillo-Cocom, L.M., Ramirez-Medrano, A. et al. (2011). Use of a Plackett–Burman statistical design to determine the effect of selected amino acids on monoclonal antibody production in CHO cells. *Biotechnol. Progr.* 27: 1709–1717.

Grabenhorst, E., Schlenke, P., Pohl, S. et al. (1999). Genetic engineering of recombinant glycoproteins and the glycosylation pathway in mammalian host cells. *Glycoconj. J.* 16: 81–97.

Grainger, R.K. and James, D.C. (2013). CHO cell line specific prediction and control of recombinant monoclonal antibody *N*-glycosylation. *Biotechnol. Bioeng.* 110: 2970–2983.

Gramer, M.J., Eckblad, J.J., Donahue, R. et al. (2011). Modulation of antibody galactosylation through feeding of uridine, manganese chloride, and galactose. *Biotechnol. Bioeng.* 108: 1591–1602.

Grammatikos, S.I., Valley, U., Nimtz, M. et al. (1998). Intracellular UDP-N-acetylhexosamine pool affects *N*-glycan complexity: a mechanism of ammonium action on protein glycosylation. *Biotechnol. Progr.* 14: 410–419.

Grey, C., Edebrink, P., Krook, M., and Jacobsson, S.P. (2009). Development of a high performance anion exchange chromatography analysis for mapping of oligosaccharides. *J. Chromatogr. B Anal. Technol. Biomed. Life Sci.* 877: 1827–1832.

Gu, X. and Wang, D.I. (1998). Improvement of interferon-gamma sialylation in Chinese hamster ovary cell culture by feeding of N-acetylmannosamine. *Biotechnol. Bioeng.* 58: 642–648.

Guile, G.R., Rudd, P.M., Wing, D.R. et al. (1996). A rapid high-resolution high-performance liquid chromatographic method for separating glycan mixtures and analyzing oligosaccharide profiles. *Anal. Biochem.* 240: 210–226.

Hamm, M., Wang, Y., and Rustandi, R.R. (2013). Characterization of N-linked glycosylation in a monoclonal antibody produced in NS0 cells using capillary electrophoresis with laser-induced fluorescence detection. *Pharmaceuticals (Basel)* 6: 393–406.

Han, L. and Costello, C.E. (2013). Mass spectrometry of glycans. *Biochemistry (Moscow)* 78: 710–720.

Hang, H.C. and Bertozzi, C.R. (2005). The chemistry and biology of mucin-type O-linked glycosylation. *Bioorg. Med. Chem.* 13: 5021–5034.

Harvey, D.J. (2000). Postsource decay fragmentation of N-linked carbohydrates from ovalbumin and related glycoproteins. *J. Am. Soc. Mass Spectrom.* 11: 572–577.

Harvey, D.J. (2015). Analysis of carbohydrates and glycoconjugates by matrix-assisted laser desorption/ionization mass spectrometry: an update for 2009–2010. *Mass Spectrom. Rev.* 34: 268–422.

Harvey, D.J., Crispin, M., Bonomelli, C., and Scrivens, J.H. (2015). Ion mobility mass spectrometry for ion recovery and clean-up of MS and MS/MS spectra obtained from low abundance viral samples. *J. Am. Soc. Mass Spectrom.* 26: 1754–1767.

Haselberg, R., de Jong, G.J., and Somsen, G.W. (2011). Capillary electrophoresis-mass spectrometry for the analysis of intact proteins 2007–2010. *Electrophoresis* 32: 66–82.

Haselberg, R., de Jong, G.J., and Somsen, G.W. (2013). Low-flow sheathless capillary electrophoresis-mass spectrometry for sensitive glycoform profiling of intact pharmaceutical proteins. *Anal. Chem.* 85: 2289–2296.

Hayes, C.A., Doohan, R., Kirkley, D. et al. (2012). Cross validation of liquid chromatography–mass spectrometry and lectin array for monitoring glycosylation in fed-batch glycoprotein production. *Mol. Biotechnol.* 51: 272–282.

Hayter, P.M., Curling, E.M., Baines, A.J. et al. (1992). Glucose-limited chemostat culture of Chinese hamster ovary cells producing recombinant human interferon-gamma. *Biotechnol. Bioeng.* 39: 327–335.

Heidemann, R., Lutkemeyer, D., Buntemeyer, H., and Lehmann, J. (1998). Effects of dissolved oxygen levels and the role of extra- and intracellular amino acid concentrations upon the metabolism of mammalian cell lines during batch and continuous cultures. *Cytotechnology* 26: 185–197.

Higel, F., Demelbauer, U., Seidl, A. et al. (2013). Reversed-phase liquid-chromatographic mass spectrometric *N*-glycan analysis of biopharmaceuticals. *Anal. Bioanal. Chem.* 405: 2481–2493.

Hills, A.E., Patel, A., Boyd, P., and James, D.C. (2001). Metabolic control of recombinant monoclonal antibody *N*-glycosylation in GS-NS0 cells. *Biotechnol. Bioeng.* 75: 239–251.

Hirabayashi, J., Kuno, A., and Tateno, H. (2011). Lectin-based structural glycomics: a practical approach to complex glycans. *Electrophoresis* 32: 1118–1128.

Hirabayashi, J., Kuno, A., and Tateno, H. (2015). Development and applications of the lectin microarray. *Top. Curr. Chem.* 367: 105–124.

Hirano, M., Adachi, Y., Ito, Y., and Totani, K. (2015). Calreticulin discriminates the proximal region at the *N*-glycosylation site of Glc1Man9GlcNAc2 ligand. *Biochem. Biophys. Res. Commun.* 466: 350–355.

Hodoniczky, J., Zheng, Y.Z., and James, D.C. (2005). Control of recombinant monoclonal antibody effector functions by Fc *N*-glycan remodeling in vitro. *Biotechnol. Progr.* 21: 1644–1652.

Hong, J.K., Cho, S.M., and Yoon, S.K. (2010). Substitution of glutamine by glutamate enhances production and galactosylation of recombinant IgG in Chinese hamster ovary cells. *Appl. Microbiol. Biotechnol.* 88: 869–876.

von Horsten, H.H., Ogorek, C., Blanchard, V. et al. (2010). Production of non-fucosylated antibodies by co-expression of heterologous GDP-6-deoxy-D-lyxo-4-hexulose reductase. *Glycobiology* 20: 1607–1618.

Horvath, B., Mun, M., and Laird, M.W. (2010). Characterization of a monoclonal antibody cell culture production process using a quality by design approach. *Mol. Biotechnol.* 45: 203–206.

Hossler, P., Goh, L.T., Lee, M.M., and Hu, W.S. (2006). GlycoVis: visualizing glycan distribution in the protein *N*-glycosylation pathway in mammalian cells. *Biotechnol. Bioeng.* 95: 946–960.

Hossler, P., McDermott, S., Racicot, C. et al. (2014). Cell culture media supplementation of uncommonly used sugars sucrose and tagatose for the targeted shifting of protein glycosylation profiles of recombinant protein therapeutics. *Biotechnol. Progr.* 30: 1419–1431.

Hu, Y., Desantos-Garcia, J.L., and Mechref, Y. (2013). Comparative glycomic profiling of isotopically permethylated *N*-glycans by liquid chromatography/electrospray ionization mass spectrometry. *Rapid Commun. Mass Spectrom.* 27: 865–877.

Huang, W., Giddens, J., Fan, S.Q. et al. (2012). Chemoenzymatic glycoengineering of intact IgG antibodies for gain of functions. *J. Am. Chem. Soc.* 134: 12308–12318.

Ijiri, S., Todoroki, K., Yoshida, H. et al. (2011). Highly sensitive capillary electrophoresis analysis of N-linked oligosaccharides in glycoproteins following fluorescence derivatization with rhodamine 110 and laser-induced fluorescence detection. *Electrophoresis* 32: 3499–3509.

Imai-Nishiya, H., Mori, K., Inoue, M. et al. (2007). Double knockdown of alpha1,6-fucosyltransferase (FUT8) and GDP-mannose 4,6-dehydratase (GMD) in antibody-producing cells: a new strategy for generating fully non-fucosylated therapeutic antibodies with enhanced ADCC. *BMC Biotechnol.* 7: 84.

Irani, Z.A., Kerkhoven, E., Shojaosadati, S.A., and Nielsen, J. (2015). Genome-scale metabolic model of *Pichia pastoris* with native and humanized glycosylation of recombinant proteins. *Biotechnol. Bioeng.* 113: 961–969.

Jan, D.C., Petch, D.A., Huzel, N., and Butler, M. (1997). The effect of dissolved oxygen on the metabolic profile of a murine hybridoma grown in serum-free medium in continuous culture. *Biotechnol. Bioeng.* 54: 153–164.

Jang, K.S., Park, H.M., Kim, Y.G. et al. (2015). An on-line platform for the analysis of permethylated oligosaccharides using capillary liquid chromatography-electrospray ionization-tandem mass spectrometry. *J. Nanosci. Nanotechnol.* 15: 3962–3966.

Jassal, R., Jenkins, N., Charlwood, J. et al. (2001). Sialylation of human IgG-Fc carbohydrate by transfected rat alpha2,6-sialyltransferase. *Biochem. Biophys. Res. Commun.* 286: 243–249.

Jedrzejewski, P.M., del Val, I.J., Constantinou, A. et al. (2014). Towards controlling the glycoform: a model framework linking extracellular metabolites to antibody glycosylation. *Int. J. Mol. Sci.* 15: 4492–4522.

Jenkins, N. and Curling, E.M. (1994). Glycosylation of recombinant proteins: problems and prospects. *Enzyme Microb. Technol.* 16: 354–364.

Jenkins, N., Parekh, R.B., and James, D.C. (1996). Getting the glycosylation right: implications for the biotechnology industry. *Nat. Biotechnol.* 14: 975–981.

Jeong, Y.T., Choi, O., Lim, H.R. et al. (2008). Enhanced sialylation of recombinant erythropoietin in CHO cells by human glycosyltransferase expression. *J. Microbiol. Biotechnol.* 18: 1945–1952.

Jimenez del Val, I., Nagy, J.M., and Kontoravdi, C. (2011). A dynamic mathematical model for monoclonal antibody N-linked glycosylation and nucleotide sugar donor transport within a maturing Golgi apparatus. *Biotechnol. Progr.* 27: 1730–1743.

Jones, D., Kroos, N., Anema, R. et al. (2003). High-level expression of recombinant IgG in the human cell line per.c6. *Biotechnol. Progr.* 19: 163–168.

Kamoda, S. and Kakehi, K. (2006). Capillary electrophoresis for the analysis of glycoprotein pharmaceuticals. *Electrophoresis* 27: 2495–2504.

Kanda, Y., Imai-Nishiya, H., Kuni-Kamochi, R. et al. (2007a). Establishment of a GDP-mannose 4,6-dehydratase (GMD) knockout host cell line: a new strategy for generating completely non-fucosylated recombinant therapeutics. *J. Biotechnol.* 130: 300–310.

Kanda, Y., Yamada, T., Mori, K. et al. (2007b). Comparison of biological activity among nonfucosylated therapeutic IgG1 antibodies with three different N-linked Fc oligosaccharides: the high-mannose, hybrid, and complex types. *Glycobiology* 17: 104–118.

Kandzia, S., Grammel, N., Grabenhorst, E., and Conradt, H. (2010). *Quantitative N-Glycan Mapping of Glycoprotein Therapeutics by HPAEC-PAD: Glycosylation Characteristics of Different Recombinant Human EPO Products*. Springer Science and Business Media B.V.

Kaneko, Y., Nimmerjahn, F., and Ravetch, J.V. (2006). Anti-inflammatory activity of immunoglobulin G resulting from Fc sialylation. *Science* 313: 670–673.

Kannicht, C., Ramstrom, M., Kohla, G. et al. (2013). Characterisation of the post-translational modifications of a novel, human cell line-derived recombinant human factor VIII. *Thromb. Res.* 131: 78–88.

Karas, M. and Hillenkamp, F. (1988). Laser desorption ionization of proteins with molecular masses exceeding 10,000 daltons. *Anal. Chem.* 60: 2299–2301.

Kim, D.Y., Chaudhry, M.A., Kennard, M.L. et al. (2013). Fed-batch CHO cell t-PA production and feed glutamine replacement to reduce ammonia production. *Biotechnol. Progr.* 29: 165–175.

Ko, K., Brodzik, R., and Steplewski, Z. (2009). Production of antibodies in plants: approaches and perspectives. *Curr. Top. Microbiol. Immunol.* 332: 55–78.

Kochanowski, N., Blanchard, F., Cacan, R. et al. (2008). Influence of intracellular nucleotide and nucleotide sugar contents on recombinant interferon-gamma glycosylation during batch and fed-batch cultures of CHO cells. *Biotechnol. Bioeng.* 100: 721–733.

Konno, Y., Kobayashi, Y., Takahashi, K. et al. (2012). Fucose content of monoclonal antibodies can be controlled by culture medium osmolality for high antibody-dependent cellular cytotoxicity. *Cytotechnology* 64: 249–265.

Kornfeld, R. and Kornfeld, S. (1985). Assembly of asparagine-linked oligosaccharides. *Annu. Rev. Biochem.* 54: 631–664.

Kovacik, V., Hirsch, J., Kovac, P. et al. (1995). Oligosaccharide characterization using collision-induced dissociation fast-atom-bombardment mass-spectrometry – evidence for internal monosaccharide residue loss. *J. Mass Spectrom.* 30: 949–958.

Krambeck, F.J., Bennun, S.V., Narang, S. et al. (2009). A mathematical model to derive *N*-glycan structures and cellular enzyme activities from mass spectrometric data. *Glycobiology* 19: 1163–1175.

Krapp, S., Mimura, Y., Jefferis, R. et al. (2003). Structural analysis of human IgG-Fc glycoforms reveals a correlation between glycosylation and structural integrity. *J. Mol. Biol.* 325: 979–989.

Krokhin, O., Ens, W., Standing, K.G. et al. (2004). Site-specific *N*-glycosylation analysis: matrix-assisted laser desorption/ionization quadrupole-quadrupole time-of-flight tandem mass spectral signatures for recognition and identification of glycopeptides. *Rapid Commun. Mass Spectrom.* 18: 2020–2030.

Kunkel, J.P., Jan, D.C., Jamieson, J.C., and Butler, M. (1998). Dissolved oxygen concentration in serum-free continuous culture affects N-linked glycosylation of a monoclonal antibody. *J. Biotechnol.* 62: 55–71.

Kuno, A., Uchiyama, N., Koseki-Kuno, S. et al. (2005). Evanescent-field fluorescence-assisted lectin microarray: a new strategy for glycan profiling. *Nat. Methods* 2: 851–856.

Laine, R.A., Yoon, E., Mahier, T.J. et al. (1991). Non-reducing terminal linkage position determination in intact and permethylated synthetic oligosaccharides having a penultimate amino sugar: fast atom bombardment ionization, collisional-induced dissociation and tandem mass spectrometry. *Biol. Mass Spectrom.* 20: 505–514.

Lam, M.P., Lau, E., Siu, S.O. et al. (2011). Online combination of reversed-phase/reversed-phase and porous graphitic carbon liquid chromatography for multicomponent separation of proteomics and glycoproteomics samples. *Electrophoresis* 32: 2930–2940.

Lareau, N.M., May, J.C., and McLean, J.A. (2015). Non-derivatized glycan analysis by reverse phase liquid chromatography and ion mobility-mass spectrometry. *Analyst* 140: 3335–3338.

Laroy, W., Contreras, R., and Callewaert, N. (2006). Glycome mapping on DNA sequencing equipment. *Nat. Protoc.* 1: 397–405.

Lattova, E. and Perreault, H. (2013). The usefulness of hydrazine derivatives for mass spectrometric analysis of carbohydrates. *Mass Spectrom. Rev.* 32: 366–385.

Lattova, E., Perreault, H., and Krokhin, O. (2004). Matrix-assisted laser desorption/ionization tandem mass spectrometry and post-source decay fragmentation study of phenylhydrazones of N-linked oligosaccharides from ovalbumin. *J. Am. Soc. Mass Spectrom.* 15: 725–735.

Lattova, E., Snovida, S., Perreault, H., and Krokhin, O. (2005). Influence of the labeling group on ionization and fragmentation of carbohydrates in mass spectrometry. *J. Am. Soc. Mass Spectrom.* 16: 683–696.

Lennarz, W.J. (1987). Protein glycosylation in the endoplasmic reticulum: current topological issues. *Biochemistry* 26: 7205–7210.

Li, H. and D'Anjou, M. (2009). Pharmacological significance of glycosylation in therapeutic proteins. *Curr. Opin. Biotechnol.* 20: 678–684.

Liu, L., Stadheim, A., Hamuro, L. et al. (2011). Pharmacokinetics of IgG1 monoclonal antibodies produced in humanized *Pichia pastoris* with specific glycoforms: a comparative study with CHO produced materials. *Biologicals* 39: 205–210.

Liu, B., Spearman, M., Doering, J. et al. (2014a). The availability of glucose to CHO cells affects the intracellular lipid-linked oligosaccharide distribution, site occupancy and the *N*-glycosylation profile of a monoclonal antibody. *J. Biotechnol.* 170: 17–27.

Liu, J., Wang, F., Zhu, J. et al. (2014b). Highly efficient N-glycoproteomic sample preparation by combining C_{18} and graphitized carbon adsorbents. *Anal. Bioanal. Chem.* 406: 3103–3109.

Lomino, J.V., Naegeli, A., Orwenyo, J. et al. (2013). A two-step enzymatic glycosylation of polypeptides with complex *N*-glycans. *Bioorg. Med. Chem.* 21: 2262–2270.

Loos, A. and Steinkellner, H. (2012). IgG-Fc glycoengineering in non-mammalian expression hosts. *Arch. Biochem. Biophys.* 526: 167–173.

Mabashi-Asazuma, H., Kuo, C.W., Khoo, K.H., and Jarvis, D.L. (2015). Modifying an insect cell *N*-glycan processing pathway using CRISPR-Cas technology. *ACS Chem. Biol.* 10: 2199–2208.

Maley, F., Trimble, R.B., Tarentino, A.L., and Plummer, T.H. Jr. (1989). Characterization of glycoproteins and their associated oligosaccharides through the use of endoglycosidases. *Anal. Biochem.* 180: 195–204.

Marasco, D.M., Gao, J., Griffiths, K. et al. (2014). Development and characterization of a cell culture manufacturing process using quality by design (QbD) principles. *Adv. Biochem. Eng. Biotechnol.* 139: 93–121.

McCracken, N.A., Kowle, R., and Ouyang, A. (2014). Control of galactosylated glycoforms distribution in cell culture system. *Biotechnol. Progr.* 30: 547–553.

McDonald, A.G., Hayes, J.M., Bezak, T. et al. (2014). Galactosyltransferase 4 is a major control point for glycan branching in N-linked glycosylation. *J. Cell Sci.* 127: 5014–5026.

Mechref, Y., Hu, Y., Desantos-Garcia, J.L. et al. (2013). Quantitative glycomics strategies. *Mol. Cell. Proteomics* 12: 874–884.

Melmer, M., Stangler, T., Schiefermeier, M. et al. (2010). HILIC analysis of fluorescence-labeled *N*-glycans from recombinant biopharmaceuticals. *Anal. Bioanal. Chem.* 398: 905–914.

Melmer, M., Stangler, T., Premstaller, A., and Lindner, W. (2011). Comparison of hydrophilic-interaction, reversed-phase and porous graphitic carbon chromatography for glycan analysis. *J. Chromatogr. A* 1218: 118–123.

Meuris, L., Santens, F., Elson, G. et al. (2014). GlycoDelete engineering of mammalian cells simplifies N-glycosylation of recombinant proteins. *Nat. Biotechnol.* 32: 485–489.

Michael, C. and Rizzi, A.M. (2015). Quantitative isomer-specific *N*-glycan fingerprinting using isotope coded labeling and high performance liquid chromatography-electrospray ionization-mass spectrometry with graphitic carbon stationary phase. *J. Chromatogr. A* 1383: 88–95.

Moremen, K.W., Tiemeyer, M., and Nairn, A.V. (2012). Vertebrate protein glycosylation: diversity, synthesis and function. *Nat. Rev. Mol. Cell Biol.* 13: 448–462.

Moseley, H.N., Lane, A.N., Belshoff, A.C. et al. (2011). A novel deconvolution method for modeling UDP-*N*-acetyl-D-glucosamine biosynthetic pathways based on ^{13}C mass isotopologue profiles under non-steady-state conditions. *BMC Biol.* 9: 37.

Nakano, M., Kakehi, K., Taniguchi, N., and Kondo, A. (2011). Capillary electrophoresis and capillary electrophoresis-mass spectrometry for structural analysis of *N*-glycans derived from glycoproteins. In: *Capillary Electrophoresis of Carbohydrates: From Monosaccharides to Complex Polysaccharides* (ed. N. Volpi). Springer Science+Business.

Nam, J.H., Zhang, F., Ermonval, M. et al. (2008). The effects of culture conditions on the glycosylation of secreted human placental alkaline phosphatase produced in Chinese hamster ovary cells. *Biotechnol. Bioeng.* 100: 1178–1192.

Nargund, S., Qiu, J., and Goudar, C.T. (2015). Elucidating the role of copper in CHO cell energy metabolism using ^{13}C metabolic flux analysis. *Biotechnol. Progr.* 31: 1179–1186.

Naso, M.F., Tam, S.H., Scallon, B.J., and Raju, T.S. (2010). Engineering host cell lines to reduce terminal sialylation of secreted antibodies. *MAbs* 2: 519–527.

Neususs, C. and Pelzing, M. (2009). Capillary zone electrophoresis-mass spectrometry for the characterization of isoforms of intact glycoproteins. *Methods Mol. Biol.* 492: 201–213.

Nilsson, J. and Larson, G. (2013). Sialic acid capture-and-release and LC–MS^n analysis of glycopeptides. *Methods Mol. Biol.* 951: 79–100.

Noguchi, A., Mukuria, C.J., Suzuki, E., and Naiki, M. (1995). Immunogenicity of *N*-glycolylneuraminic acid-containing carbohydrate chains of recombinant human erythropoietin expressed in Chinese hamster ovary cells. *J. Biochem.* 117: 59–62.

Nose, M. and Wigzell, H. (1983). Biological significance of carbohydrate chains on monoclonal antibodies. *Proc. Natl. Acad. Sci. U.S.A.* 80: 6632–6636.

Nyberg, G.B., Balcarcel, R.R., Follstad, B.D. et al. (1999). Metabolic effects on recombinant interferon-gamma glycosylation in continuous culture of Chinese hamster ovary cells. *Biotechnol. Bioeng.* 62: 336–347.

Ogorek, C., Jordan, I., Sandig, V., and von Horsten, H.H. (2012). Fucose-targeted glycoengineering of pharmaceutical cell lines. *Methods Mol. Biol.* 907: 507–517.

Okamoto, M., Takahashi, K., Doi, T., and Takimoto, Y. (1997). High-sensitivity detection and postsource decay of 2-aminopyridine-derivatized oligosaccharides with matrix-assisted laser desorption/ionization mass spectrometry. *Anal. Chem.* 69: 2919–2926.

O'Neill, R.A. (1996). Enzymatic release of oligosaccharides from glycoproteins for chromatographic and electrophoretic analysis. *J. Chromatogr. A* 720: 201–215.

Onitsuka, M., Kim, W.D., Ozaki, H. et al. (2012). Enhancement of sialylation on humanized IgG-like bispecific antibody by overexpression of alpha2,6-sialyltransferase derived from Chinese hamster ovary cells. *Appl. Microbiol. Biotechnol.* 94: 69–80.

Oyama, T., Yodohsi, M., Yamane, A. et al. (2011). Rapid and sensitive analyses of glycoprotein-derived oligosaccharides by liquid chromatography and laser-induced fluorometric detection capillary electrophoresis. *J. Chromatogr. B Anal. Technol. Biomed. Life Sci.* 879: 2928–2934.

Pacis, E., Yu, M., Autsen, J. et al. (2011). Effects of cell culture conditions on antibody N-linked glycosylation – what affects high mannose 5 glycoform. *Biotechnol. Bioeng.* 108: 2348–2358.

Pande, S., Rahardjo, A., Livingston, B., and Mujacic, M. (2015). Monensin, a small molecule ionophore, can be used to increase high mannose levels on monoclonal antibodies generated by Chinese hamster ovary production cell-lines. *Biotechnol. Bioeng.* 112: 1383–1394.

Park, J.H., Wang, Z., Jeong, H.J. et al. (2012). Enhancement of recombinant human EPO production and glycosylation in serum-free suspension culture of CHO cells through expression and supplementation of 30Kc19. *Appl. Microbiol. Biotechnol.* 96: 671–683.

Park, H.M., Hwang, M.P., Kim, Y.W. et al. (2015). Mass spectrometry-based N-linked glycomic profiling as a means for tracking pancreatic cancer metastasis. *Carbohydr. Res.* 413: 5–11.

Patel, T.P. and Parekh, R.B. (1994). Release of oligosaccharides from glycoproteins by hydrazinolysis. *Methods Enzymol.* 230: 57–66.

van Patten, S.M., Hughes, H., Huff, M.R. et al. (2007). Effect of mannose chain length on targeting of glucocerebrosidase for enzyme replacement therapy of Gaucher disease. *Glycobiology* 17: 467–478.

Pels Rijcken, W.R., Overdijk, B., van den Eijnden, D.H., and Ferwerda, W. (1995). The effect of increasing nucleotide-sugar concentrations on the incorporation of sugars into glycoconjugates in rat hepatocytes. *Biochem. J.* 305 (Pt 3): 865–870.

Pompach, P., Chandler, K.B., Lan, R. et al. (2012). Semi-automated identification of N-Glycopeptides by hydrophilic interaction chromatography, nano-reverse-phase LC-MS/MS, and glycan database search. *J. Proteome Res.* 11: 1728–1740.

Prater, B.D., Connelly, H.M., Qin, Q., and Cockrill, S.L. (2009). High-throughput immunoglobulin G N-glycan characterization using rapid resolution reverse-phase chromatography tandem mass spectrometry. *Anal. Biochem.* 385: 69–79.

Prime, S., Dearnley, J., Ventom, A.M. et al. (1996). Oligosaccharide sequencing based on exo- and endoglycosidase digestion and liquid chromatographic analysis of the products. *J. Chromatogr. A* 720: 263–274.

Rademacher, T.W., Jaques, A., and Williams, P.J. (1996). The defining characteristics of immunoglobulin glycosylation. In: *Abnormalities of IgG Glycosylation and Immunological Disorders* (ed. D. Isenberg and T. Rademacher). New York: Wiley.

Rathore, A.S., Kumar Singh, S., Pathak, M. et al. (2015). Fermentanomics: relating quality attributes of a monoclonal antibody to cell culture process variables and raw materials using multivariate data analysis. *Biotechnol. Progr.* 31: 1586–1599.

Raymond, C., Robotham, A., Spearman, M. et al. (2015). Production of alpha2,6-sialylated IgG1 in CHO cells. *MAbs* 7: 571–583.

Rearick, J.I., Chapman, A., and Kornfeld, S. (1981). Glucose starvation alters lipid-linked oligosaccharide biosynthesis in Chinese hamster ovary cells. *J. Biol. Chem.* 256: 6255–6261.

Reinhold, V.N., Reinhold, B.B., and Costello, C.E. (1995). Carbohydrate molecular weight profiling, sequence, linkage, and branching data: ES–MS and CID. *Anal. Chem.* 67: 1772–1784.

Restelli, V., Wang, M.D., Huzel, N. et al. (2006). The effect of dissolved oxygen on the production and the glycosylation profile of recombinant human erythropoietin produced from CHO cells. *Biotechnol. Bioeng.* 94: 481–494.

Reusch, D., Haberger, M., Kailich, T. et al. (2013). High-throughput glycosylation analysis of therapeutic immunoglobulin G by capillary gel electrophoresis using a DNA analyzer. *MAbs* 6: 185–196.

Reusch, D., Haberger, M., Maier, B. et al. (2015). Comparison of methods for the analysis of therapeutic immunoglobulin G Fc-glycosylation profiles – Part 1: Separation-based methods. *MAbs* 7: 167–179.

Richter, W.J., Muller, D.R., and Domon, B. (1990). Tandem mass spectrometry in structural characterization of oligosaccharide residues in glycoconjugates. *Methods Enzymol.* 193: 607–623.

Rodig, J., Hennig, R., Schwarzer, J. et al. (2009). *Optimized CGE-LIF-Based Glycan Analysis for High-Throughput Applications.* Springer Science+Business Media.

Rodig, J., Hennig, R., Schwarzer, J. et al. (2012). Optimized CGE-LIF-based glycan analysis for high-throughput applications. Proceedings of the 21st Annual Meeting of the European Society for Animal Cell Technology (ESACT) Dublin, Volume 5, Ireland (7–10 June 2009), pp. 599–603.

Rodrigues, J.A., Taylor, A.M., Sumpton, D.P. et al. (2007). Mass spectrometry of carbohydrates: newer aspects. *Adv. Carbohydr. Chem. Biochem.* 61: 59–141.

Rodriguez, J., Spearman, M., Huzel, N., and Butler, M. (2005). Enhanced production of monomeric interferon-beta by CHO cells through the control of culture conditions. *Biotechnol. Prog.* 21: 22–30.

Rodriguez, J., Spearman, M., Tharmalingam, T. et al. (2010). High productivity of human recombinant beta-interferon from a low-temperature perfusion culture. *J. Biotechnol.* 150: 509–518.

Rothman, R.J., Perussia, B., Herlyn, D., and Warren, L. (1989a). Antibody-dependent cytotoxicity mediated by natural killer cells is enhanced by castanospermine-induced alterations of IgG glycosylation. *Mol. Immunol.* 26: 1113–1123.

Rothman, R.J., Warren, L., Vliegenthart, J.F., and Hard, K.J. (1989b). Clonal analysis of the glycosylation of immunoglobulin G secreted by murine hybridomas. *Biochemistry* 28: 1377–1384.

Royle, L., Radcliffe, C.M., Dwek, M., and Rudd, P. (2007). *Detailed Structural Analysis of N-Glycans Released from Glycoproteins in SDS-PAGE Gel Bands Using HPLC Combined with Exoglycosidase Array Digestions.* Totowa, NJ: Humana Press.

Royle, L., Campbell, M.P., Radcliffe, C.M. et al. (2008). HPLC-based analysis of serum *N*-glycans on a 96-well plate platform with dedicated database software. *Anal. Biochem.* 376: 1–12.

Rudd, P.M., Colominas, C., Royle, L. et al. (2001). A high-performance liquid chromatography based strategy for rapid, sensitive sequencing of N-linked oligosaccharide modifications to proteins in sodium dodecyl sulphate polyacrylamide electrophoresis gel bands. *Proteomics* 1: 285–294.

Ruhaak, L.R., Zauner, G., Huhn, C. et al. (2010). Glycan labeling strategies and their use in identification and quantification. *Anal. Bioanal. Chem.* 397: 3457–3481.

Rustandi, R.R., Anderson, C., and Hamm, M. (2013). Application of capillary electrophoresis in glycoprotein analysis. *Methods Mol. Biol.* 988: 181–197.

Ryll, T. (2003). Mammalian cell culture process for producing glycoproteins. US Patent 6528286 B1.

Sareneva, T., Pirhonen, J., Cantell, K., and Julkunen, I. (1995). *N*-glycosylation of human interferon-gamma: glycans at Asn-25 are critical for protease resistance. *Biochem. J.* 308 (Pt 1): 9–14.

Scallon, B.J., Tam, S.H., McCarthy, S.G. et al. (2007). Higher levels of sialylated Fc glycans in immunoglobulin G molecules can adversely impact functionality. *Mol. Immunol.* 44: 1524–1534.

Schachter, H. (1986). Biosynthetic controls that determine the branching and microheterogeneity of protein-bound oligosaccharides. *Adv. Exp. Med. Biol.* 205: 53–85.

Schaub, J., Clemens, C., Schorn, P. et al. (2010). CHO gene expression profiling in biopharmaceutical process analysis and design. *Biotechnol. Bioeng.* 105: 431–438.

Sealover, N.R., Davis, A.M., Brooks, J.K. et al. (2013). Engineering Chinese hamster ovary (CHO) cells for producing recombinant proteins with simple glycoforms by zinc-finger nuclease (ZFN)—mediated gene knockout of mannosyl (alpha-1,3-)-glycoprotein beta-1,2-*N*-acetylglucosaminyltransferase (Mgat1). *J. Biotechnol.* 167: 24–32.

Sellick, C.A., Croxford, A.S., Maqsood, A.R. et al. (2011). Metabolite profiling of recombinant CHO cells: designing tailored feeding regimes that enhance recombinant antibody production. *Biotechnol. Bioeng.* 108: 3025–3031.

Sellick, C.A., Croxford, A.S., Maqsood, A.R. et al. (2015). Metabolite profiling of CHO cells: molecular reflections of bioprocessing effectiveness. *Biotechnol. J.* 10: 1434–1445.

Shahrokh, Z., Royle, L., Saldova, R. et al. (2011). Erythropoietin produced in a human cell line (Dynepo) has significant differences in glycosylation compared with erythropoietins produced in CHO cell lines. *Mol. Pharm.* 8: 286–296.

Shields, R.L., Lai, J., Keck, R. et al. (2002). Lack of fucose on human IgG1 N-linked oligosaccharide improves binding to human Fcgamma RIII and antibody-dependent cellular toxicity. *J. Biol. Chem.* 277: 26733–26740.

Shinkawa, T., Nakamura, K., Yamane, N. et al. (2003). The absence of fucose but not the presence of galactose or bisecting *N*-acetylglucosamine of human IgG1 complex-type oligosaccharides shows the critical role of enhancing antibody-dependent cellular cytotoxicity. *J. Biol. Chem.* 278: 3466–3473.

Shubhakar, A., Reiding, K.R., Gardner, R.A. et al. (2015). High-throughput analysis and automation for glycomics studies. *Chromatographia* 78: 321–333.

Sinclair, A.M. and Elliott, S. (2005). Glycoengineering: the effect of glycosylation on the properties of therapeutic proteins. *J. Pharm. Sci.* 94: 1626–1635.

Sola, R.J. and Griebenow, K. (2010). Glycosylation of therapeutic proteins: an effective strategy to optimize efficacy. *BioDrugs* 24: 9–21.

Son, Y.D., Jeong, Y.T., Park, S.Y., and Kim, J.H. (2011). Enhanced sialylation of recombinant human erythropoietin in Chinese hamster ovary cells by combinatorial engineering of selected genes. *Glycobiology* 21: 1019–1028.

Sou, S.N., Sellick, C., Lee, K. et al. (2015). How does mild hypothermia affect monoclonal antibody glycosylation? *Biotechnol. Bioeng.* 112: 1165–1176.

Spahn, P.N., Hansen, A.H., Hansen, H.G. et al. (2015). A Markov chain model for N-linked protein glycosylation – towards a low-parameter tool for model-driven glycoengineering. *Metab. Eng.* 33: 52–66.

St. Amand, M.M., Tran, K., Radhakrishnan, D. et al. (2014). Controllability analysis of protein glycosylation in CHO cells. *PLoS ONE* 9, e87973.

Stadheim, T.A., Li, H., Kett, W. et al. (2008). Use of high-performance anion exchange chromatography with pulsed amperometric detection for O-glycan determination in yeast. *Nat. Protoc.* 3: 1026–1031.

Stadlmann, J., Pabst, M., Kolarich, D. et al. (2008). Analysis of immunoglobulin glycosylation by LC-ESI-MS of glycopeptides and oligosaccharides. *Proteomics* 8: 2858–2871.

Stanley, P. (2011). Golgi glycosylation. *Cold Spring Harbor Perspect. Biol.* 3: a005199.

Stanley, P., Schachter, H., and Taniguchi, N. (2009). *Essentials of Glycobiology*, 2, Chapter 8e. Cold Spring Harbor, NY: Cold Spring Harbor Laboratory Press.

Starr, C.M., Masada, R.I., Hague, C. et al. (1996). Fluorophore-assisted carbohydrate electrophoresis in the separation, analysis, and sequencing of carbohydrates. *J. Chromatogr. A* 720: 295–321.

Stavenhagen, K., Plomp, R., and Wuhrer, M. (2015). Site-specific protein N- and O-glycosylation analysis by a C18-porous graphitized carbon–liquid chromatography-electrospray ionization mass spectrometry approach using pronase treated glycopeptides. *Anal. Chem.* 87: 11691–11699.

Stockmann, H., Adamczyk, B., Hayes, J., and Rudd, P.M. (2013). Automated, high-throughput IgG-antibody glycoprofiling platform. *Anal. Chem.* 85: 8841–8849.

Surve, T. and Gadgil, M. (2015). Manganese increases high mannose glycoform on monoclonal antibody expressed in CHO when glucose is absent or limiting: implications for use of alternate sugars. *Biotechnol. Progr.* 31: 460–467.

Suzuki, S., Kakehi, K., and Honda, S. (1996). Comparison of the sensitivities of various derivatives of oligosaccharides in LC/MS with fast atom bombardment and electrospray ionization interfaces. *Anal. Chem.* 68: 2073–2083.

Swiech, K., de Freitas, M.C., Covas, D.T., and Picanco-Castro, V. (2015). Recombinant glycoprotein production in human cell lines. *Methods Mol. Biol.* 1258: 223–240.

Szabo, Z., Guttman, A., Rejtar, T., and Karger, B.L. (2010). Improved sample preparation method for glycan analysis of glycoproteins by CE-LIF and CE-MS. *Electrophoresis* 31: 1389–1395.

Szabo, Z., Guttman, A., Bones, J., and Karger, B.L. (2011). Rapid high-resolution characterization of functionally important monoclonal antibody N-glycans by capillary electrophoresis. *Anal. Chem.* 83: 5329–5336.

Szekrenyes, A., Park, S.S., Santos, M. et al. (2016). Multi-site N-glycan mapping study 1: capillary electrophoresis - laser induced fluorescence. *MAbs* 8 (1): 56–64.

Tachibana, H., Kim, J.Y., and Shirahata, S. (1997). Building high affinity human antibodies by altering the glycosylation on the light chain variable region in N-acetylglucosamine-supplemented hybridoma cultures. *Cytotechnology* 23: 151–159.

Tarentino, A.L., Gomez, C.M., and Plummer, T.H. Jr. (1985). Deglycosylation of asparagine-linked glycans by peptide:N-glycosidase F. *Biochemistry* 24: 4665–4671.

Taschwer, M., Hackl, M., Hernandez Bort, J.A. et al. (2012). Growth, productivity and protein glycosylation in a CHO EpoFc producer cell line adapted to glutamine-free growth. *J. Biotechnol.* 157: 295–303.

Tayi, V. and Butler, M. (2014). Methods to produce single glycoform monoclonal antibodies. U.S. Provisional Patent February 18 Patent Application US20170051328A1, pp. 1887–216.

Templeton, N., Dean, J., Reddy, P., and Young, J.D. (2013). Peak antibody production is associated with increased oxidative metabolism in an industrially relevant fed-batch CHO cell culture. *Biotechnol. Bioeng.* 110: 2013–2024.

Tharmalingam, T., Adamczyk, B., Doherty, M.A. et al. (2013). Strategies for the profiling, characterisation and detailed structural analysis of N-linked oligosaccharides. *Glycoconj. J.* 30: 137–146.

Tharmalingam, T., Wu, C.H., Callahan, S., and Goudar, C.T. (2015). A framework for real-time glycosylation monitoring (RT-GM) in mammalian cell culture. *Biotechnol. Bioeng.* 112: 1146–1154.

Tian, E. and Ten Hagen, K.G. (2009). Recent insights into the biological roles of mucin-type O-glycosylation. *Glycoconj. J.* 26: 325–334.

Umana, P., Jean-Mairet, J., Moudry, R. et al. (1999). Engineered glycoforms of an antineuroblastoma IgG1 with optimized antibody-dependent cellular cytotoxic activity. *Nat. Biotechnol.* 17: 176–180.

Valley, U., Nimtz, M., Conradt, H.S., and Wagner, R. (1999). Incorporation of ammonium into intracellular UDP-activated *N*-acetylhexosamines and into carbohydrate structures in glycoproteins. *Biotechnol. Bioeng.* 64: 401–417.

Villacres, C., Tayi, V.S., Lattova, E. et al. (2015). Low glucose depletes glycan precursors, reduces site occupancy and galactosylation of a monoclonal antibody in CHO cell culture. *Biotechnol. J.* 10: 1051–1066.

Wang, L.X. and Lomino, J.V. (2012). Emerging technologies for making glycan-defined glycoproteins. *ACS Chem. Biol.* 7: 110–122.

Wang, Z., Hilder, T.L., van der Drift, K. et al. (2013). Structural characterization of recombinant alpha-1-antitrypsin expressed in a human cell line. *Anal. Biochem.* 437: 20–28.

Wang, Q., Stuczynski, M., Gao, Y., and Betenbaugh, M.J. (2015). Strategies for engineering protein *N*-glycosylation pathways in mammalian cells. *Methods Mol. Biol.* 1321: 287–305.

Webster, D.E. and Thomas, M.C. (2012). Post-translational modification of plant-made foreign proteins; glycosylation and beyond. *Biotechnol. Adv.* 30: 410–418.

Weitzhandler, M., Rohrer, J., Thayer, J.R., and Avdalovic, N. (1998). HPAEC-PAD analysis of monosaccharides released by exoglycosidase digestion using the CarboPac MA1 column. *Methods Mol. Biol.* 76: 71–78.

Wildt, S. and Gerngross, T.U. (2005). The humanization of *N*-glycosylation pathways in yeast. *Nat. Rev. Microbiol.* 3: 119–128.

Wong, N.S., Yap, M.G., and Wang, D.I. (2006). Enhancing recombinant glycoprotein sialylation through CMP-sialic acid transporter over expression in Chinese hamster ovary cells. *Biotechnol. Bioeng.* 93: 1005–1016.

Wong, D.C., Wong, N.S., Goh, J.S. et al. (2010a). Profiling of *N*-glycosylation gene expression in CHO cell fed-batch cultures. *Biotechnol. Bioeng.* 107: 516–528.

Wong, N.S., Wati, L., Nissom, P.M. et al. (2010b). An investigation of intracellular glycosylation activities in CHO cells: effects of nucleotide sugar precursor feeding. *Biotechnol. Bioeng.* 107: 321–336.

Wright, A. and Morrison, S.L. (1997). Effect of glycosylation on antibody function: implications for genetic engineering. *Trends Biotechnol.* 15: 26–32.

Wright, A., Sato, Y., Okada, T. et al. (2000). In vivo trafficking and catabolism of IgG1 antibodies with Fc associated carbohydrates of differing structure. *Glycobiology* 10: 1347–1355.

Wurm, F.M. (2004). Production of recombinant protein therapeutics in cultivated mammalian cells. *Nat. Biotechnol.* 22: 1393–1398.

Xie, L. and Wang, D.I. (1997). Integrated approaches to the design of media and feeding strategies for fed-batch cultures of animal cells. *Trends Biotechnol.* 15: 109–113.

Xu, X., Nagarajan, H., Lewis, N.E. et al. (2011). The genomic sequence of the Chinese hamster ovary (CHO)-K1 cell line. *Nat. Biotechnol.* 29: 735–741.

Yagi, Y., Yamamoto, S., Kakehi, K. et al. (2011). Application of partial-filling capillary electrophoresis using lectins and glycosidases for the characterization of oligosaccharides in a therapeutic antibody. *Electrophoresis* 32: 2979–2985.

Yamane-Ohnuki, N., Kinoshita, S., Inoue-Urakubo, M. et al. (2004). Establishment of FUT8 knockout Chinese hamster ovary cells: an ideal host cell line for producing completely defucosylated antibodies with enhanced antibody-dependent cellular cytotoxicity. *Biotechnol. Bioeng.* 87: 614–622.

Yang, M. and Butler, M. (2000). Effects of ammonia on CHO cell growth, erythropoietin production, and glycosylation. *Biotechnol. Bioeng.* 68: 370–380.

Yang, Z., Wang, S., Halim, A. et al. (2015). Engineered CHO cells for production of diverse, homogeneous glycoproteins. *Nat. Biotechnol.* 33: 842–844.

Yu, M., Brown, D., Reed, C. et al. (2012). Production, characterization, and pharmacokinetic properties of antibodies with N-linked mannose-5 glycans. *MAbs* 4: 475–487.

Zaia, J. (2013). Capillary electrophoresis-mass spectrometry of carbohydrates. *Methods Mol. Biol.* 984: 13–25.

Zanghi, J.A., Mendoza, T.P., Knop, R.H., and Miller, W.M. (1998). Ammonia inhibits neural cell adhesion molecule polysialylation in Chinese hamster ovary and small cell lung cancer cells. *J. Cell. Physiol.* 177: 248–263.

Zauner, G., Deelder, A.M., and Wuhrer, M. (2011). Recent advances in hydrophilic interaction liquid chromatography (HILIC) for structural glycomics. *Electrophoresis* 32: 3456–3466.

Zhang, P., Zhang, Y., Xue, X. et al. (2011). Relative quantitation of glycans using stable isotopic labels 1-(d0/d5) phenyl-3-methyl-5-pyrazolone by mass spectrometry. *Anal. Biochem.* 418: 1–9.

Zhang, P., Chan, K.F., Haryadi, R. et al. (2013). CHO glycosylation mutants as potential host cells to produce therapeutic proteins with enhanced efficacy. *Adv. Biochem. Eng. Biotechnol.* 131: 63–87.

Zhao, Y., Szeto, S.S., Kong, R.P. et al. (2014). Online two-dimensional porous graphitic carbon/reversed phase liquid chromatography platform applied to shotgun proteomics and glycoproteomics. *Anal. Chem.* 86: 12172–12179.

Zheng, K., Bantog, C., and Bayer, R. (2011). The impact of glycosylation on monoclonal antibody conformation and stability. *MAbs* 3: 568–576.

Zhong, X., Cooley, C., Seth, N. et al. (2012). Engineering novel Lec1 glycosylation mutants in CHO-DUKX cells: molecular insights and effector modulation of *N*-acetylglucosaminyltransferase I. *Biotechnol. Bioeng.* 109: 1723–1734.

Zhong, X., Chen, Z., Snovida, S. et al. (2015). Capillary electrophoresis-electrospray ionization-mass spectrometry for quantitative analysis of glycans labeled with multiplex carbonyl-reactive tandem mass tags. *Anal. Chem.* 87: 6527–6534.

Zhou, Q., Shankara, S., Roy, A. et al. (2008). Development of a simple and rapid method for producing non-fucosylated oligomannose containing antibodies with increased effector function. *Biotechnol. Bioeng.* 99: 652–665.

Zhu, Z., Su, X., Clark, D.F. et al. (2013). Characterizing O-linked glycopeptides by electron transfer dissociation: fragmentation rules and applications in data analysis. *Anal. Chem.* 85: 8403–8411.

Part II

Bioreactors

4

Bioreactors for Stem Cell and Mammalian Cell Cultivation

Ana Fernandes-Platzgummer*, Sara M. Badenes*, Cláudia L. da Silva, and Joaquim M. S. Cabral

Department of Bioengineering and Institute for Bioengineering and Biosciences, Insituto Superior Técnico, Universidade de Lisboa, Lisboa, Portugal

4.1 Overview of (Mammalian and Stem) Cell Culture Engineering

The use of cultured cells for the production of viral vaccines was the first application of mammalian cell culture technology. In the mid-1950s, the need of mass cultivation techniques for vaccine production led to the development of large-scale cell culture bioreactors (Fenge and Lüllau 2006). The hybridoma technology had been established in 1975 (Kohler and Milstein 1975), and in the 1980s, several companies started producing monoclonal antibodies for research and diagnostic purposes, using small hollow-fiber bioreactors (Papoutsakis 2009). Recombinant DNA technology enabled the production of several protein therapeutics in mammalian cell culture in the 1980s. Tolbert et al. (1980, 1982) pioneered the development of large-scale mammalian cell culture technology identifying several important scale-up issues. Since the mid-1990s, while mammalian cell culture was maturing, the available knowledge has also been applied to the design of bioreactors for stem cell culture. In recent years, the potential of stem cell research for Tissue Engineering and Cellular Therapies has become established, which will require cell production on a large-scale using bioreactors.

In a biological system, extensive studies are required to understand cell needs concerning cell growth, metabolism, as well as focusing genetic manipulation, protein, or other product expression. Then, when choosing and scaling-up the appropriate bioreactor, strategy for a specific biological system is essential for the understanding of the influence of the complexity of the fluid-mechanical,

**These authors equally contribute to this chapter.

Bioprocessing Technology for Production of Biopharmaceuticals and Bioproducts, First Edition.
Edited by Claire Komives and Weichang Zhou.
© 2019 John Wiley & Sons, Inc. Published 2019 by John Wiley & Sons, Inc.

nutritional, and physicochemical environment in the bioreactors in the cells, which depend on their intrinsic characteristics (Zhong 2010). Importantly, cells must be cultured in conditions compliant with current good manufacturing practice (cGMP) standards. There are several differences between biopharmaceutical production using mammalian cell culture and stem cell production schemes that are relevant for the bioreactor design. Of notice, in the latter case, the target product is a cell, rather than a protein or a vector.

4.1.1 Cell Products for Therapeutics

The rapid growth in biopharmaceutical industry, with a surge of protein, antibody, and peptide drugs expected in the next 10–20 years, has been led by various products produced by mammalian cells. Different therapeutic and diagnostic products from mammalian cells culture are been produced since 1996, as examples, *Herceptin*™, a monoclonal antibody (mAb) for treating breast cancer; *Enbrel*™, an immunoglobulin- tumor necrosis factor (TNF) receptor fusion protein for treating rheumatoid arthritis; and *Vaqta*™, an inactivated hepatitis A vaccine (Chu and Robinson 2001). Between January 2008 and June 2011, U. S. Food and Drug Administration (FDA) approved 12 recombinant proteins produced using mammalian expression systems (Table 4.1). Additionally, more than 50% of the therapeutic proteins on the market are produced using those systems since mammalian cells are able to synthetize proteins similar to those naturally occurring in humans, regarding molecular structures and biochemical properties (Zhu 2012).

A consequence of the increasing demand for these products has been the expansion of the capacity of those industries, which has been accomplished with large-scale cell culture production in reactors. Large-scale cultivation of mammalian cells is usually performed in conventional stirred tanks with homogeneous mixing of cells, gases, and nutrients, where issues such as adaption of cells to suspension culture, shear sensitivity, and oxygen supply are mostly solved. Fed-batch feeding strategies, where concentrated key nutrients are added to the culture during the process, are the most common processes, which can be operated at scales up to 20 000 l working volume (Birch and Racher 2006). There are more diversified production systems for low-volume and specific applications, like roller bottles (RBs), stacked plates, and hollow fibers, among others. Research and development on traditional bioreactor designs have considerably diminished after the remarkable advances in the 1980s and early 1990s and have been focused on the increase of productivity ensuring product quality and consistency, the optimization of feeding strategies, the improvement of process control, the removal of animal-derived components from cell culture medium and the control of carbon dioxide (CO_2) accumulation (Hu and Aunins 1997). On the other hand, during the past 10 years, the disposable manufacturing industry has grown exponentially,

Table 4.1 Biopharmaceutical products from mammalian expression systems approved by FDA, between January 2008 and June 2011.

Product	Year approved	Manufacturer	Indication
Belatacept (CTL4-Ig fusion)	2011	BMS	Prevention of acute rejection in adult kidney transplant patients
Yervoy (Ipilimumab)	2011	BMS	Metastatic melanoma
Benlysta (Belimumab)	2011	HGS	Sytemic lupus erythematosus
Prolia (denosumab)	2010	Amgen	Osteoporosis
Vpriv (Velaglucerase)	2010	Shire	Gaucher disease
Lumizyme (Alglucosidase alfa)	2010	Genzyme	Pompe disease
Actemra (Tocilizamab)	2010	Genentech	Systemic juvenile idiopathic arthritis
Stelama (ustekinumab)	2009	Centocor/JandJ	Plaque psoriasis
Arzerra (Ofatumumab)	2009	Genmab	Chronic Lymphocytic Leukemia
Simponi (Golimumab)	2009	Centocor/JandJ	Immune dysfunction-related arthritis
Ilaris (Canakinumab)	2009	Novartis	Cryopyrin-associated periodic syndromes
Arcalyst (Rilonacept)	2008	Regeneron	CAPS and FCAS

CAPS, Cryopyrin-Associated Periodic Syndromes; FCAS, Familial Cold Autoinflammatory Syndrome
Source: Zhu (2012). Adapted from Elsevier.

and disposable bioreactors have been widely utilized for preclinical, clinical, and commercial-scale biotechnological applications (Loffelholz et al. 2014). This rapid shift from traditional bioreactors to single-use systems due to their simplicity, safety, and flexibility is revolutionizing how biotherapeutics are manufactured today.

Suspension cell culture processes are usually the chosen system for large-scale production because of the grounded knowledge of the principles of scale-up and an easier process control compared with the adherent cell systems. The most common systems being used are the stirred tank reactor (STR), the airlift reactor (ALR), and more recently, the disposable bioreactors such as the Wave bioreactor (Rodrigues et al. 2010a). On the other hand, adherent culture systems have been developed and optimized due to the intrinsic anchorage-dependent nature of most mammalian cells and to solve problems inherent of suspension culture systems, such as cell damage due to hydrody-namic shear forces aroused by aeration and agitation. The most well-know system to cultivate adherent cells on a larger scale is the microcarrier culture in which cells are immobilized on particles that are suspended in the culture medium. Microcarrier culture can be performed in different bioreactor config-urations, the STR being the most common. Alternative bioreactors have been tested for anchorage-dependent cell culture such as hollow-fiber bioreactors and packed bed reactors. These are systems where mammalian cells are immobilized in the membrane capillaries or on carriers matrix, respectively, and are perfused with culture medium, contained in an external reservoir (Rodrigues et al. 2010a). However, these two systems are difficult to scale up.

Different approaches are emerging in the scale-up of bioreactors strategy, whether to increase the size or the number of reactors when moving to indus-trial scale. Advantages of multiple reactors strategy are the operational flexibil-ity, ease of start-up, maintenance, and sterilization, among others. Rouf et al. reported an economic analysis of single versus multiple bioreactor scale-up strategy for the Tissue Plasminogen activator (t-PA) production from Chinese hamster ovary (CHO) cells (Rouf et al. 2000). They concluded that the benefits of increasing the size of the bioreactor diminishes when the integrated process is considered, even more when the product purification is a major part of the overall cost. The multiple bioreactor approach increases the return of invest-ment of the base process by 122%.

The CHO cells and the murine myeloma lines SP2/0 and Nonsecreting null (NS0) are the most commonly used mammalian cell lines for the production of the approved biopharmaceutical products (Chu and Robinson 2001; Wurm 2004; Zhu 2012), from which several are industrially produced in bioreactors (Table 4.2). In the case of recombinant therapeutic products, which are mainly recombinant proteins or antibodies and require significant production quanti-ties, at least 70% of the licensed processes use stirred tank bioreactors.

4.1.2 Cell as a Product: Stem Cells

Stem cells have been envisioned as an unlimited cell source for Regenerative Medicine applications due to their properties of self-renewal, differentiation, and *in vivo* engraftment (Kirouac and Zandstra 2008), and also an important tool for developing *in vitro* model systems to test drugs and chemicals and predict toxicity in humans (Davila et al. 2004; McNeish 2007). Both applica-tions, however, still represent a challenge, mainly due to technical limitations in scaling-up stem cell cultures. In addition, the controlled expansion and differ-entiation of stem cells present technical challenges due to the complex kinetics

Table 4.2 Industrial products with biological license approval (BLA), generated in mammalian culture in bioreactors, listed on the Current Food and Drug Administration website (fda.gov).

Product type	Mammalian cell culture	Bioreactor type	BLA name
Recombinant therapeutics	CHO	Stirred-tank bioreactor, suspension	Enbrel™, Herceptin™, Benefix™, Avonex™, Rituxan™
		Continuous perfusion	ReFacto™
	NSO, recombinant	Stirred-tank bioreactor, suspension, fed-batch	Synagis™
	Murine myeloma cell line, recombinant	Stirred-tank bioreactor, suspension, perfusion	Simulect™
Vaccines	Human MRC-5 diploid fibroblasts	Continuous fed-batch culture	Vaqta™
Diagnostic products	Murine hybridoma cell line	Hollow fiber	ProstaScint™

MRC, Medical Research Council

of the heterogeneous starting culture population and the multiple interactions between culture parameters as growth factor concentrations, dissolved oxygen (DO) tensions and cell–cell interactions, among others (Rodrigues et al. 2011a; Serra et al. 2012).

As stem cell-based therapeutics progress toward clinical trials and drug discovery platforms are developed, the robust large-scale production of well-characterized human stem cells in tightly controlled conditions is required to meet the doses needed, with attested quality. Several challenges exist across the scale-up process for obtaining large batches of stem cells and/or their progeny, where the cells themselves are the target product. Contrary to the biotherapeutic and vaccine markets in which the product is contained in the culture supernatant, when cells are the product, these should be harvested without cell damage, and the yield of harvesting depends mainly on cell dissociation and separation methods (Rowley et al. 2012). Recently, Rodrigues and collaborators (Rodrigues et al. 2011a) published an overview of the progress in cultivating different stem cell populations in different bioreactor systems, describing the recent developments and technologies established. During the last few years, the expansion and/or differentiation of embryonic stem cells (ESCs) and adult stem cells, namely mesenchymal stem cell (MSC), hematopoietic stem cell (HSC), and neural stem cell (NSC), have been evaluated using several types of bioreactors (Table 4.3).

Stirred and perfused bioreactors have significant advantages when compared to static systems routinely used in stem cell culture: a controlled environment

Table 4.3 Stem cell cultivation in bioreactors.

Stem cell culture	Bioreactor type	References
Mouse and human ESC		
As aggregates	Stirred suspension bioreactor	Zandstra et al., (2003); Kehoe et al., (2008); Singh et al., (2010); Abbasalizadeh et al. (2012); Hunt et al. (2014)
	Rotating wall vessel	Fridley et al. (2010)
Encapsulation	Stirred suspension bioreactor	Jing et al. (2010)
On microcarriers	Stirred suspension bioreactor	Oh et al. (2009); Fernandes-Platzgummer et al., (2011); Chen et al., (2014); Fernandes-Platzgummer et al., (2014); Lam et al., (2015a,b)
NSC		
As aggregates	Stirred suspension bioreactor	Kallos et al. (1999); Baghbaderani et al. (2008)
On microcarriers	Stirred suspension bioreactor	Rodrigues et al. (2011b)
Encapsulation	Rotating wall vessel	Lin et al., (2004)
HSC		
Suspension culture	Stirred suspension bioreactor	Zandstra et al., (1994)
	Rotating wall vessel	Liu et al., (2006)
	Flat-bed single/multi-step perfusion in parallel plates bioreactor	Koller et al., (1993); Palsson et al., (1993)
	Hollow-fiber bioreactor	Sardonini and Wu (1993)
	Packed-bed bioreactor	Wang et al. (1995)
	Fluidized-bed bioreactor	Meissner et al. (1999)
MSC		
Suspension culture	Stirred suspension bioreactor	Frith et al. (2010)
	Rotating wall vessel	Chen et al. (2006)
	Flat-bed single-step perfusion in parallel plates bioreactor	Dennis et al. (2007)
	Packed-bed bioreactor	Weber et al. (2010a)
On microcarriers	Stirred suspension bioreactor	Yang et al. (2007a); Dos Santos et al. (2010); Eibes et al. (2010); Dos Santos et al. (2014); Carmelo et al. (2015); Shekaran et al. (2015); Rafiq et al. (n.d.)

Source: Rodrigues et al. (2011a). Adapted from Elsevier.

Figure 4.1 Number of active clinical trials testing clinical interventions for mesenchymal, hematopoietic, embryonic and neural stem cells, listed on the U. S. NIH website (clinicaltrials.gov).

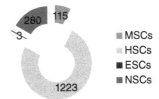

leading to homogeneous culture, simplicity of handling, the ability to control and monitor crucial culture parameters, and a less susceptibility to contamination. However, challenges remain on determining the optimal values of dissolved oxygen tension, pH, temperature, and nutrient concentrations for cell expansion or differentiation, as well as concerning the hydrodynamic shear stress developed in stirred and gas-sparged reactors (King and Miller 2007; Rodrigues et al. 2011a).

Scalability can be also implemented horizontally, a scale-out as a replication of controlled technologies in many units, instead of the vertical scalability (the scale-up as larger volumes). In autologous settings, in which cells are derived from the patient, the culture processes are usually based on scale-out platforms, growing large numbers of small-volume batches. On the other hand, in allogeneic applications (i.e. cells derived from a donor) processes are typically scaled-up into larger-volume batches (Hampson et al. 2008; Lapinskas 2010). Single-use, closed, easy-to-use, disposable cell production "kits" may represent a desired design strategy for patient-specific cell therapy manufacturing protocols (Kirouac and Zandstra 2008).

The number of clinical trials with stem cell-based therapies has surged in recent years, establishing the clinical pathways for an emergent new medicine. The majority of active clinical trials in public databases (Figure 4.1) is based on decades of research and clinical experience in hematopoietic transplantation, including strategies to expand the suboptimal dose of HSC through viral transgene delivery or to engineer T cells to attack malignancy via adoptive immunotherapy. The clinical trials with MSC target bone and cartilage repair, immunological applications due to their ability for cytoprotection and immunosuppression, and cardiac tissue repair, among others. In the case of NSC, there are hundreds of investigator-initiated clinical trials occurring in the academic setting, and several companies have forged efforts to develop novel therapies through intracerebral or spinal transplantation of NSC in order to treat neurodegenerative diseases, catastrophic stroke, traumatic brain injury and spinal paralysis, among others. Human ESC have begun to make their way through clinical trials, with hESC-derived oligodendrocyte progenitor cells for safety studies on thoracic spinal cord injuries and with retinal pigment epithelial progenitor cells derived from hESC for transplantation to treat retinal blindness (Trounson et al. 2011; Daley 2012).

4.2 Bioprocess Characterization

4.2.1 Cell Cultivation Methods

Mammalian cells can be grown as suspension cells or as anchorage-dependent cells. For mammalian cells that grow individually in suspension, the range of fermentation equipment developed for bacteria in the past has been adapted and scaled up. On the other hand, for those cells that cannot be easily adapted to proliferate as suspension cells and only grow when attached to a substrate, the scale-up is far more difficult to achieve.

A wide range of alternative culture systems are available to immobilize adherent cells under dynamic culture conditions. One of the most effective consists of using microcarriers. The development of carriers to support cell growth was initially driven by the large-scale production systems for adherent cells for vaccine production. After the development of the first microcarrier in 1967 by van Wezel (1967), all adherent cell cultures are potentially scalable to stirred bioreactors and other bioreactor configurations. The great success of this technology led to the development of a large number of different microcarriers for use in different reactors systems (see Section 4.3). Microcarriers are basically cell-supporting particles, which can be made of many different materials (e.g. dextran, polystyrene, collagen, gelatin, and glass) of various shapes (e.g. spherical and disk shapes). Besides the advantage of allowing the cultivation of adherent cells as suspension cultures, microcarriers also facilitate culture medium exchange due to the fast sedimentation of cell-containing microcarriers.

Microcarriers for suspension systems can be nonporous and microporous if they allow cell adhesion and growth only on the external surface of the support. Nonporous microcarrieres are nonpermeable even for low molecular components (e.g. *2D MicroHex*™, Nunc) and microporous microcarriers allow diffusion of macromolecules up to 100 kDa (e.g. *Cytodex 3*, GE Healthcare). Finally, there are macroporous microcarriers that allow cells to use the interior of the bead for cell adhesion and proliferation, since pores are typically 30–400 µm diameter (e.g. *Cultispher S*, Sigma) (Markvicheva and Grandfils 2004). This internal matrix in the macroporous beads may protect cells from any disruptive forces caused by bead interactions or turbulence in the system, allowing culture of both adherent and suspension cells; however, this type of matrix usually leads to mass transfer limitations. In contrast, cells grown on the bead surface of nonporous/microporous microcarriers are vulnerable to damage due to bead collision or hydrodynamic forces in the system (Ng et al. 1996).

One of the important aspects in cell culture with microcarriers concerns the initial cell adhesion, which is essential for further cell proliferation, and is influenced by different factors, such as the ratio of inoculated cells to microcarriers. According to Forestell et al. (1992), although inoculum density does not interfere with environmental conditions controlling cell adhesion, the cell-to-carrier ratio is an important parameter since it influences the distribution of cells on the microcarrier's surface and determines the proportion of microcarriers that do not become occupied with cells (Ng et al. 1996). According to literature, the key to obtain high cell yields in microcarrier cell cultures is to assure that the majority of microcarriers is inoculated with cells, since bead-to-bead transfer is not likely to occur along cultivation. Nevertheless, Ohlson et al. (1994) and later on Melero-Martin et al. (2006), reported bead-to-bead transfer of CHO cells and chondro-progenitor cells, respectively, on porous *Cultispher G* microcarriers. One potential strategy to improve cell adhesion is to customize the microcarrier surface according to the cell-type specific needs, by attaching specific synthetic peptides or extracellular matrix molecules (Varani et al. 1993; Kato et al. 2003). In addition, it should be mentioned that the type and concentration of proteins in the culture medium influence cell adhesion and, consequently, cell proliferation (Mukhopadhyay et al. 1993; Ng et al. 1996).

Adherent cells can also be cultivated under suspension in the form of aggregates in suspension and furthermore, to protect cells against harmful environmental conditions (e.g. high agitation and aeration rates), these aggregates can be encapsulated (Dang et al. 2004; Bauwens et al. 2005). With the cell encapsulation strategy, it is possible to create suitable microenvironments for cell growth, since the environment of the capsule can be customized and designed using specific biomaterials. Several materials have been described for suspension cell encapsulation, such as Ca-alginate, Na-alginate, agarose, cellulose sulfate and collagen (Fenge and Lüllau 2006). The cell leakage could be reduced using polyethylene glycol (PEG)-alginate composite beads, and the mechanical strength could be improved using photosensitive polymers. In the specific case of stem cell culture, this strategy is even more advantageous in order to create a microenvironment suitable for directing their differentiation along with promoting a 3D configuration, similar to those in native tissues. Nevertheless, the structure instability of the cell-containing capsules is detrimental to scale up and long-term operation (Eibl et al. 2009). Also, the diffusion-controlled transport of nutrients, waste products, and oxygen, which may lead to cell necrosis in the center of larger aggregates/capsules, is a relevant drawback of this strategy. Therefore, as it will be discussed on Section 4.2.4, the control of the aggregate/capsule size is important.

4.2.2 Cell Metabolism

A deep understanding of cell metabolism is crucial for designing feeding strategies to improve productivity and decrease culture costs. The metabolic pathways of glutamine and glucose have been reported to be quite similar to all mammalian cells including stem cells (Lao and Toth 1997).

Glucose is utilized either through the pentose phosphate pathway to provide nucleotides for biosynthesis or through the glycolysis pathway to provide metabolic intermediates, such as pyruvate, and energy. Pyruvate is preferably converted to lactate and fatty acids, but it can also become oxidized into the tricarboxylic acid (TCA) (Butler 1985; Williams et al. 1988) cycle originating water (H_2O) and carbon dioxide (CO_2), with oxygen (O_2) consumption and adenosine triphosphate (ATP) production.

On the other hand, glutamine is oxidized to CO_2 and is a key amino acid since it is consumed at an order of magnitude higher than other amino acids. Glutamine can be used to produce other amino acids, or incorporated into biomass; however, the majority is converted to glutamate. Glutamate can then be converted to α-ketoglutarate and enter the TCA via one of the two pathways: a less efficient pathway involving the production of alanine, and a pathway that results in the production of an additional molecule of ammonia (Cruz et al. 1999). Although the glutamine and glucose catabolic pathways have been reported to be comparable for different cell types (Zielke et al. 1984; Glacken et al. 1988; Fitzpatrick et al. 1993), the tolerance levels toward the two major metabolic waste products, lactate and ammonia, seem to be cell line-specific and vary widely (Hassell et al. 1991). These differences may be due to the sensitivities of key enzymes in the glucose and glutamine metabolic pathways, and metabolic shifts in response to the adverse environment. In addition, there are several reports on the effects of ammonia and lactate on two of the most commonly used cell lines, i.e. hybridoma and CHO cells, focusing their impacts on growth and productivity. However, few groups discussed the metabolic changes, especially amino acid metabolism, induced by these two waste products and the possible mechanisms that cells use to counteract these effects. Since the metabolism of glucose and glutamine closely interact, control strategies addressing one of these nutrients may yield unexpected results on the other (Hu et al. 1987) and, therefore, understanding such metabolic changes may be essential in designing suitable culture control strategies (Lao and Toth 1997).

High initial levels of glucose (higher than 14 mM) inhibit oxidative metabolism resulting in lactate production, which will cause premature cellular stress or apoptosis (Whitford 2006; Karra et al. 2010). Omasa et al. reported the concentration effect of certain amino acids on cell growth and antibody productivity. It was observed that the specific growth rate decreased at higher glutamine concentrations (10–35 mM), while the antibody production rate increased with the increase of glutamine concentration in the range of less than about 25 mM (Omasa et al. 1992).

Of notice, it has been reported that for mammalian cells in general, cell growth might be inhibited by lactate and ammonia at concentrations above 20 and 4 mM, respectively (Ozturk et al. 1992; Patel et al. 2000). Gagnon et al. developed a method for controlling lactate accumulation in suspensions of

CHO cell cultures. Since, when glucose drops to a low level (generally below 1 mM), cells begin to take up lactic acid from the medium rising the pH, a glucose-feeding method based on pH control was performed in this work (Gagnon et al. 2011).

4.2.3 Culture Medium Design

The medium to cultivate mammalian cells has to provide all the necessary nutrients (e.g. glucose, amino acids, vitamins, mineral salts, and trace elements) required for growth and product formation. In addition, to maintain the pH value between 7.0 and 7.3, the medium should have a buffer capacity in the range of 15–40 mM and should provide an appropriate osmolality (approximately 350 mOsm l^{-1}) to avoid damage to the sensitive cell membrane (Hu and Oberg 1990).

Presently, there are many available sources of basal media for mammalian cell culture (e.g. Eagle's minimal essential medium (MEM), Dulbecco's enriched (modified) Eagle's medium (DMEM), Ham's F12, and Roswell Park Memorial Institute (RPMI) 1640) and supplements, crucial to achieve successful and predictable experimental results within cell culture research. However, there is a certain variability in most culture media formulations typically used, which results primarily from the addition of serum or serum substitutes (e.g. transferin, insulin, ethanolamine, or albumin). Indeed, serum is a biological product and, therefore, has an intrinsic high batch-to-batch variability, being poorly characterized (Mukhopadhyay et al. 1993). In addition, animal serum is a potential factor of infectious agent transmission (Ying et al. 2003). Fetal bovine serum, typically used in mammalian cell culture, is generally produced in large batches originating from several thousand individual animals. This pooling method minimizes the variability of cell culture performance; however, raw serum samples can vary in animal age at collection, animal diet, country of origin and several other factors. With large lots, the likelihood of consistent cell culture performance increases.

Serum substitutes (e.g. Knockout Serum Replacement, Invitrogen) that are more defined components may improve cell culture performance by providing a xeno-free and more predictable culture medium composition and eliminating the risk of undesired biological contamination. Furthermore, in the last years, especially in the area of stem cells, much effort is being made to discover synthetic compounds that can control signaling pathways, and subsequently stem cell behavior, in order to develop chemically defined, xeno-free media (Wiles and Johansson 1999; Ying et al. 2003). In addition, the use of small molecules as substitutes of recombinant, cost-expensive growth factors and cytokines that are currently obligatory components of many medium formulations might also contribute for the cost-effectiveness of stem cell bioprocesses (Zweigerdt 2009).

4.2.4   Culture Parameters

In addition to the composition of the culture medium, it is clear that physico-chemical parameters such as pH and oxygen levels have significant effects on mammalian cell responses. pH variations can affect many cellular functions and parameters such as intracellular pH, metabolism, glucose transport, and the ATP/adenosine diphosphate (ADP) ratio (Borys et al. 1993). Häggström reported that changes in intracellular pH could affect the activity of certain key enzymes of the metabolism (e.g. phosphofructokinase) (Häggström 2003). Moreover, concerning stem cells, extracellular pH variations can lead to significant changes in cell phenotype, as reported by Kohn et al. In this study, the authors observed that small shifts in extracellular pH led to significant changes in the ability of bone marrow stromal cells to express markers of the osteoblast phenotype (Kohn et al. 2002).

The decrease of culture pH is normally associated with lactate accumulation in culture medium. In controlled bioreactors, this acidification of the culture medium is counterbalanced by using one of two strategies, depending on the presence or not of sodium bicarbonate in the medium (Thilly 1986): in the absence of sodium bicarbonate, a base (e.g. sodium hydroxide (NaOH)) should be added, which can ultimately lead to the dilution of the medium and growth factors (Langheinrich and Nienow 1999).

Oxygen is usually used by mammalian cell cultures for the production of energy from organic carbon sources and is often the first component to become limiting at high cell densities because of the low solubility of oxygen in culture medium at $37\,^{\circ}C$ (approximately, $0.2\,mmol\,l^{-1}$ for RPMI medium (Oller et al. 1989)). Therefore, it is crucial to measure and control O_2 levels in *in vitro* cell culture. Of notice, in aggregate cultures, it is important to control the aggregate size especially since, as oxygen transfer to cells within the aggregate occurs through diffusion, it is possible to predict the maximum diffusion distance that allows all cells (including those at the center) to receive an adequate supply of oxygen (Kallos and Behie 1999).

Concerning stem cells, mouse blastocysts were found to grow better when *in vitro* oxygen concentration was between 2.5 and 5% (Quinn and Harlow 1978), compared to atmospheric conditions (21% O_2). The beneficial effects of a hypoxia environment might be associated with the lower levels of free radicals, which are highly toxic comparing to normoxic conditions. In this context, several studies have been performed to study the effect of hypoxic conditions in the expansion and differentiation of stem cells (Bauwens et al. 2005; Niebruegge et al. 2008; Dos Santos et al. 2010; Fernandes et al. 2010; Rodrigues et al. 2010b). For example, Fernandes et al. reported that under a 2% oxygen tension, mouse ESC proliferation and viability were reduced when compared with cells grown under normoxia using serum-free (SF) medium supplemented with Leukemia inhibitory factor (LIF) (Fernandes et al. 2010).

Oxygen also influence cell metabolism since a greater conversion of glucose to lactate was observed in cultures of myocytes (Silverman et al. 1997), hybridoma cells (Ozturk and Palsson 1990), and mouse embryonic stem cells (mESC) (Fernandes et al. 2010) under lower oxygen concentrations. Furthermore, the oxygen consumption rate can also serve as a rapid and sensitive indicator for culture monitoring, since a sudden decrease can occur in response to changes in pH, temperature, or depletion of essential medium components.

4.2.5 Culture Modes

For the majority of the biotechnological processes, strategies for the operation of bioreactors can be classified as discontinuous (batch, repeated batch, fed-batch, and medium exchange) or continuous. The different culture operating modes are presented in Figure 4.2.

In batch culture (Figure 4.2a), all nutrients are supplied in the beginning of the culture. The cells grow until these nutrients become limiting, and afterward, the cells and/or product can be harvested. It cannot be considered a truly "closed system" since during time in culture O_2, antifoam, and base, acid or CO_2 for pH control are added. Typically, an initial lag phase, with reduced cell growth, is followed by a phase of exponential growth until either substrates are consumed

Figure 4.2 Culture operating modes.
(a) Batch, (b) fed-batch, (c) repeated batch, (d) medium exchange, (e) continuous, and (f) perfusion.

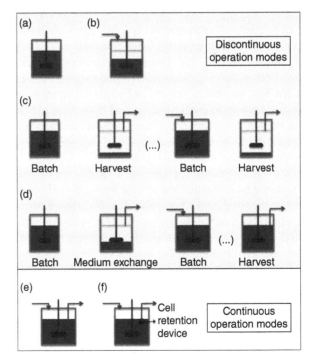

or inhibiting metabolites have accumulated (decline and death phases). One of the disadvantages of batch systems is the limited productivity due to the initial concentration of substrates.

One way to overcome limitations of batch operation is by operating the system either in a fed-batch (Figure 4.2b) mode or in a repeated fed-batch (Figure 4.2c). In the first, fresh medium or nutrients are added in increments, as soon as substrates are consumed or have reached growth-limiting values, as the culture progresses up to the maximum working volume. The main advantages of a fed-batch system compared to a batch are a longer growth phase, higher cell and/or product concentrations and a higher product yield.

In the repeated fed-batch mode, 85–95% of the exhausted medium and cells are harvested at the end of the exponential growth phase, and the remaining part is used as inoculum for the next run. The culture system is filled with fresh medium and the process is restarted. This procedure can be repeated until final harvest. However, if the objective is to reach high cell densities (i.e. by cell retention), another strategy can be used, namely medium exchange (Figure 4.2d). This operation mode, commonly used for microcarrier stem cell cultures, starts with a batch culture. After a certain period of time, the agitation is stopped allowing the cell-containing microcarriers to settle down and a defined percentage of the exhausted medium is exchanged by fresh medium. After the medium exchange, agitation is turned on, another batch culture is initiated, and the whole process repeated until the final harvest.

In continuous culture (Figure 4.2e), fresh medium is continuously pumped to the culture system at the same flow rate as the exhausted medium, including the cells, is removed. Therefore, nutrients are continuously replenished and metabolites diluted, minimizing time-varying culture conditions. The culture is initiated as a batch culture and when substrate concentrations reach critical values, feeding of fresh medium and harvest of exhausted medium (plus cells) are started. Although continuous cultures are a valuable tool for research (e.g. kinetic studies), for production scale, they are limited due to the low cell (and product) concentrations attained.

In perfusion cultures (Figure 4.2f) under continuous operation, the cells are separated from the outflow stream and retained inside the bioreactor. The culture is initiated, and when substrate concentrations reach critical values, fresh medium is added to the culture system at the same flow rate as the exhausted medium is removed. One of the advantages of perfusion operation is that, as the cells are retained in the bioreactor, high flow rates can be applied leading to more efficient mass transfer, which results in higher cell concentrations. A major challenge for perfusion bioreactors design and operation is the reliability of the cell retention device (Woodside et al. 1998; Voisard et al. 2003). Therefore, when choosing the appropriate cell retention several aspects have to be taken in account. For example, the device should function effectively during the expected period of time, without any maintenance or replacement (e.g. in order to minimize additional risk of contamination). To maintain high cell concentration in the bioreactor, it is also important that cell viability and/or productivity are not affected. For that reason, it is expected that the device has the ability to separate nearly 100% of the cells from the outflow stream, under any working cell concentration and flow rate conditions (Woodside et al. 1998).

Overall, fed-batch is the culture mode of choice in the pharmaceutical sector due to the high product titers (e.g. antibodies) and high cell densities attained, as well as its simplicity, among other features. Concerning the stem cell research field, since the cells are the product itself, the operation modes likely to be used are medium exchange and continuous perfusion. Continuous mode in particular presents several advantages, when compared to discontinuous cultures, such as the ability to operate at very high cell densities, the ease of handling medium exchanges, the easy removal of metabolites and other inhibitors, and the prospect of easy scale-up.

4.3   Cell Culture Systems

Cell culture systems from small to large scale have been developed over the past 50 years for mammalian cell culture-based applications, and their primary role is to provide sustainable conditions for cell growth and/or product formation. Concerning stem cell expansion, in addition to increase cell number, the maintenance of the self-renewing and multilineage differentiation capabilities is crucial. Therefore, a quality control panels either based on surface markers or morphology are an important consideration in culture system selection (Portner et al. 2005; Ulloa-Montoya et al. 2005). In Table 4.4, some bench-scale culture systems for use in incubators or warm room at 37 °C and 5% CO_2 are described.

4.3.1   Static Culture Systems

Static culture systems such as multi well-plates, T-flasks, or tissue culture Petri dishes are the most widely used options for expanding mammalian cells (Table 4.4). However, despite their widespread usage, static environments are limited in their productivity by the number of cells that can be supported by a given surface area (Cabrita et al. 2003).

In order to overcome this limitation, other configurations aiming the scale-up of static systems were developed, such as TripleFlask, HyperFlask vessel, CellCube module, and CellStack culture chamber, which consist in multilayer vessels. In fact, the latter represents a valuable alternative to the RBs, which are the most common culture systems for cultivating adherent mammalian cells (e.g. for vaccine production). A higher cultivation surface area can be attained with this new system, reducing significantly the labor

Table 4.4 Examples of available culture systems for bench-scale for use in incubators or warm rooms at 37°C and 5% CO_2.

	Culture system	Suspension cells	Adherent cells	Maximum volume (ml)	Maximum S/A (cm²)	Advantages	Disadvantages
Static culture systems	T-Flask 175	Yes Maximum: 1.5×10^8	Yes Maximum: 10^7	150	175	Cheap, disposable, no cleaning/sterilization required.	Small scale. Multiples required for larger batches.
	PetakaG3™	No	Yes Maximum: 3×10^7	20	150	No additional CO_2 and humidity required. Easy to handle. Excellent for shipping live cells without requiring freezing and dry ice.	New product, still in trial stage.
	Triple flask	Yes Maximum: 1.5×10^8	Yes Maximum: 3×10^7	150	525	Same as T-Flasks.	Difficult to harvest attached cells. Multiple systems required for larger batches.
	Hyperflask™	N/A	Yes Maximum: 2×10^8	560	1720	Disposable Small footprint	Large volumes of medium needed.
	Multiple parallel plates (e.g. CellSTACK®)	No	Yes Maximum: 1.5×10^{10}	8 000	25 280	Not expensive, no cleaning/sterilization required. Potentially used at a production scale.	Requires additional systems. Difficult to harvest attached cells.

Dynamic culture systems						
	Shake-flasks	Yes Maximum: 1×10^9	N/A	1 000	N/A	Some disposables available.	Suspension only. Glass vessels need to be cleaned and sterilized. Requires shaker incubation.
	Roller bottles	Yes Maximum: 1×10^9	Yes Maximum: 1×10^8	2 000	1700	Not expensive, no cleaning/sterilization required. Versatile, automated systems available.	Requires appropriate systems to assure proper rotation. Automation very costly.
	Rotating membrane flask (MiniPerm)	Yes Maximum: 1×10^8	Yes Maximum: 1×10^8	35 + 400	240	The dialysis membrane and rotation of the bioreactor facilitates exchange of dissolved gases, nutrients, and waste metabolites.	Requires appropriate systems to assure proper rotation.
	Spinner flasks	Yes Max. 1×10^9	Only with the use of 3-D scaffolds	1 000	N/A	Simple stirred vessels	Glass vessels need to be cleaned and sterilized. Requires Stirrer-base and incubator.

Source: Adapted from da Silva et al. (2006).

requirement and the contamination risk. Nevertheless, the inhomogeneous nature of a static system results in concentration gradients (e.g. pH, dissolved oxygen, and metabolites) in the culture medium (Portner et al. 2005).

In an attempt to overcome some of the limitations of static systems, innovative configurations were developed. For example, concerning stem cells, Oh et al. (2005) developed an alternative simple and scalable bioprocess, which combines automated perfusion feeding and culture of mESC on petriperm dishes. Unlike ordinary Petri dishes, petriperm dishes have a gas-permeable base. This means that the cells grow at the junction of the gas and liquid phases, mimicking more closely physiological conditions. Two mESC lines were tested in these petriperm dishes and a 64-fold increase in cell density was obtained compared to Petri dish cultures, which typically yielded a ninefold-increase (Oh et al. 2005). Nevertheless, for the vast majority of static systems, the environmental conditions are not readily monitored or controlled online and these systems require repeated handling in order to feed cultures or obtain data on culture performance.

4.3.2 Roller Bottles

RBs are cylindrical screw-capped bottles mostly made of disposable plastic with a total volume of 1–1.5 l. The cells grow on the internal surface of the bottle, which are rolled in their axis in such a way that the culture medium inside the device is agitated axially forming a thin layer of media on the upper part of the bottle. Even though a single RB is considered laboratory-scale equipment, there are systems used for long at the industrial scale for vaccine production consisting of hundreds of RB. In order to overcome the labor and time-consuming handling of RB for industrial scale production, it has been developed a robotic handling system.

RB is one of the simplest systems for achieving high cell densities of adherent and suspended cells. The slow rotation of the RB (1–5 rpm) provides a low shear mixing environment ideal for sensitive cells, while assuming proper mass transfer. In addition, they can be operated without specialized training, are easily scalable for clinical purposes, and involve very low capital investment when compared with a fully controlled bioreactor.

Although it has been extensively used for mammalian cell culture for vaccine production (Eibl et al. 2009), only recently Andrade-Zaldívar et al. reported the expansion of stem cells, particularly mononuclear cells from the umbilical cord blood, in RB (Andrade-Zaldivar et al. 2011).

4.3.3 Spinner Flask

Spinner flasks are glass or plastic vessels with a central magnetic stirrer shaft for agitation, and side arms for the addition and removal of cells and medium.

For mammalian cell culture, the impeller speed is normally kept between 30 and 100 rpm and oxygen is supplied via the headspace through slightly opened caps. These small-scale laboratory (50–1000 ml) bioreactors offer attractive advantages of readily scalability and relative simplicity. Their relatively homogeneous nature makes them uniquely suited for investigations of different culture parameters (e.g. DO, cytokine concentration, antibody titers, serum components, medium exchange rates) that may influence the viability and turnover of cells (Portner et al. 2005).

Spinner flask has been the culture system of choice for growing suspension cells including hybridomas, CHO cells, and adherent cells that have been adapted to growth in suspension. Although its original form does not allow any kind of parameter control (except agitation speed), alternative configurations of spinner flasks were developed recently to allow online monitoring and control of pH, dissolved oxygen, cytokine and nutrient concentrations, in an effort to assess more rigorously their effects on cell proliferation.

Concerning stem cells, in the last years, several methods of expanding adherent mESC in homogeneous suspension in spinner flasks in the form of aggregates (Fok and Zandstra 2005; Cormier et al. 2006; Storm et al. 2010), encapsulated (Dang et al. 2004; Bauwens et al. 2005) or using microcarriers as substrate for cell attachment (Fok and Zandstra 2005 ; Abranches et al. 2007; Fernandes et al. 2007; Marinho et al. 2010; Storm et al. 2010; Alfred et al. 2011; Dos Santos et al. 2011; Rodrigues et al. 2011b) have been reported.

4.3.4 Airlift Bioreactor

ALRs (Figure 4.3a) are structurally very simple and because of the low level and homogeneous distribution of hydrodynamic shear, they are considered to have vast potential in industrial bioprocesses. Their main advantages, compared to stirred reactors, include low capital investments and energy costs, lack of moving parts (e.g. mechanical agitators) and low shear characteristics. However, the ALR are not flexible to accommodate variations of operating conditions, for example a minimum liquid volume is required for the reactor to operate. The use of ALR revealed to be extremely satisfactory in several cases such as suspension cultures of baby hamster kidney 21 (BHK21), human lymphoblastoid, CHO, and hybridomas (Eibl et al. 2009).

In the ALR, the major patterns of fluid circulation are determined by the design of the reactor, which has a section for gas–liquid upflow (Eibl et al. 2010), where the sparger is located, and other section for the bubble-free downflow (downcomer). Therefore, as the riser section is not as dense as the downcomer section, the difference in hydrostatic pressure between the two sections induces the liquid circulation upward. The hydrodynamic behavior within the reactor depends on the geometry, gas flow rate, gas hold up, and circulation velocity but only gas flow rate is autonomously controllable. In terms of operation, ALR have been used in a fed-batch mode to produce antibodies for therapeutic and diagnosis use. In addition, they can be equipped with cell retention devices to operate in a perfusion mode.

Concerning stem cell cultivation, Sardonini and Wu reported the expansion and differentiation of human hematopoietic cells in an ALR (Sardonini and Wu 1993) and in 1996, Highfill et al. developed a large-scale production of murine bone marrow cells in an airlift-packed bed bioreactor (Highfill et al. 1996).

4.3.5 Fixed/Fluidized-Bed Bioreactor

Fixed-bed or packed-bed and fluidized-bed bioreactors (Figure 4.3b) have been used for the cultivation of a wide range of anchorage-dependent cell lines for the production of several biological products (e.g. antibodies, recombinant proteins) (Kang et al. 2000; Kaufman et al. 2000). In both systems, cells are immobilized within porous or nonporous carriers (e.g. ceramic or glass porous beads) in a "bed" through which culture medium is recirculated. Whereas in fixed (packed)-bed bioreactors the stationary bed of carriers is not disturbed, in fluidized-bed bioreactors, the carriers are kept floating ("fluidization") and the height of the bed will increase as the fluid flow increases. The advantages of fixed/fluidized-bed reactors, concerning the production of biopharmaceuticals, include high volume-specific cell density, high volume-specific productivity during long-term cultures, simple medium exchange, cell/product separation, and scalability. On the other hand, the potential disadvantages of these culture systems are the concentration gradients, nonhomogeneous cell distribution, impossibility to directly count cells, and the costs, if used in perfusion cultivating mode.

Concerning stem cells, it was already reported that the culture of human hematopoietic progenitor cells (Meissner et al. 1999) and MSCs (Weber et al. 2010a, 2010b) using a fixed-bed bioreactor is based on porous and nonporous glass carriers, respectively.

4.3.6 Wave Bioreactor

The Wave bioreactor (Figure 4.3c) is one of the disposable cell culture systems that is gaining increasing popularity for clinical applications (e.g. immunotherapy). In this bioreactor, cells are only in contact with the plastic disposable cultivation chamber, so there is no risk of cross-contamination, and it is not necessary to perform the cleaning and sterilization steps. The cultivation is performed in disposable bags that are placed on a rocking platform, which imparts the wave motion to the suspension inside the culture bag and provides the necessary mixing and oxygen transfer. Bags can be operated with suspension cells or adherent cells on carriers, and rocking can be performed at different angle and rocking speeds for culture mixing. Oxygen supply is assumed by headspace aeration, and it has been shown that the oxygen transfer in Wave bioreactors

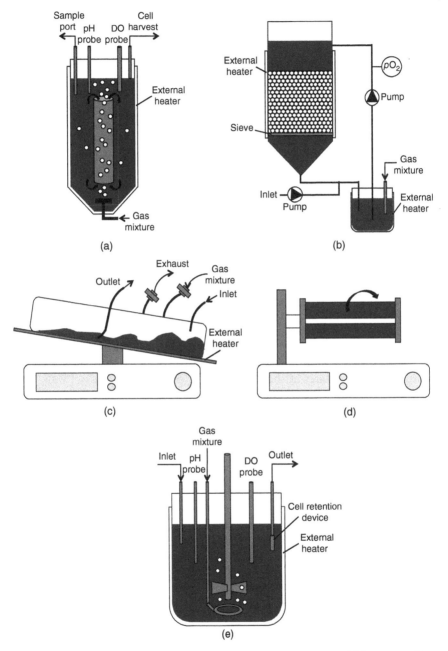

Figure 4.3 Cell cultivation systems applicable to mammalian cells. (a) Airlift bioreactor, (b) packed-bed bioreactor, (c) wave bioreactor, (d) rotating-wall vessel bioreactor, and (e) stirred tank reactor.

is not limiting and the dissolved oxygen levels remains above 50% saturation. It has attained a comparable performance to STR for working volumes between 1 and 100 l in terms of NS0 cell growth and antibody production (Singh 1999). Standard formats or customized bags are now available for several applications. These can be equipped with disposable pH and dissolved oxygen electrodes and operate in batch or perfusion modes.

Wave bioreactor has been used to culture suspended SF9 insect cells (Urabe et al. 2002) to produce adeno-associated virus, one of the most promising vectors for gene delivery in clinical trials. More recently, Sadeghi et al. reported the expansion of tumor-infiltrating lymphocytes from four melanoma patients (Sadeghi et al. 2011) and Timmins et al. reported clinical-scale manufacture of neutrophils from hematopoietic progenitor cells (Timmins et al. 2009). So far, there are very few reports on the use of Wave bioreactors for the cultivation of adherent eukaryotic cells using microcarriers (Hundt et al. 2005).

4.3.7 Rotating-Wall Vessel Bioreactor

In the 1990s, the National Aeronautics and Space Administration (NASA) developed the rotating-wall vessel reactor (RWVR, Figure 4.3d) in order to simulate microgravity effects similar to those occurring in the space shuttle (Schwarz et al. 1992). RWVR is basically a cylindrical vessel that maintains cells in suspension by slow rotation about its horizontal axis, with a coaxial tubular silicon membrane for oxygenation. When the vessel is rotated, the culture medium rotates at the same angular velocity as the vessel wall, with laminar fluid flow. The sedimentation of cells or microcarriers within the vessel is counterbalanced by the rotating fluid, creating a constant "free-fall" state of the cell suspension that facilitates nutrient exchange and localized mixing. In addition, unlike RBs for example, the vessel is completely filled with culture media, avoiding the turbulence created by a headspace and bubble formation. Then in this bubble-free environment, the cell damage and shear stress are minimized. A silicone membrane wrapped around a central cylinder provides oxygenation of the medium and avoids bubble formation as well. Due to the low shear stress, the optimal mass transfer, and the microgravity effect, mammalian cells cultured in RWVR can grow in three dimensions using different types of supports (e.g. microcarriers).

Concerning stem cell cultivation, it has been demonstrated that the simulation of microgravity can alter directly the gene expression, or indirectly the proliferation and differentiation of the cells within the bioreactor (Eibl et al. 2009). Therefore, in order to investigate how microgravity affects cell growth and differentiation of stem cells, several studies on human mesenchymal stem cells (Frith et al. 2010; Kedong et al. 2010; Sheyn et al. 2010), HSCs (Liu et al. 2006; Kedong et al. 2010) and NSCs (Low et al. 2001; Lin et al. 2004) have been reported.

4.3.8 Stirred Tank Bioreactor

Stirred tank bioreactors (STRs) are the most common type of reactors used in the pharmaceutical industry (Figure 4.3e). Important commercial products obtained from cells in STR include viral vaccines, monoclonal antibodies, and interferons, as well as recombinant therapeutic proteins. The attractiveness of STR results from their simplicity, the ease of monitoring and straight scale-up. Design of STR for cultivation of mammalian cells has to consider several aspects such as geometrical characteristics (e.g. sparger design) and operational parameters (e.g. gas flow rate, feeding, temperature, pH, and DO).

As the spinner flask system, the STRs are commonly used for growing suspension cells, but adherent cells can be grown on microcarriers or as aggregates, thus handled similarly to a suspension culture. STRs are cylindrical vessels, typically filled up to 70–80% of the total volume, only to provide sufficient headspace for sedimentation of airdrops or for removal of foam. STRs are equipped with one or more impellers fixed on a rotating shaft, which is activated by an external motor, and the aeration can be provided by the headspace (surface aeration) or by a sparger close to the bottom (sparger aeration). Although some bioreactors might contain up to four baffles to minimize the vortex associated with high impeller speed (characteristic of microbial cultures), these are generally absent since these can cause additional damage, and in mammalian cell cultures, the agitation rates (30–300 rpm) are normally lower than in microbial cultures (200–1000 rpm). Additionally, to overcome any mixing limitation due to lower agitation rates, STRs for mammalian cells often have a curved bottom instead of a flat one (used in microbial fermentation). Heating can be achieved by using an electrical heating jacket fixed to the outer vessel wall or by pumping a preheated liquid (usually tap water circulating in a loop) through a double jacket, at laboratory scale.

STRs are generally equipped with sensors to control temperature, agitation, DO, and pH, connected to a control system, which can act accordingly to the information received. Additional parameters to be monitored include cell density and viability, waste metabolites and nutrients, which have been typically measured offline. Nevertheless, technologies are becoming available that enable online measurement of those parameters (Baradez and Marshall 2011).

Disposable STR for benchtop scale (e.g. Mobius CellReady 3-l Bioreactor from Millipore) equipped with a 3-blade marine impeller, microsparger, and sensors for pH, DO, and temperature were recently developed (Eibl et al. 2010). Concerning stem cell expansion in STRs, Fernandes-Platzgummer et al. established a microcarrier-based STR platform toward the maximization of mESC production, in a controlled, reproducible, and cost-effective bioprocess (Fernandes-Platzgummer et al. 2011). In addition, Schroeder et al. reported the controlled cardiomyocyte differentiation via embryoid body (EB) formation of mESC in a 2-l instrumented and controlled STR, operating in batch

(Schroeder et al. 2005) and perfusion mode (Niebruegge et al. 2008). A study using a parallel cultivation system was also reported (Bauwens et al. 2005) consisting of four vessels of 500 ml each, working in a perfusion fed mode. This system, described by Bauwens and colleagues, implemented a process combining two steps: (i) hydrogel encapsulation to prevent ESC aggregation and (ii) genetic selection to obtain a purified population of cardiomyocytes from the whole heterogeneous cell population, with medium perfusion in a controlled stirred bioreactor environment.

Since STRs have become the system of choice in commercial biomanufacturing processes, the problems associated with this type of bioreactor will be further discussed.

4.3.8.1 Agitation/Shear Stress

Mixing is required in bioreactors to provide a homogeneous environment within the bioreactor and to increase the mass transfer rates of oxygen and other nutrients to the cells. Nevertheless, excessive agitation rates can alter surface markers (McDowell and Papoutsakis 1998) potentially impacting cell function or affect cell membrane integrity (Sun et al. 2000). On the other hand, excessive cell agglomeration and concentration gradients result from an insufficient mixing. In addition, growth factor concentration gradients, as a result of deficient mixing, were reported to be associated to stem cell differentiation (Gurdon et al. 1994). In both cases, culture viability is seriously compromised. Therefore, defining the appropriate agitation rate is of major importance when culturing mammalian cells, and more specifically stem cells, under dynamic conditions.

Of notice, agitation introduces two contradictory effects on initial cell adhesion in microcarrier cultures. For instance, a high agitation rate increases the collision frequency between cells and microcarriers, favoring the initial adhesion of cells onto the carriers (Sun et al. 2000) (namely, if the nature of cell-carrier adhesion is electrostatic). However, a minimum cell-carrier contact time is needed to ensure that cells remain adherent and proliferate. Since high agitation rates shorten cell-carrier contact time, a decrease in adhesion efficiency may potentially occur.

For stem cell culture in stirred suspension systems, hydrodynamic shear stress due to agitation is a critical factor to consider since there is a limited range of shear stress conditions that are amenable for cell proliferation and/or controlled differentiation (Kinney et al. 2011). For instances, ESC cultured as monolayer in the presence of fluid shear stress exhibited increased expression of endothelial and hematopoietic markers (Yamamoto et al. 2005; Adamo et al. 2009). More recently, Toh and Voldman reported that hydrodynamic shear stress primes mESC toward differentiation in a self-renewing environment via heparan sulfate proteoglycans transduction

Table 4.5 Main causes of shear stress in stirred bioreactors with direct sparging.

	Device/parameter	Function	Consequence	Damage	Solution
Agitation	Impeller type	Maintains the cells/MC in suspension	Vortex formation	Foam layer at the surface	Low gas flow rates
	Agitation rate				
		Provides a homogeneous environment	Collisions MC–MC, MC-impeller and MC-sensors	Deposit of MCs on the vessel walls due to bubbles bursting	Small bubble diameter for high oxygen transfer
		Control of the oxygen transfer rates		Cell death	Low-intensity agitation
Aeration	Sparger geometry (nozzle diameter)	Oxygen control	Bubble-associated damage		Use of antifoams and shear protectants
		pH control			
	Gas flow rate				
			Foam formation		

MCs, microcarriers.

(Toh and Voldman 2011). Therefore, if the objective is the maintenance of pluripotency, high agitation rates should be avoided.

On the other hand, in stirred bioreactors with direct sparging, although aeration is necessary for the delivery of O_2 to the cells and CO_2 for pH control, the shear stress caused by sparging bubbles in the culture medium proved to be detrimental to the cells at high flow rates (Cherry and Papoutsakis 1986). Then, through manipulation of the agitation rate and the aeration flow rate, the level of hydrodynamic shear stress within the bioreactor may be tuned. Hydrodynamic effects of agitation and sparging on mammalian cells grown on microcarriers are summarized in Table 4.5.

4.4 Cell Culture Modeling

A kinetic model of mammalian cells can be defined as a quantitative description of the main phenomena that have influence on the growth, death, and metabolic activities of cells, i.e. mathematical expressions that correlate the rates of cellular growth, death, and metabolism with the composition of the medium. The establishment of these kinetic models can be an efficient tool for the design and the optimization of media and reactor operational parameters, in batch or continuous operation, due to predictive capabilities. Also, it can be useful in

Table 4.6 Main equations for the kinetic analysis for mammalian cell cultures.

Parameter	Equation
Total cell production	$X_t = X_v + X_d$
Specific growth rate	$\mu = \frac{dX_t}{X_v dt}(h^{-1})$
Specific death rate	$k_d = \frac{dX_d}{X_v dt}(h^{-1})$
Specific rate of glucose consumption	$v_{Glc} = \frac{-d[Glc]}{X_v dt}(mmol\ 10^{-9}\ cells\ h^{-1})$
Specific rate of glutamine consumption	$v_{Gln} = \frac{-d[Gln]}{X_v dt}\frac{k_{deg}[Gln]}{X_v}(mmol\ 10^{-9}\ cells\ h^{-1})$
Specific rate of ammonia production	$\pi_{NH_4} = \frac{d[NH_4]}{X_v dt} - \frac{k_{deg}[Gln]}{X_v}(mmol\ 10^{-9}\ cells\ h^{-1})$
Specific rate of lactate production	$\pi_{Lac} = \frac{d[Lac]}{X_v dt}(mmol\ 10^{-9}\ cells\ h^{-1})$
Specific rate of antibody production	$\pi_{MAbs} = \frac{d[MAbs]}{X_v dt}(mmol\ 10^{-9}\ cells\ h^{-1})$

the development of software sensors to estimate the time variation of the media composition online.

Several steps should be considered when constructing a kinetic model. First, prior to the construction of the kinetic model, a detailed kinetic analysis of the experimental data have to be performed in order to identify the main rate-limiting effects and the relationships that may exist between the different kinetic variables (Doyle and Griffiths 1998). In mammalian cell cultures, the rates usually measured are the rates of cell growth and death, the rates of uptake of the main nutrients, glucose and glutamine, the rates of production of the main metabolites, lactate and ammonia, and the rate of secretion of proteins (as cell product) (Table 4.6). In the case of glutamine consumption and ammonia production, the specific rates must take into consideration the spontaneous glutamine decomposition in the medium into ammonia. This process follows first-order rate kinetics with respect to glutamine concentration, being k_{deg} the glutamine degradation constant.

After analyzing the experimental data and formulate hypotheses on the nature of the rate limiting steps, rate expressions describing the influence of this phenomena on the cellular processes are chosen. The different rate expressions presented in Table 4.7 are general equations that have been found to correctly modeling batch or continuous cultures of several mammalian cell lines (Doyle and Griffiths 1998).

In order to simulate the time-course of a culture with the rate expressions previously introduced, it is necessary to construct the mass balance equations for each of the considered species. These equations will depend on the operation strategy used (i.e. continuous or discontinuous culture). Finally, all these equations have to be integrated by a numerical method (e.g. Runge–Kutta

Table 4.7 Main equations for the construction of kinetic models for mammalian cell cultures.

Parameter	Equation
Specific rate of cell growth	$\mu = \mu_{max} \left(\frac{[Glc]}{K_{Glc}+[Glc]} \right) \left(\frac{[Gln]}{K_{Gln}+[Gln]} \right) \left(\frac{1}{1+\frac{[Lac]}{K_{Lac}}} \right) \left(\frac{1}{1+\frac{[NH_4]}{K_{NH_4}}} \right)$
Specific rate of cell death	$k_d = A_d \left\{ \left(\frac{1}{1+\frac{[Gln]}{C_1}} \right) + \left(\frac{1}{1+\frac{[Glc]}{C_2}} \right) + k_1[NH_4] + k_2[Lac] \right\}$
Specific rate of glucose consumption	$v_{Glc} = Y_{Glc/X}\,\mu + m_{Glc}$
Specific rate of glutamine consumption	$v_{Gln} = Y_{Gln/X}\,\mu + m_{Gln}$
Specific rate of ammonia production	$\pi_{NH_4} = Y_{NH_4/X}\,\mu + m_{NH_4}$
Specific rate of lactate production	$\pi_{Lac} = Y_{Lac/X}\,\mu + m_{Lac}$
Specific rate of antibody production	$\pi_{MAbs} = Y_{Mabs/X}\,\mu + m_{MAbs}$
Rate of glutamine degradation	$r_{Gln} = k_{deg}[Gln]$

method), and the values of the parameters introduced in the different rate expressions determined. In the end, the kinetic model should be validated with different experimental results.

Concerning stem cell manufacturing, it should be taken into account that the cells are the target and a term accounting for both expansion and differentiation of the cell should be introduced in the mass balance equations (da Silva et al. 2003).

4.5 Case Studies

In this section, two case studies focusing mammalian cell culture for antibody production and mouse ESC culture, in bioreactors systems, are presented.

4.5.1 Antibody Production in Bioreactor Systems

In order to obtain high product yields with high volume throughput, which are the advantages of a fed-batch and perfusion operations, Yang et al. (2000, 2007b) developed a robust bioprocess for mAb production. A recombinant mouse myeloma cell line, Sp2/0-Ag 14, was used to produce a humanized mAb hLL2 (*Epratuzumab*), against non-Hodgkin's lymphoma. In this work, authors (Yang et al. 2000) used a "controlled-fed perfusion" approach, though careful feeding of nutrient supplements in combination with toxic metabolite removal. The laboratory-scale experiments were conducted in a 3-l Bellco spinner flask bioreactor system (working volume of 1.5 l) with controlled pH by the addition of CO_2, aeration through a cylindrical sintered sparger at 0.033 vvm, and the DO controlled above 40% air saturation by intermittent sparging of O_2. Perfusion was accomplished by recirculating the culture suspension from the bioreactor through a sterile external hollow-fiber module.

The system was scaled up to a 20-l bioreactor system from BioStat C (working volume of 15 l) in a cGMP-controlled clean-room environment. Culture reached the maximum steady-state viable cell density of 1.0×10^7 cells ml with an antibody concentration of 125 U/L by day 9, comparable to those observed in the 3-l perfused culture. It was observed that the cell growth was limited by glucose starvation, and then this nutrient was supplemented to the perfused culture upon day 9. The steady-state viable cell density increased to 1.5×10^7 cells ml^{-1}, resulting in a 2.5-fold increase in antibody production. This work showed that a long-term cultivation in this system was sustainable and remained productive for five weeks.

In another study by the same group (Yang et al. 2007b), a stepwise scale-up strategy (3–2500 l scale) was presented, with careful selection of several available mixing models for a successful scale-up of agitation speed. An integrated scale-up of nutrient feeding, inoculation and set-points of the operational parameters (temperature, pH, DO, DCO_2, and aeration) were also considered. The 2500-l fed-batch process is currently in use for the routine clinical production of *Epratuzumab*.

In another mammalian cell culture system reported by Tharmalingan et al. (2011), CHO cells were grown to produce either recombinant human beta interferon (β-IFN) or recombinant human tissue-plasminogen activator (t-PA) embedded in macroporous microcarriers (Cytopore 1 or 2). Batch cultures were established in 100 ml cultures in spinner flaks. Although the CHO cells are capable of growing freely in suspension, the electrostatic charge of the matrix of these microcarriers ensures the entrapment of the entire cell population, achieving high cell densities within the microcarriers. This resulted in a significant enhancement of the specific productivity of either recombinant protein. In the microcarrier-based culture, the volumetric production of both β-IFN and t-PA was enhanced by up to 2.5-fold compared to equivalent suspension cultures of CHO cells. With Cytopore 2 (cellulose-based), the maximum cell density reached 3×10^8 cells ml^{-1}, when specific productivity increased to over five times higher than in suspension cultures.

More recently, Valasek et al. (2011) performed the scale-up of a process for PER.C6 human cell line (derived from human embryonic retinal cells) expressing the PAT-SM6 IgM antibody, which is proprietary to Patrys Ltd. and is under evaluation in a phase I melanoma study. The cell scale-up process began with fed-batch culture in shake flasks and then in the Wave bioreactor (volume up to 10 l). The growth rate and viability of SM6-expressing cells remained comparable in both systems, reaching 2×10^6 viable cells ml^{-1} after 72 hours. The next step was the SM6 cells inoculation in a 12-L Applikon glass stirred tank bioreactor where CO_2 was cascaded to pH control and sparged to the culture based on demand. A continuous-feed strategy was used for a fed-batch production with two feed streams: one of culture media and the other with a high-pH amino acid supplement. At passage 7, 1.7 l of cell suspension from the Wave bag was inoculated in the Applikon bioreactor that reached a final 7.6; of medium volume. The viable cell density reached a maximum of 30×10^6 cells ml^{-1}, with the viability dropping to a 70% at harvest (day 14), and the productivity was $0.6\, g_{IgG}\, l^{-1}$.

Finally, the process was transferred from 12-l bioreactor to the 250-l scale using a Thermo Scientific HyClone Single-Use Bioreactor (SUB). The 250-l run ended on day 14 after reaching a 70% of cell viability target, the cell growth and productivity showed to be superior to those in 12-l run, while high productivity levels were obtained ($0.8\, g_{IgG}\, l^{-1}$).

4.5.2 mESC Expansion on Microcarriers in a Stirred Tank Bioreactor

Efficient strategies to scale up mESC expansion to dynamic culture systems, such as STRs, operating under a continuous perfusion mode with cell retention, was recently developed (Fernandes-Platzgummer 2011). One important parameter in perfused cultures is the flow rate at which medium is renewed. The aim of this work was to study the influence of the residence time on the expansion of mESC on microcarrieres, using SF medium, in a STR operating under a continuous perfusion mode with cell retention.

The mESC expansion was previously established using a discontinuous feeding strategy of 50% medium exchange every day (Fernandes-Platzgummer et al. 2011). The operational parameters were set to: temperature 37 °C, pH 7.2, agitation rate 60 rpm, airflow rate 100–200 ccm, and dissolved oxygen concentration 20%.

Residence times of 12, 24, 32, and 48 hours supported mESC expansion, whereas for a residence time of 96 hours, in which only 25% of the medium was exchanged per day, the cells stopped growing after day 7 probably due to nutrient/growth factor depletion and accumulation of inhibitory metabolites. For the other residence times studied, the maximum cell numbers and specific growth rates are presented in Table 4.8.

Comparing the growth of the cells expanded with a culture medium residence time of 48 h (i.e. 50% volume medium renewal/day) and the cells fed once per day (i.e. 50% medium renewal/change) in the previously established discontinuous feeding strategy, it was observed that shifting the feeding scheme from discontinuous to continuous mode increased the cell density by twofold. In the discontinuous medium exchange protocol, a large portion of medium

Table 4.8 Growth kinetic characterization and cell yields for the different residence times (Fernandes-Platzgummer 2011).

	Residence times			
	12 h	24 h	32 h	48 h
Maximum cell number (cells)	4.0×10^9	5.5×10^9	6.4×10^9	5.5×10^9
Fold increase in total cell number	114 ± 5	156 ± 10	184 ± 8	156 ± 19
Specific growth rate (day^{-1})	1.3 ± 0.1	1.6 ± 0.2	1.3 ± 0.2	1.4 ± 0.1
Doubling time (day)	0.6 ± 0.2	0.4 ± 0.1	0.5 ± 0.1	0.5 ± 0.2
Cell yield (cells ml^{-1} of medium used)	0.4×10^6	0.8×10^6	1.2×10^6	1.4×10^6

is replaced at a time (50% every 24 hours), which might affect cell growth in two ways: if medium exchange is performed too early in culture and/or a large portion of medium is replaced at a time, a dilution of important autocrine factors can occur; if medium exchange is performed too late, an accumulation of toxic metabolic by-products can inhibit cell growth and ultimately lead to cell death. On the other hand, in the continuous perfusion mode, the addition and removal of metabolites and other inhibitors can be made in a controlled way without the dilution of the autocrine factors necessary to mESC expansion. Permanent medium exchange via perfusion is thus an important step forward the automation and standardization of mESC bioprocessing.

To evaluate which residence time would maximize cell yield, the total cell number produced was divided by the total volume of medium used. In Table 4.8, it can be seen that the maximum cell yields – 1.4×10^6 and 1.2×10^6 cells ml^{-1} of medium used – were attained with the residence times of 48 and 32 hours, respectively.

Importantly, mESC expanded in the fully controlled STR, using SF medium, retained the expression of pluripotency markers and their differentiation potential into cells of the three embryonic germ layers (ectoderm, mesoderm, and endoderm). The controlled bioprocess developed herein can be potentially adaptable to other cell types, thus representing a promising starting point for the development of novel technologies for the controlled production of differentiated derivatives from human pluripotent stem cells.

4.6 Concluding Remarks

The use of mammalian cells to produce biological products, including vaccines, growth factors, therapeutic monoclonal antibodies, and other therapeutic proteins, in large-scale industrial processes is well established, and there is an enormous potential to be further expanded. Among the existing static and dynamic bioreactor culture systems that have been described in this chapter, traditional stainless steel stirred bioreactors are still the gold standard and dominate both in research and manufacturing areas.

For the specific case of stem cell cultivation for Regenerative Medicine applications, namely stem cell-based cellular or gene therapies, where the target product is the cell itself rather than secreted product, high numbers of cells meeting strict quality control standards are required to assure safe and clinically effective cellular products. The expansion of human stem cells under fully defined conditions, namely concerning culture medium composition, in clinical-scale bioreactors, remains a challenge. The selection of the appropriate bioreactor is strongly affected by the stem cell type and its characteristics, and the purpose of cultivation (expansion and/or differentiation). Therefore, it will be important to develop optimal culture conditions for different stem cell types (either anchorage-dependent – e.g. ESCs – or suspension cells – i.e. HSCs), which implies research on fundamental issues of the cellular system biology. Presently, disposable bioreactor platforms meeting good manufacturing practice (GMP) standards can be used to produce stem cells for clinical applications and, although their engineering aspects are not yet fully characterized, these systems are expected to pave the way for "custom-made" medicine (autologous settings). However, it will be necessary to overcome the challenge of scale-up to large-volume production systems, specially in the context of allogeneic therapies where lot sizes of 10^{11}–10^{12} cells should be produced.

List of Symbols

X_t	total cell production (10^9 cells l^{-1})
X_v	viable cell concentration (10^9 cells l^{-1})
X_d	dead cell concentration (10^9 cells l^{-1})
μ	specific growth rate (h^{-1})
k_d	specific death rate (h^{-1})
μ_{max}	maximum specific growth rate (h^{-1})
v_{Glc}	specific glucose consumption rate (mmol 10^{-9} cells h^{-1})
v_{Gln}	specific glutamine consumption rate (mmol 10^{-9} cells h^{-1})
π_{NH_4}	specific ammonia production rate (mmol 10^{-9} cells h^{-1})
π_{Lac}	specific lactate production rate (mmol 10^{-9} cells h^{-1})
π_{MAbs}	specific antibody production rate (mmol 10^{-9} cells h^{-1})
[Glc]	glucose concentration (mM)
[Gln]	glutamine concentration (mM)
[NH$_4$]	ammonium ion concentration (mM)
[Lac]	lactate concentration (mM)
[MAbs]	monoclonal antibody concentration (mM)
K_{Glc}	glucose activation constant in specific growth rate expression (mM)

K_{Gln} glutamine activation constant in specific growth rate expression (mM)

K_{NH_4} ammonia inhibition constant in specific growth rate expression (mM)

K_{Lac} lactate inhibition constant in specific growth rate expression (mM)

A_d death constant (h^{-1})

C_1 glutamine activation constant in specific death rate expression (mM)

C_2 glucose activation constant in specific death rate expression (mM)

k_1 lactate activation constant in specific death rate expression (mM)

k_2 ammonia activation constant in specific death rate expression (mM)

Y growth-associated yields (mmol 10^{-9} cells)

m non-growth-associated specific rates (mmol 10^{-9} cells h^{-1})

r_{Gln} glutamine degradation rate (mmol l^{-1} h^{-1})

k_{deg} glutamine degradation constant (h^{-1})

References

Abbasalizadeh, S. et al. (2012). Bioprocess development for mass production of size-controlled human pluripotent stem cell aggregates in stirred suspension bioreactor. *Tissue Eng. Part C Methods* 18 (11): 831–851.

Abranches, E. et al. (2007). Expansion of mouse embryonic stem cells on microcarriers. *Biotechnol. Bioeng.* 96 (6): 1211–1221.

Adamo, L. et al. (2009). Biomechanical forces promote embryonic haematopoiesis. *Nature* 459 (7250): 1131–1135.

Alfred, R. et al. (2011). Efficient suspension bioreactor expansion of murine embryonic stem cells on microcarriers in serum-free medium. *Biotechnol. Progr.* 27 (3): 811–823.

Andrade-Zaldivar, H. et al. (2011). Expansion of human hematopoietic cells from umbilical cord blood using roller bottles in CO_2 and CO_2-free atmosphere. *Stem Cells Dev.* 20 (4): 593–598.

Baghbaderani, B.A. et al. (2008). Expansion of human neural precursor cells in large-scale bioreactors for the treatment of neurodegenerative disorders. *Biotechnol. Progr.* 24 (4): 859–870.

Baradez, M.O. and Marshall, D. (2011). The use of multidimensional image-based analysis to accurately monitor cell growth in 3D bioreactor culture. *PLoS One* 6 (10): e26104.

Bauwens, C. et al. (2005). Development of a perfusion fed bioreactor for embryonic stem cell-derived cardiomyocyte generation: oxygen-mediated enhancement of cardiomyocyte output. *Biotechnol. Bioeng.* 90 (4): 452–461.

Birch, J.R. and Racher, A.J. (2006). Antibody production. *Adv. Drug Delivery Rev.* 58 (5–6): 671–685.

Borys, M.C., Linzer, D.I., and Papoutsakis, E.T. (1993). Culture pH affects expression rates and glycosylation of recombinant mouse placental lactogen proteins by Chinese hamster ovary (CHO) cells. *Biotechnology (N Y)* 11 (6): 720–724.

Butler, M. (1985). Growth limitations in high density microcarrier cultures. *Dev Biol. Stand.* 60: 269–280.

Cabrita, G.J. et al. (2003). Hematopoietic stem cells: from the bone to the bioreactor. *Trends Biotechnol.* 21 (5): 233–240.

Carmelo, J.G. et al. (2015). A xeno-free microcarrier-based stirred culture system for the scalable expansion of human mesenchymal stem/stromal cells isolated from bone marrow and adipose tissue. *Biotechnol. J.* 10 (8): 1235–1247.

Chen, X. et al. (2006). Bioreactor expansion of human adult bone marrow-derived mesenchymal stem cells. *Stem Cells* 24 (9): 2052–2059.

Chen, A.K. et al. (2014). Inhibition of ROCK-myosin II signaling pathway enables culturing of human pluripotent stem cells on microcarriers without extracellular matrix coating. *Tissue Eng. Part C Methods* 20 (3): 227–238.

Cherry, R.S. and Papoutsakis, E.T. (1986). Hydrodynamic effects on cells in agitated tissue culture reactors. *Bioprocess. Eng.* 1: 29–41.

Chu, L. and Robinson, D.K. (2001). Industrial choices for protein production by large-scale cell culture. *Curr. Opin. Biotechnol.* 12 (2): 180–187.

Cormier, J.T. et al. (2006). Expansion of undifferentiated murine embryonic stem cells as aggregates in suspension culture bioreactors. *Tissue Eng.* 12 (11): 3233–3245.

Cruz, H.J., Moreira, J.L., and Carrondo, M.J. (1999). Metabolic shifts by nutrient manipulation in continuous cultures of BHK cells. *Biotechnol. Bioeng.* 66 (2): 104–113.

Daley, G.Q. (2012). The promise and perils of stem cell therapeutics. *Cell Stem Cell* 10 (6): 740–749.

Dang, S.M. et al. (2004). Controlled, scalable embryonic stem cell differentiation culture. *Stem Cells* 22 (3): 275–282.

Davila, J.C. et al. (2004). Use and application of stem cells in toxicology. *Toxicol. Sci.* 79 (2): 214–223.

Dennis, J.E. et al. (2007). Clinical-scale expansion of a mixed population of bone-marrow-derived stem and progenitor cells for potential use in bone-tissue regeneration. *Stem Cells* 25 (10): 2575–2582.

Dos Santos, F. et al. (2010). Ex vivo expansion of human mesenchymal stem cells: a more effective cell proliferation kinetics and metabolism under hypoxia. *J. Cell. Physiol.* 223 (1): 27–35.

Dos Santos, F. et al. (2011). Ex vivo expansion of human mesenchymal stem cells on microcarriers. *Methods Mol. Biol.* 698: 189–198.

Dos Santos, F. et al. (2014). A xenogeneic-free bioreactor system for the clinical-scale expansion of human mesenchymal stem/stromal cells. *Biotechnol. Bioeng.* 111 (6): 1116–1127.

Doyle, A. and Griffiths, J. (1998). *Cell and Tissue Culture: Laboratory Procedures in Biotechnology*. Wileyand.

Eibes, G. et al. (2010). Maximizing the ex vivo expansion of human mesenchymal stem cells using a microcarrier-based stirred culture system. *J. Biotechnol.* 146 (4): 194–197.

Eibl, R. et al. (2009). *Cell and Tissue Reaction Engineering*. Berlin, Heidelberg: Springer-Verlag.

Eibl, R. et al. (2010). Disposable bioreactors: the current state-of-the-art and recommended applications in biotechnology. *Appl. Microbiol. Biotechnol.* 86 (1): 41–49.

Fenge, C. and Lüllau, E. (2006). Cell culture bioreactors. In: *Cell Culture Technology for Pharmaceutical and Cell-Based Therapies* (ed. S.S. Ozturk and W.S. Hu). New York: Taylor and Francis.

Fernandes, A.M. et al. (2007). Mouse embryonic stem cell expansion in a microcarrier-based stirred culture system. *J. Biotechnol.* 132 (2): 227–236.

Fernandes, T.G. et al. (2010). Different stages of pluripotency determine distinct patterns of proliferation, metabolism, and lineage commitment of embryonic stem cells under hypoxia. *Stem Cell Res.* 5 (1): 76–89.

Fernandes-Platzgummer, A. (2011). *Bioreactor culture systems for the expansion of mouse embryonic stem cells*. Lisbon: Department of Bioengineering, Instituto Superior Técnico, Technical University of Lisbon.

Fernandes-Platzgummer, A. et al. (2011). Scale-up of mouse embryonic stem cell expansion in stirred bioreactors. *Biotechnol. Progr.* 27 (5): 1421–1432.

Fernandes-Platzgummer, A. et al. (2014). Maximizing mouse embryonic stem cell production in a stirred tank reactor by controlling dissolved oxygen concentration and continuous perfusion operation. *Biochem. Eng. J.* 82: 81–90.

Fitzpatrick, L., Jenkins, H.A., and Butler, M. (1993). Glucose and glutamine metabolism of a murine B-lymphocyte hybridoma grown in batch culture. *Appl. Biochem. Biotechnol.* 43 (2): 93–116.

Fok, E.Y. and Zandstra, P.W. (2005). Shear-controlled single-step mouse embryonic stem cell expansion and embryoid body-based differentiation. *Stem Cells* 23 (9): 1333–1342.

Forestell, S.P. et al. (1992). Development of the optimal inoculation conditions for microcarrier cultures. *Biotechnol. Bioeng.* 39 (3): 305–313.

Fridley, K.M. et al. (2010). Unique differentiation profile of mouse embryonic stem cells in rotary and stirred tank bioreactors. *Tissue Eng. Part A* 16 (11): 3285–3298.

Frith, J.E., Thomson, B., and Genever, P.G. (2010). Dynamic three-dimensional culture methods enhance mesenchymal stem cell properties and increase therapeutic potential. *Tissue Eng. Part C Methods* 16 (4): 735–749.

Gagnon, M. et al. (2011). High-end pH-controlled delivery of glucose effectively suppresses lactate accumulation in CHO fed-batch cultures. *Biotechnol. Bioeng.* 108 (6): 1328–1337.

Glacken, M.W., Adema, E., and Sinskey, A.J. (1988). Mathematical descriptions of hybridoma culture kinetics: I. Initial metabolic rates. *Biotechnol. Bioeng.* 32 (4): 491–506.

Gurdon, J.B. et al. (1994). Activin signalling and response to a morphogen gradient. *Nature* 371 (6497): 487–492.

Häggström, L. (2003). Cell Metabolism, Animal. In: *Encyclopedia of Cell Technology*. R. E. Spier (Ed.). doi:10.1002/0471250570.spi040.

Hampson, B., Rowley, J., and Venturi, N. (2008). Manufacturing patient-specific cell therapy products. *Bioprocess Int.* September: 60–72.

Hassell, T., Gleave, S., and Butler, M. (1991). Growth inhibition in animal cell culture. The effect of lactate and ammonia. *Appl. Biochem. Biotechnol.* 30 (1): 29–41.

Highfill, J.G., Haley, S.D., and Kompala, D.S. (1996). Large-scale production of murine bone marrow cells in an airlift packed bed bioreactor. *Biotechnol. Bioeng.* 50 (5): 514–520.

Hu, W.S. and Aunins, J.G. (1997). Large-scale mammalian cell culture. *Curr. Opin. Biotechnol.* 8 (2): 148–153.

Hu, W.S. and Oberg, M.G. (1990). Monitoring and control of animal cell bioreactors: biochemical engineerings considerations. *Bioprocess Technol.* 10: 451–481.

Hu, W.S. et al. (1987). Effect of glucose on the cultivation of mammalian cells. *Dev. Biol. Stand.* 66: 279–290.

Hundt, B., Schänzler, A., and Reichl, U. (2005). Serum free cultivation of primary chicken embryo fibroblasts in microcarrier systems for vaccine production. In: *Animal Cell Technology Meets Genomics*, ESACT Proceedings, vol. 2 (ed. F. Gòdia and M. Fussenegger). Dordrecht: Springer.

Hunt, M.M. et al. (2014). Factorial experimental design for the culture of human embryonic stem cells as aggregates in stirred suspension bioreactors reveals the potential for interaction effects between bioprocess parameters. *Tissue Eng. Part C Methods* 20 (1): 76–89.

Jing, D., Parikh, A., and Tzanakakis, E.S. (2010). Cardiac cell generation from encapsulated embryonic stem cells in static and scalable culture systems. *Cell Transplant.* 19 (11): 1397–1412.

Kallos, M.S. and Behie, L.A. (1999). Inoculation and growth conditions for high-cell-density expansion of mammalian neural stem cells in suspension bioreactors. *Biotechnol. Bioeng.* 63 (4): 473–483.

Kallos, M.S., Behie, L.A., and Vescovi, A.L. (1999). Extended serial passaging of mammalian neural stem cells in suspension bioreactors. *Biotechnol. Bioeng.* 65 (5): 589–599.

Kang, S.H., Kim, B.G., and Lee, G.M. (2000). Justification of continuous packed-bed reactor for retroviral vector production from amphotropic PsiCRIP murine producer cell. *Cytotechnology* 34 (1–2): 151–158.

Karra, S., Sager, B., and Karim, M.N. (2010). Multi-scale modeling of heterogeneities in mammalian cell culture processes. *Ind. Eng. Chem. Res.* 49 (17): 7990–8006.

Kato, D. et al. (2003). The design of polymer microcarrier surfaces for enhanced cell growth. *Biomaterials* 24 (23): 4253–4264.

Kaufman, J.B. et al. (2000). Continuous production and recovery of recombinant Ca^{2+} binding receptor from HEK 293 cells using perfusion through a packed bed bioreactor. *Cytotechnology* 33 (1–3): 3–11.

Kedong, S. et al. (2010). Simultaneous expansion and harvest of hematopoietic stem cells and mesenchymal stem cells derived from umbilical cord blood. *Journal of Materials Science. Materials in Medicine* 21 (12): 3183–3193.

Kehoe, D.E. et al. (2008). Propagation of embryonic stem cells in stirred suspension without serum. *Biotechnol. Progr.* 24 (6): 1342–1352.

King, J.A. and Miller, W.M. (2007). Bioreactor development for stem cell expansion and controlled differentiation. *Curr. Opin. Chem. Biol.* 11 (4): 394–398.

Kinney, M.A., Sargent, C.Y., and McDevitt, T.C. (2011). The multiparametric effects of hydrodynamic environments on stem cell culture. *Tissue Eng. Part B Rev.* 17 (4): 249–262.

Kirouac, D.C. and Zandstra, P.W. (2008). The systematic production of cells for cell therapies. *Cell Stem Cell* 3 (4): 369–381.

Kohler, G. and Milstein, C. (1975). Continuous cultures of fused cells secreting antibody of predefined specificity. *Nature* 256 (5517): 495–497.

Kohn, D.H. et al. (2002). Effects of pH on human bone marrow stromal cells in vitro: implications for tissue engineering of bone. *J. Biomed. Mater. Res.* 60 (2): 292–299.

Koller, M.R. et al. (1993). Expansion of primitive human hematopoietic progenitors in a perfusion bioreactor system with IL-3, IL-6, and stem cell factor. *Biotechnology (N Y)* 11 (3): 358–363.

Lam, A.T. et al. (2015a). Improved human pluripotent stem cell attachment and spreading on xeno-free Laminin-521-coated microcarriers results in efficient growth in agitated cultures. *Biores. Open Access* 4 (1): 242–257.

Lam, A.T. et al. (2015b). Integrated processes for expansion and differentiation of human pluripotent stem cells in suspended microcarriers cultures. *Biochem. Biophys. Res. Commun.* .

Langheinrich, C. and Nienow, A.W. (1999). Control of pH in large-scale, free suspension animal cell bioreactors: alkali addition and pH excursions. *Biotechnol. Bioeng.* 66 (3): 171–179.

Lao, M.S. and Toth, D. (1997). Effects of ammonium and lactate on growth and metabolism of a recombinant Chinese hamster ovary cell culture. *Biotechnol. Progr.* 13 (5): 688–691.

Lapinskas, E. (2010. November). Scaling up research to commercial manufacturing. *Chem. Eng. Prog.* S44–S55.

Lin, H.J. et al. (2004). Neural stem cell differentiation in a cell-collagen-bioreactor culture system. *Brain Res. Dev. Brain Res.* 153 (2): 163–173.

Liu, Y. et al. (2006). Ex vivo expansion of hematopoietic stem cells derived from umbilical cord blood in rotating wall vessel. *J. Biotechnol.* 124 (3): 592–601.

Loffelholz, C. et al. (2014). Dynamic single-use bioreactors used in modern liter- and m(3)-scale biotechnological processes: engineering characteristics and scaling up. *Adv. Biochem. Eng. Biotechnol.* 138: 1–44.

Low, H.P., Savarese, T.M., and Schwartz, W.J. (2001). Neural precursor cells form rudimentary tissue-like structures in a rotating-wall vessel bioreactor. *In Vitro Cell. andDev. Biol. Anim.* 37 (3): 141–147.

Marinho, P.A. et al. (2010). Maintenance of pluripotency in mouse embryonic stem cells cultivated in stirred microcarrier cultures. *Biotechnol. Progr.* 26 (2): 548–555.

Markvicheva, E. and Grandfils, C. (2004). Microcarriers for animal cell culture. In: *Fundamentals of Cell Immobilization Biotechnology* (ed. V. Nedovic and R. Willaert). Dordrecht, The Netherlands: Klumer Academic Publishers.

McDowell, C.L. and Papoutsakis, E.T. (1998). Increased agitation intensity increases CD13 receptor surface content and mRNA levels, and alters the metabolism of HL60 cells cultured in stirred tank bioreactors. *Biotechnol. Bioeng.* 60 (2): 239–250.

McNeish, J.D. (2007). Stem cells as screening tools in drug discovery. *Curr. Opin. Pharmacol.* 7 (5): 515–520.

Meissner, P. et al. (1999). Development of a fixed bed bioreactor for the expansion of human hematopoietic progenitor cells. *Cytotechnology* 30 (1–3): 227–234.

Melero-Martin, J.M. et al. (2006). Expansion of chondroprogenitor cells on macroporous microcarriers as an alternative to conventional monolayer systems. *Biomaterials* 27 (15): 2970–2979.

Mukhopadhyay, A., Mukhopadhyay, S.N., and Talwar, G.P. (1993). Influence of serum proteins on the kinetics of attachment of Vero cells to cytodex microcarriers. *J. Chem. Technol. Biotechnol.* 56 (4): 369–374.

Ng, Y.C., Berry, J.M., and Butler, M. (1996). Optimization of physical parameters for cell attachment and growth on macroporous microcarriers. *Biotechnol. Bioeng.* 50 (6): 627–635.

Niebruegge, S. et al. (2008). Cardiomyocyte production in mass suspension culture: embryonic stem cells as a source for great amounts of functional cardiomyocytes. *Tissue Eng. Part A* 14 (10): 1591–1601.

Oh, S.K. et al. (2005). High density cultures of embryonic stem cells. *Biotechnol. Bioeng.* 91 (5): 523–533.

Oh, S.K. et al. (2009). Long-term microcarrier suspension cultures of human embryonic stem cells. *Stem Cell Res.* 2 (3): 219–230.

Ohlson, S., Branscomb, J., and Nilsson, K. (1994). Bead-to-bead transfer of Chinese hamster ovary cells using macroporous microcarriers. *Cytotechnology* 14 (1): 67–80.

Oller, A.R. et al. (1989). Growth of mammalian cells at high oxygen concentrations. *J. Cell Sci.* (94 (Pt 1): 43–49.

Omasa, T. et al. (1992). The enhancement of specific antibody production rate in glucose- and glutamine-controlled fed-batch culture. *Cytotechnology* 8 (1): 75–84.

Ozturk, S.S. and Palsson, B.O. (1990). Effects of dissolved oxygen on hybridoma cell growth, metabolism, and antibody production kinetics in continuous culture. *Biotechnol. Progr.* 6 (6): 437–446.

Ozturk, S.S., Riley, M.R., and Palsson, B.O. (1992). Effects of ammonia and lactate on hybridoma growth, metabolism, and antibody production. *Biotechnol. Bioeng.* 39 (4): 418–431.

Palsson, B.O. et al. (1993). Expansion of human bone marrow progenitor cells in a high cell density continuous perfusion system. *Biotechnology (N Y)* 11 (3): 368–372.

Papoutsakis, E.T. (2009). From CHO-cell to stem-cell biotechnology, oxygenation, and mixing in animal-cell culture: bioreactors, bubbles, and cell injury. *Biotechnol. Bioeng.* 102 (4): 976–979.

Patel, S.D. et al. (2000). The lactate issue revisited: novel feeding protocols to examine inhibition of cell proliferation and glucose metabolism in hematopoietic cell cultures. *Biotechnol. Progr.* 16 (5): 885–892.

Portner, R. et al. (2005). Bioreactor design for tissue engineering. *J. Biosci. Bioeng.* 100 (3): 235–245.

Quinn, P. and Harlow, G.M. (1978). The effect of oxygen on the development of preimplantation mouse embryos in vitro. *J. Exp. Zool.* 206 (1): 73–80.

Rafiq, Q.A. et al. Systematic microcarrier screening and agitated culture conditions improves human mesenchymal stem cell yield in bioreactors. *Biotechnol. J.* 2015.

Rodrigues, M.E. et al. (2010a). Technological progresses in monoclonal antibody production systems. *Biotechnol. Progr.* 26 (2): 332–351.

Rodrigues, C.A. et al. (2010b). Hypoxia enhances proliferation of mouse embryonic stem cell-derived neural stem cells. *Biotechnol. Bioeng.* 106 (2): 260–270.

Rodrigues, C.A. et al. (2011a). Stem cell cultivation in bioreactors. *Biotechnol. Adv.* 29 (6): 815–829.

Rodrigues, C.A. et al. (2011b). Microcarrier expansion of mouse embryonic stem cell-derived neural stem cells in stirred bioreactors. *Biotechnol. Appl. Biochem.* 58 (4): 231–242.

Rouf, S.A. et al. (2000). Single versus multiple bioreactor scale-up: economy for high-value products. *Biochem. Eng. J.* 6 (1): 25–31.

Rowley, J. et al. (2012). Meeting lot-size challenges of manufacturing adherent cells for therapy. *Bioprocess Int.* 10 (3): S16–S22.

Sadeghi, A. et al. (2011). Large-scale bioreactor expansion of tumor-infiltrating lymphocytes. *J. Immunol. Methods* 364 (1–2): 94–100.

Sardonini, C.A. and Wu, Y.J. (1993). Expansion and differentiation of human hematopoietic cells from static cultures through small-scale bioreactors. *Biotechnol. Progr.* 9 (2): 131–137.

Schroeder, M. et al. (2005). Differentiation and lineage selection of mouse embryonic stem cells in a stirred bench scale bioreactor with automated process control. *Biotechnol. Bioeng.* 92 (7): 920–933.

Schwarz, R.P., Goodwin, T.J., and Wolf, D.A. (1992). Cell culture for three-dimensional modeling in rotating-wall vessels: an application of simulated microgravity. *J. Tissue Culture Meth.* 14 (2): 51–57.

Serra, M. et al. (2012). Process engineering of human pluripotent stem cells for clinical application. *Trends Biotechnol.* 30 (6): 350–359.

Shekaran, A. et al. (2015). Enhanced in vitro osteogenic differentiation of human fetal MSCs attached to 3D microcarriers versus harvested from 2D monolayers. *BMC Biotech.* 15 (1): 102.

Sheyn, D. et al. (2010). The effect of simulated microgravity on human mesenchymal stem cells cultured in an osteogenic differentiation system: a bioinformatics study. *Tissue Eng. Part A* 16 (11): 3403–3412.

da Silva, C.L. et al. (2003). Modelling of ex vivo expansion/maintenance of hematopoietic stem cells. *Bioprocess. Biosyst. Eng.* 25 (6): 365–369.

da Silva, C.L., Ferreira, B.S., and Cabral, J.M.S. (2006). Biorreactores para cultura de células animais. In: *Reactores Biológicos* (ed. M.N.R.D. Fonseca and J.A. Teixeira). LiDEL-Edições Técnica Lda.

Silverman, H.S. et al. (1997). Myocyte adaptation to chronic hypoxia and development of tolerance to subsequent acute severe hypoxia. *Circ. Res.* 80 (5): 699–707.

Singh, V. (1999). Disposable bioreactor for cell culture using wave-induced agitation. *Cytotechnology* 30 (1–3): 149–158.

Singh, H. et al. (2010). Up-scaling single cell-inoculated suspension culture of human embryonic stem cells. *Stem Cell Res.* 4 (3): 165–179.

Storm, M.P. et al. (2010). Three-dimensional culture systems for the expansion of pluripotent embryonic stem cells. *Biotechnol. Bioeng.* 107 (4): 683–695.

Sun, L., Sun, T.T., and Lavker, R.M. (2000). CLED: a calcium-linked protein associated with early epithelial differentiation. *Exp. Cell. Res.* 259 (1): 96–106.

Tharmalingam, T. et al. (2011). Enhanced production of human recombinant proteins from CHO cells grown to high densities in macroporous microcarriers. *Mol. Biotechnol.* 49 (3): 263–276.

Thilly, W.G. (1986). *Mammalian Cell Technology*. Butterworth-Heinemann.

Timmins, N.E. et al. (2009). Clinical scale ex vivo manufacture of neutrophils from hematopoietic progenitor cells. *Biotechnol. Bioeng.* 104 (4): 832–840.

Toh, Y.C. and Voldman, J. (2011). Fluid shear stress primes mouse embryonic stem cells for differentiation in a self-renewing environment via heparan sulfate proteoglycans transduction. *FASEB J.* 25 (4): 1208–1217.

Tolbert, W.R., Hitt, M.M., and Feder, J. (1980). Cell aggregate suspension culture for large-scale production of biomolecules. *In Vitro* 16 (6): 486–490.

Tolbert, W.R. et al. (1982). Large-scale mammalian cell culture: design and use of an economical batch suspension system. *Biotechnol. Bioeng.* 24 (7): 1671–1679.

Trounson, A. et al. (2011). Clinical trials for stem cell therapies. *BMC Medicine* 9: 52.

Ulloa-Montoya, F., Verfaillie, C.M., and Hu, W.S. (2005). Culture systems for pluripotent stem cells. *J. Biosci. Bioeng.* 100 (1): 12–27.

Urabe, M., Ding, C., and Kotin, R.M. (2002). Insect cells as a factory to produce adeno-associated virus type 2 vectors. *Hum. Gene Ther.* 13 (16): 1935–1943.

Valasek, C. et al. (2011). Production and purification of a PER.C6-expressed IgM antibody therapeutic. *Bioprocess Int.* 9 (11): 28–37.

Varani, J. et al. (1993). Use of recombinant and synthetic peptides as attachment factors for cells on microcarriers. *Cytotechnology* 13 (2): 89–98.

Voisard, D. et al. (2003). Potential of cell retention techniques for large-scale high-density perfusion culture of suspended mammalian cells. *Biotechnol. Bioeng.* 82 (7): 751–765.

Wang, T.Y., Brennan, J.K., and Wu, J.H. (1995). Multilineal hematopoiesis in a three-dimensional murine long-term bone marrow culture. *Exp. Hematol.* 23 (1): 26–32.

Weber, C. et al. (2010a). Expansion of human mesenchymal stem cells in a fixed-bed bioreactor system based on non-porous glass carrier–part A: inoculation, cultivation, and cell harvest procedures. *Int. J. Artif. Organs* 33 (8): 512–525.

Weber, C. et al. (2010b). Expansion of human mesenchymal stem cells in a fixed-bed bioreactor system based on non-porous glass carrier – Part B: modeling and scale-up of the system. *Int. J. Artif. Organs* 33 (11): 782–795.

van Wezel, A.L. (1967). Growth of cell-strains and primary cells on micro-carriers in homogeneous culture. *Nature* 216 (5110): 64–65.

Whitford, W.G. (2006). Fed-batch mammalian cell culture in bioproduction. *Bioprocess Int.* April: 30–40.

Wiles, M.V. and Johansson, B.M. (1999). Embryonic stem cell development in a chemically defined medium. *Exp. Cell. Res.* 247 (1): 241–248.

Williams, R.L. et al. (1988). Myeloid leukaemia inhibitory factor maintains the developmental potential of embryonic stem cells. *Nature* 336 (6200): 684–687.

Woodside, S.M., Bowen, B.D., and Piret, J.M. (1998). Mammalian cell retention devices for stirred perfusion bioreactors. *Cytotechnology* 28 (1–3): 163–175.

Wurm, F.M. (2004). Production of recombinant protein therapeutics in cultivated mammalian cells. *Nat. Biotechnol.* 22 (11): 1393–1398.

Yamamoto, K. et al. (2005). Fluid shear stress induces differentiation of Flk-1-positive embryonic stem cells into vascular endothelial cells in vitro. *Am. J. Physiol. Heart Circ. Physiol.* 288 (4): H1915–H1924.

Yang, J.D. et al. (2000). Achievement of high cell density and high antibody productivity by a controlled-fed perfusion bioreactor process. *Biotechnol. Bioeng.* 69 (1): 74–82.

Yang, Y., Rossi, F.M., and Putnins, E.E. (2007a). Ex vivo expansion of rat bone marrow mesenchymal stromal cells on microcarrier beads in spin culture. *Biomaterials* 28 (20): 3110–3120.

Yang, J.D. et al. (2007b). Fed-batch bioreactor process scale-up from 3-L to 2,500-L scale for monoclonal antibody production from cell culture. *Biotechnol. Bioeng.* 98 (1): 141–154.

Ying, Q.L. et al. (2003). BMP induction of Id proteins suppresses differentiation and sustains embryonic stem cell self-renewal in collaboration with STAT3. *Cell* 115 (3): 281–292.

Zandstra, P.W., Eaves, C.J., and Piret, J.M. (1994). Expansion of hematopoietic progenitor cell populations in stirred suspension bioreactors of normal human bone marrow cells. *Biotechnology (N Y)* 12 (9): 909–914.

Zandstra, P.W. et al. (2003). Scalable production of embryonic stem cell-derived cardiomyocytes. *Tissue Eng.* 9 (4): 767–778.

Zhong, J.J. (2010). Recent advances in bioreactor engineering. *Korean J. Chem. Eng.* 27 (4): 1035–1041.

Zhu, J. (2012). Mammalian cell protein expression for biopharmaceutical production. *Biotechnol. Adv.* 30 (5): 1158–1170.

Zielke, H.R., Zielke, C.L., and Ozand, P.T. (1984). Glutamine: a major energy source for cultured mammalian cells. *Fed. Proc.* 43 (1): 121–125.

Zweigerdt, R. (2009). Large scale production of stem cells and their derivatives. *Adv. Biochem. Eng. Biotechnol.* 114: 201–235.

5

Model-Based Technologies Enabling Optimal Bioreactor Performance

Rimvydas Simutis[1], Marco Jenzsch[2], and Andreas Lübbert[1]

[1] *Martin Luther University, Halle-Wittenberg, Halle/Salle, Germany*
[2] *Roche Diagnostics Operations, Nonnenwald 2, Penzberg, Germany*

5.1   Introduction

Bioprocess engineers aim in using mechanistic knowledge to optimize their processes. Quantitatively exploitable models are used to formulate this knowledge and to exploit it subsequently with numerical optimization procedures. Models are goal-specific tools employed for optimizing design and operational procedures and must be tightly restricted to concrete tasks. Before they can be used for decision-making, they must be validated. For that purpose, accurate experimental process data are required that allow discriminating between possible model variants. Modeling without concrete validation data is at best an academic exercise.

Where not all important aspects of the process are understood mechanistically, experimental data can be used to formulate data-driven models. The engineering correlations traditionally used for this purpose have now been replaced by advanced modeling techniques. In many practical cases, however, combinations of data-driven and mechanistic approaches, referred to as hybrid models, are the methods of choice (Psichogios and Ungar 1992; Thompson and Kramer 1994; von Stosch et al. 2012, 2014).

Unfortunately, in industrial practice, trial-and-error methods still prevail. Even if this is supported by statistical methods of experimental design, this approach requires many experiments and does not really exploit the available a priori knowledge. This chapter presents some simple practical examples illustrating how model-based optimization of industrial bioprocesses becomes easily accessible to the bioprocess engineer responsible for manufacturing bioreactors in industry. First, we describe the general procedure of model-based optimization and then we explain details by means of realistic case studies.

Bioprocessing Technology for Production of Biopharmaceuticals and Bioproducts, First Edition.
Edited by Claire Komives and Weichang Zhou.
© 2019 John Wiley & Sons, Inc. Published 2019 by John Wiley & Sons, Inc.

5.2 Basics

5.2.1 Balances

Commonly mechanistic models are based on balances of conserved extensive variables, such as mass and energy, around some control volume. Our simplest models assume that the culture is well mixed, i.e. that we do not need distinguishing different volume elements in the culture. Then the control volume comprises the entire culture, and the dynamic models can be expressed by ordinary differential equations.

As we are less familiar with masses, mass balances are usually transformed into equations in terms of concentrations c (e.g. Shuler and Kargi 2002)

$$\frac{\mathrm{d}c}{\mathrm{d}t} = R + \frac{F}{W}(c_F - c) \tag{5.1}$$

It describes changes $\mathrm{d}c/\mathrm{d}t$ with time in the concentrations c of the relevant substances. These are either due to reactions taking place at rate R within the reactor volume W or by transport across the boundaries of the culture, e.g. by feeding with a solution of concentration c_F at rate F. In simple biomass growth processes, the most important elements of the vector c are the concentrations of biomass, X, substrate, S, and product, P. Most industrial cultivations, however, are operated in the fed-batch mode. Then the liquid volume W increases, and we need an additional, albeit very simple, equation

$$\frac{\mathrm{d}W}{\mathrm{d}t} = F \tag{5.2}$$

The critical part of such balances is the volumetric conversion rate vector R for the components of c. It is usually expressed in terms of the specific conversion rate vectors q. In this respect, specific means relative to the biomass concentration X.

These specific rates can be described by a Monod-type function, for instance the specific substrate uptake rate σ by

$$\sigma = \sigma_{\max} \frac{S}{K_s + S} \tag{5.3}$$

where K_s is a saturation constant, and $\sigma_{\max}$ the maximal specific substrate uptake rate the cells can make. When we assume that the conversion process can be described by a simple net reaction equation, then the specific biomass growth and product formation rates, μ and π, are proportional to σ, where $Y_{xs} = \mu/\sigma$ and $Y_{ps} = \pi/\sigma$ are the proportionality constants referred to as biomass yield on the substrate consumed by the cells and product formation rate per rate of substrate consumed. The volumetric conversion rate vector R is related to the specific rates vector $q = [\mu;\sigma;\pi]$ by $R = q\,X$.

In most cases, balances in the form of Eq. (5.1) can be established; however, simple expressions for the conversion rates are often not known to a comparable level of accuracy. Hence more complex expressions for R must be found either by mechanistic reasoning or by exploiting process data (Simutis and Lübbert 1997a). Much experience has been made with artificial neural networks (ANNs) to replace Monod-type expressions for the conversion rates in the balance equations (e.g. Gnoth et al. 2010).

When the assumption of homogeneous conditions within the entire culture is not justified at all, the models may become quite complex. In the general case, we must then describe the turbulent mixing process within the bioreactor by fluid dynamical models. At the first glance, such models also look quite simple, as the fluid flow can be described by the Navier–Stokes equation, but its solution under realistic boundary conditions in a multiphase system is nearly impossible within a process optimization project: a single solution of these equations may take days even on fast computers. Hence, rough but realistic approximations must be taken, e.g. compartment models in which we assume that the culture can be divided into parts that can be treated individually as ideally stirred tanks.

Attempts to explain the nonlinear behavior of cellular conversion processes mechanistically by means of detailed metabolic modes are widespread in literature (e.g. Ramkrishna 2003; Wiechert and Noack 2011). Such models gain our insights, for instance into details of the conversion processes (Kim et al. 2012), by showing that not only stoichiometry and biochemical kinetics are responsible for their nonlinearities and constraints but also cell-internal regulation processes. However, although these investigations make valuable contributions to our mechanistic knowledge, they cannot yet be employed for quantitatively optimizing the operational procedures in industrial production reactors. These models are far too complex and their many free parameters cannot simultaneously be identified and validated in situations relevant to industrial manufacturing processes. Since, such models usually cannot be maintained in industrial manufacturing plants, we do not further consider them in this review.

5.2.2 Model Identification

Before we can employ a model for process optimization, we must determine its parameters. This parameter identification is already an optimization task, where optimal model parameters are found by fitting the model to available process data.

For a new process, one can start this model identification using data from similar processes. Then, in an iterative procedure, the model with its current parameters, i.e. the model that represents our current process knowledge, is used to plan an experiment by which the process is probed. This is a further optimization procedure where optimal operational trajectories are proposed. The data measured during such an experiment are then, in the next iteration cycle, used to improve the model parameters, and the improved model is again used to lay out a further experiment that provides data from situations that are closer to the final optimum.

Parameter identification is performed numerically using nonlinear optimization procedures. A wide variety of optimization routines are available for that purpose. Most often, software packages, such as Matlab®, are used, which offer routines of all complexity levels. Identification procedures for hybrid models combining ANNs and mass balance equations require a little bit more efforts. They can be identified using the sensitivity equation approach together with gradient-based methods (Psichogios and Ungar 1992; Schubert et al. 1994; Kadlec et al. 2009). For identification of more complicated, e.g. not differentiable models, random search algorithms, including genetic and evolutionary approaches, can be employed (Simutis and Lübbert 1997b).

The important point to make here is that neither optimization software nor computational power is decisively limiting model development for process optimization. More important is a deep understanding of the process dynamics and informative precise measurement data from the process under consideration.

Three critical aspects must be considered with respect to data when creating models for industrial bioprocess optimization:

1) Data are required from tightly controlled experiments, e.g. at $\Delta p\text{H} = \pm 0.01$, $\Delta p\text{O}_2 = \pm 2\%$, $\Delta T = \pm 0.02\,°\text{C}$, and $\Delta F = \pm 1\%$ from their set-point profiles. Larger variations can lead to some unpredictable process changes and consequently to inaccurate models that are inadequate for prediction and thus useless for process optimization. It is not alone the preciseness but also the accuracy that is of importance. This must be checked by simple mass, C-mol, or electron balances. These must be closed before any model identification based on their values makes sense.
2) Because the most important process variables (biomass, substrate, and product concentrations) cannot be measured to sufficient accuracy at sufficiently small time increments, special soft-sensor techniques must be applied to generate reliable on-line estimates of these variables.
3) The process itself must be run in a highly reproducible way in order to produce useful data for process modeling. Data from a single run only are not sufficient.

5.2.3   Model-Based Process Optimization

Modeling must be goal-specific and tightly restricted to solving the special problem under consideration in order to keep the number of process data needed for model identification and validation of optimization results within reasonable limits.

The main goals of modeling for bioreactor performance optimizations are the following:

(i) First main goal of modeling for process optimization is supporting estimates of the key quantities governing the process. As most of them cannot be measured directly in a sufficiently accurate way they must be derived indirectly in a multivariate way. This requires models that interrelate available measurement signals and the quantities to be estimated. Without accurate and precise data, modeling remains a play with computers.

(ii) Currently, most processes in biochemical production systems are operated in an open-loop fashion, where they are run along predefined trajectories of action variables such as the main substrate feed rate profile $F(t)$. Although one is generally interested in high product yields, the higher-ranking optimization objective is batch-to-batch reproducibility. When there is no feedback, these trajectories must be chosen so that the process becomes robust with respect to commonly appearing disturbances. This must usually be paid for by cutbacks in the volumetric productivity.

(iii) Process optimization with respect to product quality is number one priority in many companies manufacturing biologics. Model-supported optimization techniques help reducing the number of experiments required to solve these problems. As everywhere in science and technology, utilization of a priori knowledge reduces the work load as compared to pure statistical approaches.

(iv) When the state of the process can be estimated online, this data can be used to correct for deviations from the desired paths by feedback control. Then the process can be run closer to its optimum. Model-supported control is most effective in this sense, particularly in nonlinear chemical and biochemical productions systems where the process dynamics is often considerably changing with time. With respect to modeling, this closed-loop approach has the advantage in comparison with open-loop control, that it does not only correct for process disturbances but also for small model insufficiencies so that small inaccuracies of the models are compensated for as well.

(v) Most often quasi-optimal process-operational procedures and process control can be derived from unstructured unsegregated models. However, some problems require consideration of structured models. One then needs solving partial differential equations to determine sufficiently accurate process trajectories. However, in some cases, solutions can be obtained easier simply by considering process model components on different time scales.

(vi) In order to really optimize an industrial bioreactor, one needs process supervision techniques. These must be distinguished from process control in the following sense: a controller reacts on deviations from its set-point profile in a clearly defined way to keep the process on track. A supervisor, however, examines whether or not the measurement information on which the controller acts is correct at all. Hence, it looks for data consistency and accuracy, which, of course, can only be decided on the basis of process know-how and thus reliable process models.

(vii) Finally, in industrial applications, the models must be compatible with the control structures at the production plants. A modern way to make sure that the process models can properly work within the industrial automation systems is developing, testing, and finally validating them directly within the process automation systems. This led to the development of virtual plants (McMillan et al. 2008). A well-established virtual plant can then be used as a powerful process simulator not only for comparing different control designs and operational techniques but also for the training of personnel such as we know it from flight simulators.

5.3   Examples

5.3.1   Model-Based State Estimation

Ab initio modeling is not possible in bioprocess engineering. In all cases, models must be identified using data from the processes under consideration and can thus not be more accurate than the data. As most key quantities of bioprocesses such as biomass and product concentration cannot be measured accurately enough at small time increments, indirect measurements are necessary to make them measurable at all or at least to significantly increase the measurement accuracy. This means multivariate, model-supported measurements and thus sufficiently accurate models that interrelate the available signals to obtain reliable data that enable at all modeling for bioreactor optimization (Mandenius 2004; Luttmann et al. 2012).

Two main levels of sophistication can be distinguished for state and parameter estimation: (i) using static and (ii) using dynamic models. Both will be discussed at the example of the biomass estimation.

5.3.1.1   Static Model Approach

Static models use fixed relationships between the easily measurable process variables and the desired ones. Linear relations are the simplest approaches for biomass estimation.

Figure 5.1 shows that the easy measurable volume fractions of CO_2 and O_2 in the fermenter vent line, which are generally used to determine carbon dioxide production rates (CPRs), carry much information about the biomass. A linear

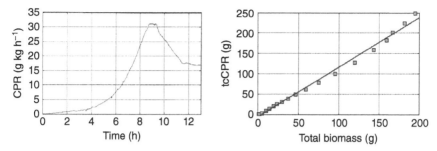

Figure 5.1 Correlation between biomass and the total mass of carbon dioxide derived from the volume fraction measurements of CO_2 in the vent line.

correlation of the total cumulative CPR provides a fairly good first estimate for the biomass. The same applies for the oxygen uptake rate (OUR). As both can be measured reliably at bioreactors of all sizes, they can be used as a rich information source for biomass estimation.

As this correlation is not truly linear, improvements are necessary. The simplest improvement is taking both signals and to form simple polynomial relationships between biomass x and the tcCPR as well as the corresponding tcOUR data. tcCPR and tcOUR are abbreviations for the total cumulative CPR and OUR signals (Jenzsch et al. 2006b). However, better estimation performance is obtained with ANNs.

Simple feedforward ANNs with a single hidden and a linear output layer are sufficient for most purposes in bioprocess engineering. They use internal non-linear auxiliary functions ϕ

$$h_j(x_i) = \phi\left(\sum^{D_{;i=0}} V_{ji}\, x_i + V_{j0}\right) \tag{5.4}$$

to formulate the nonlinear parametric mapping

$$y_k(x_i) = \sum^{M_{;j=0}} U_{kj}\, h_j(x_i) + U_{k0} \tag{5.5}$$

U and V are coefficient matrices, referred to as weights. An often-used example for the nonlinear internal function ϕ is the hyperbolic tangent (tanh()).

Hence, ANNs cannot be considered black art. They are just parametric functional representations of data. Their parameters must be fitted to measurement data. For that purpose, general nonlinear optimization routines can be taken.

A typical example obtained with such a simple ANN is shown in Figure 5.2. It depicts an estimation of the viable cell density in a production reactor at the 12 000 scale. The four input variables are oxygen/pressurized air ratio and sparging rate from the pO_2 controller as well as base consumption and finally reactor weight. The ANN was trained/validated on data sets from 49 production runs.

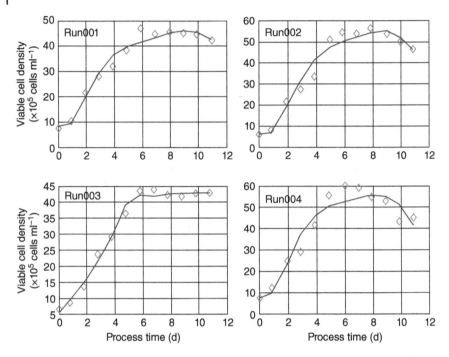

Figure 5.2 Viable cell density estimate for a MAB production cell culture. The lines are the online estimates; the symbols depict the offline measured data available afterward. These offline data were not used during the network training.

ANNs are replacing simple polynomial relationships or power laws conventionally used in engineering to interrelate variables (Glassey et al. 1997; Norgaard et al. 2003). Although they usually lead to much higher estimation performance as the correlations, their essential disadvantage is that the number of free parameters (weights U and V) is considerably high. Hence, more data are required to reproducibly determine them. Moreover, ANN models sensitively depend on the first guesses for their weights in the training, consequently for the same data, a group of models with very similar characteristics can be formed. This may lead to confusions during the validation process of control procedures based on these models. To cope with this problem, new black-box models have been developed in the machine learning community. The most important variants are Vapnik's (1998) support vector machines (SVMs) and Tipping's (2000) relevance vector machines (RVMs). Both are based on linear combinations of nonlinear functions $K(c,c_n)$ referred to as kernel functions (Schölkopf and Smola 2002):

$$x(c) = \sum_{n=1}^{N;n=1} w_n K(c, c_n) \qquad (5.6)$$

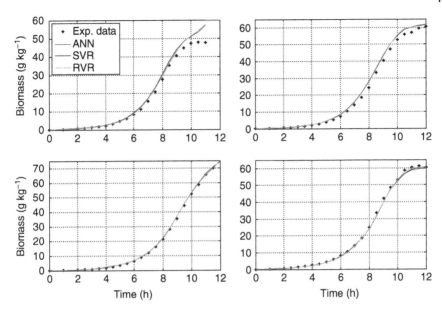

Figure 5.3 Comparison of different black-box models in estimating the biomass x (g) in *E. coli* cultures. Here validation data are shown that have not been used for model fitting. The various estimators were trained on cumulative OUR and CPR data.

A typical kernel function is the Gaussian

$$K(c, c_n) = \frac{1}{\sqrt{(2\pi\sigma^2)^N}} \cdot e^{\left(-\frac{(c-c_n)^2}{2\sigma^2}\right)} \tag{5.7}$$

These representations perform well with a relatively small number of terms. Tipping's RVM usually needs less terms than Vapnik's SVM. A typical example is shown in Figure 5.3. As can be seen, the estimation performances of these techniques do not differ much from each other, and the decisive difference is the number of free parameters. This commonly improves generalization as well as the extrapolation quality and implementation issues of these kernel models.

Where a single black box model (ANN, SVM, of RVM) cannot cover the entire process, it is straightforward to divide the process into different phases and to train different static models to each phase. The transition between the phased can be made smoothly by means of a few simple fuzzy rules (Simutis et al. 1993).

5.3.1.2 Dynamic Alternatives

Where static models are not considered sufficient, the relationships can be formulated dynamically by means of ordinary differential equations. The technique most widely distributed across the entire engineering is the Kalman filter and its extensions. A clear advantage in our context is that the Kalman filter and its extensions can be used not only to estimate biomass, substrate, and product concentrations but also to estimate specific biomass growth, substrate consumption, and product formation rates. Their conceptual advantage is that they combine measurement data with its corresponding model-based estimates and weights their mutual influence on the final estimate by their relative uncertainties, expressed by covariance matrices.

The original Kalman filter is restricted to linear models. Hence, in engineering, one usually takes its nonlinear extension, the so-called extended Kalman filter (EKF). However, EKF is restricted to slightly nonlinear models. The reason is that the model nonlinearity is approximated by the first two terms of its Taylor expansion. This requires the computation of the Jacobian matrix, which restricts the models (structures) that can be employed in practice. The most important difficulty, however, is the tuning of the covariances, i.e. the uncertainties in the measurements and the model predictions.

Most disadvantages are removed with a new development (Julier and Uhlmann 2004), the unscented Kalman filters (UKFs), which no longer need model approximations but can directly employ a full nonlinear model. An example of biomass and specific growth estimates is shown in Figure 5.4.

Lifting the restrictions for the models opens a large number of new Kalman filter applications in bioprocess engineering.

Another very efficient modeling technique that can be used not only for state estimation is hybrid modeling combining data-driven ANNs and dynamic differential equations as shown in Figure 5.5 (Schubert et al. 1994; Aehle et al. 2010; Gnoth et al. 2010).

As compared with Kalman filters, this approach has the advantage that measurement and modeling errors are weighted automatically during ANN training. This makes the difficult tuning procedures of Kalman filters obsolete. Estimates for the amount of viable cells and the product mass are very good as shown in Figure 5.6.

5.3.2 Optimizing Open Loop-Controlled Cultivations

5.3.2.1 Robust Cultivation Profiles

For open loop-controlled fermentations, trajectories of the action variables must be found that make the culture robust against unavoidable randomly appearing distortions. Missing robustness leads to different outcomes at every production run which is unacceptable.

In industrial recombinant protein productions, an important source of process variability is variations in the inoculum size. The simple model formulated in Eq.(5.1) allows simulating its influence on biomass propagation: most production processes are started in the batch mode with a relatively high initial substrate concentration leading to biomass growth at maximal growth

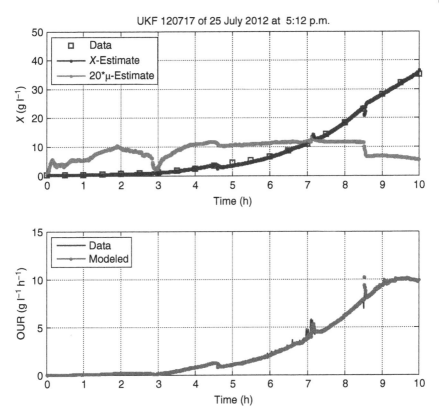

Figure 5.4 Unscented Kalman filter, simplest case of estimating biomass concentrations and specific growth rates from OUR data. After an initial growth phase, the filter gives quite reliable estimates.

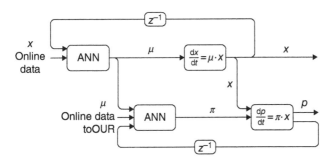

Figure 5.5 Hybrid model used for estimating viable cell count together with specific growth rates and product mass with and product formation rates in a CHO-culture producing erythropoietin (EPO).

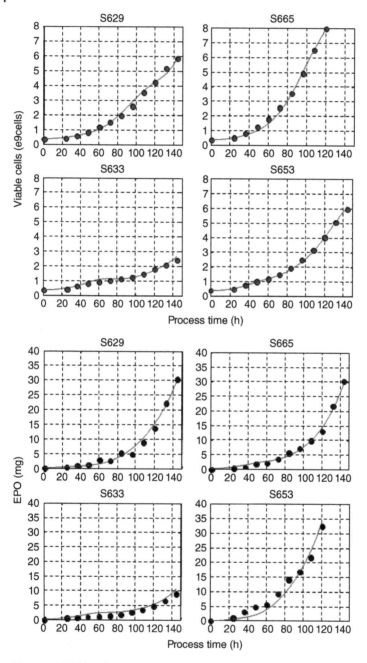

Figure 5.6 Viable cell counts and product mass in a CHO culture measured (symbols, validation data) and estimated (lines) online by means of the hybrid model depicted in Figure 5.5.

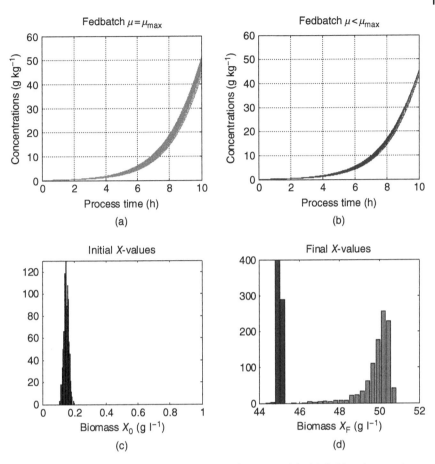

Figure 5.7 Comparison of fed-batch operation with $\mu = \mu_{max}$ (a) with fed-batch operation with $\mu_{set} < \mu_{max}$ (b). For both, the same X_0-variability (c) of 10% was taken. The variability of the final biomass X_F is depicted in (d).

rate until the substrate becomes depleted. The biomass concentration in this situation very sensitively depends of the initial biomass concentration.

Robust operation can be obtained by starting the process in the fed-batch mode with a constant specific growth rate $\mu_{set} < \mu_{max}$ as shown in Figure 5.7.

In the case that the biomass is higher than the nominal one for which the feed was computed, the cells will see less substrate than expected, hence, their growth rate will become lower. When on the other hand, the inoculum is smaller, there is more substrate available for the cells, and hence, they grow faster. In this way, the initial errors are automatically corrected. This, however, is not the case when the cells are already run at maximal speed, then, if the inoculum is smaller, their growth rate cannot be increased any more. Hence, only at a smaller growth rate robustness and thus reproducibility can be obtained (Jenzsch et al. 2006a). Obviously, robust operation is obtained at the cost of productivity. The more precisely the initial biomass can be adjusted, the smaller the difference between μ_{set} and μ_{max} must be chosen, and the smaller is the cost of robustness in terms of productivity.

This way of process operation is of particular importance to the biomass growth phase of recombinant protein production processes as in this initial phase the accuracies by which state variables can be estimated is known to generally be low and hence, the alternative to guaranteeing reproducibility, namely closed loop control, cannot successfully be used. In later process phases, where the estimates work perfect, the process can be kept very close to its real optimum by feedback control.

Consequently, to achieve robust and reproducible processes, the cultivations should entirely be operated in the fed-batch mode using feeding strategies that keep the specific biomass growth rates slightly below their maxima.

5.3.2.2 Evolutionary Modeling Approach

In the beginning of any process design, only a few quantitative data is available from the process under consideration. Missing data must be compensated for by possibly relevant a priori knowledge taken from literature and by experiences with similar production processes. This initial knowledge or hypothesis about the process dynamics, specified in the form of a first model, should directly be used to compute the best feed rate profile, which is possible under the assumption that the model is accurate. Then, not earlier, first experiments should be conducted along the computed trajectories. The data measured during this exploratory experiment provides a basis for a decision about the accuracy of the initial model. It is then employed for improving the model structure and correction of the model parameters in such a way that the next model outputs and this measured data match as good as possible.

This procedure, schematically shown on the left of Figure 5.8, can be repeated iteratively: at each stage of this design procedure, the currently available quantitative process knowledge in form of a numerically exploitable model is used to find the best process control trajectories by numerical optimization techniques. In this way, two goals are approached simultaneously: (i) the model becomes better in each iteration step and (ii) the productivity becomes better as well.

The efficiency of this evolutionary development strategy was demonstrated at the example of the production of the light chain of the antibody MAK33 in *Escherichia coli* (Galvanauskas et al. 2004). This protein was expressed by pUBS520 p12023 bacteria under the control of the tac promoter, where lactose was used for induction. During only four experiments displayed on the right of Figure 5.8, the quasi-optimal feed profile was determined. This evolutionary design procedure allowed increasing the target recombinant protein production by approximately 300%. Figure 5.8 illustrates the development from the initial experiment to the target product concentration profile.

(a)

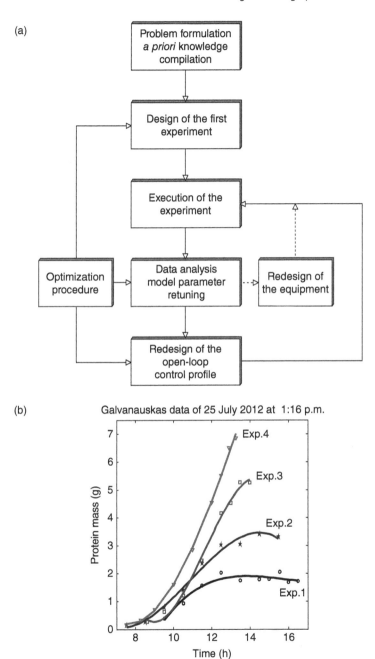

(b)

Galvanauskas data of 25 July 2012 at 1:16 p.m.

Figure 5.8 (a) Scheme of the evolutionary procedure. (b) Evolutionary improvement of the model for the formation of the short chain of the antibody MAK33 in an *E. coli* culture.

5.3.3 Optimization by Model-Aided Feedback Control

5.3.3.1 Improving the Basic Control

Industrial cultures of microbial and animal cells cannot be considered capricious creatures that behave different every day. As already stated by Sonnleitner and Fiechter (1992), variations in experimental results can essentially be traced back to inexactly controlled process operation. Temperature, pH, and pO_2 should tightly be controlled in order to keep the variability low enough. In most industrial cultivation processes, these basic environmental quantities are controlled by simple proportional integral derivative (PID) controllers with fixed parameters. However, as the dynamics of cellular growth and product formation processes change significantly with time, the controller parameters change as well and so that these controllers cannot perform well across the entire fermentation. Fortunately, closed loop controllers do not need highly exact models for the adaptation of the controller parameters; here fairly rough models are sufficient (Åström and Hägglund 2005; Dochain 2008).

As shown by Gnoth et al. (2009) and Kuprijanov et al. (2009), the variations (errors) of conventional (PID) pH and pO_2 controllers are roughly proportional to the CPR, which coarsely reflects the change in the activity of the culture during the production run. Hence, changes in the controller parameters can simply be related to CPR. A relatively simple gain-scheduling controller base on the measured CPR solves the adaptation problem. Results are shown in Figure 5.9.

Instead of the CPR signal, one can also take the OUR signal for gain scheduling. The latter was preferred in animal cell cultures (Aehle et al. 2011a).

5.3.3.2 Optimizing the Amount of Soluble Product

In *E. coli* cultures, the recombinant protein may appear in its soluble form or as an inclusion body. When refolding is not possible in an efficient way, as much soluble protein as possible is desired. Often, it helps optimizing the temperature and the specific growth rate in the product formation phase to change the relative fraction of these product forms.

As such a temperature dependency cannot easily be described analytically; the kinetics of the process was described by means of ANNs as shown schematically in Figure 5.10. Two different ANNs were taken for the specific biomass growth rate μ and the specific product formation rate π. These were used in the basic mass balance equations for biomass and product. Hence, we are dealing with a simple hybrid model (Gnoth et al. 2010). A visualization of the soluble product formation rate is also shown in the figure. It shows a clear maximum over the specific growth rate μ and temperature T, which allows for numerical optimization of the operational conditions. The maximum position is usually changing during the cultivation process. Using the hybrid model and

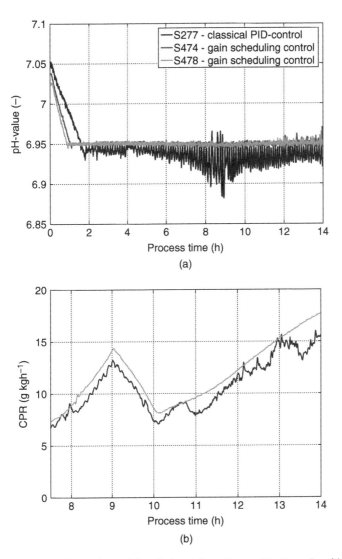

Figure 5.9 Comparison of the pH signals from three cultivations. Panel (a) was conventionally PID-controlled with fixed parameters, the two others with a gain scheduling controller that used the measured CPR signal. (b) The result in the CPR signal is shown. It depicts a much higher signal-to-noise ratio. That is, its information content is much higher.

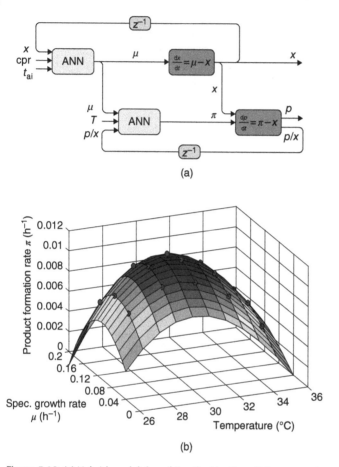

Figure 5.10 (a) Hybrid model describing the kinetics of biomass and product formation by ANNs making use of the measured CPR and the culture temperature after induction. (b) Product formation rate described by the model as a function of temperature and the specific biomass growth rate. Note that this relationship is usually changing during the cultivation process.

numerical optimization procedures, it is straightforward to determine optimal time profiles of the specific growth rate $\mu(t)$ and the temperature $T(t)$.

In order to make sure that the process will consistently run at this optimum, feedback control along this optimal $\mu(t)$-profile seems to be straightforward as most advanced control techniques published in literature took this physiological key quantity μ as the control variable. This, however, has two disadvantages: the first is that μ cannot be measured but must be estimated from the measurable quantities. The second is more important as every

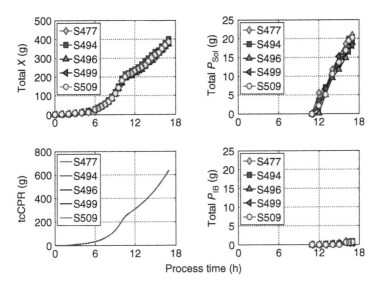

Figure 5.11 Five repetitions of the *E. coli* culture that is controlled along the tcCPR profile. Biomass, as well as both product fractions, is kept on their predefined profiles.

random deviation from the set point for the specific growth rate leads to differences in the desired biomass profile that cannot be corrected by later controller actions and thus leads to variations in the final biomass. Hence, where the objective is batch-to-batch reproducibility, direct μ-control is not advisable. An indirect μ-control by following the desired biomass profile could be used instead. However, this would require a good biomass estimator. As already shown, this is possible; however, an even simpler control variable that can be measured quite directly is the profile of the total cumulative carbon dioxide profile, i.e. the mass of CO_2 produced up to time t (Jenzsch et al. 2007). This can be used to indirectly keep $\mu(t)$ under tight control and at the same time ensuring a high batch-to-batch reproducibility. Experimental evidence for the perfect reproducibility with this model-based approach is provided in Figure 5.11: five validation runs impressively show that the batch-to-batch reproducibility is perfect not only for the controlled variable but also for the key variables biomass and the two product mass profiles. This shows that the process can reliably run on its model-aided optimized paths. The final variability in biomass and product are not caused by model uncertainties but result from the measurements. Biomass and particularly recombinant protein concentrations cannot be measured as accurate as CPRs.

At the same time, this example shows that all product that initially was roughly equally distributed among inclusion bodies and soluble protein, nearly quantitatively appeared in its soluble form.

5.3.4 CO_2-Removal in Large-Scale Cell Cultures

Oxygen supply and CO_2 removal and mixing are generally recognized bottlenecks of bioreactors (Noorman 2011), particularly in animal cell cultures. As well known in industrial practice with large cell culture reactors, CO_2 concentrations accumulate in the culture until it reaches levels at which cell growth and product formation can significantly become inhibited (e.g. Zhu et al. 2005).

The key to solving this problem is understanding the reason for CO_2 accumulation within production-scale cell cultures (Sieblist et al. 2011). Dissolved CO_2 continuously produced by the cells must be stripped from the culture by gas sparging. In small reactors, this does not make any problem, the normal aeration will do the job. In larger reactors, however, the gas bubbles pass the culture on longer ways and then, importantly, they may become saturated with CO_2 before they leave the culture. Once saturated, the bubbles can no longer take up further carbon dioxide. For CO_2, this takes only some seconds. When the reactor is large enough, all gas bubbles will be saturated. Then the stripping rate does no longer depend on the mass transfer rates between liquid and gas, and the only remaining influence variable is the gas throughput. This does not apply to the oxygen mass transfer in cell cultures, as with respect to oxygen a gas bubble can take part in mass transfer for several minutes instead of seconds.

From the modeling point of view, we can no longer assume the reactor to be well mixed with respect to the gas phase. Fortunately, we do not need complex partial differential equations to provide a sufficiently accurate model describing the O_2/CO_2 mass transfer process. A simple estimation shows that the time constant at which the dissolved CO_2 concentration in the bulk of the culture changes is much longer than that for accumulation of CO_2 within the rising bubble. Hence, for the individual rising bubble, we can assume that the bulk concentration does not change significantly during its residence time in the reactor. Hence, we can consider both effects independent from each other.

The amount n_{CO_2} is transferred into a representative rising bubble of surface area A at a rate that follows Fick's first law:

$$J = \frac{dn_{CO_2}}{A\,dt} = k_L \cdot (CO_2{}^* - CO_2) \tag{5.8}$$

The flux J is driven by the difference between concentration CO_2 in the bulk liquid and concentration $CO_2{}^*$ directly at the gas liquid interface. k_L is the mass transfer coefficient. Initially, the bubble contains air only and thus practically no CO_2. During its rise in the culture, its CO_2 content, n_{CO_2}, rises, while pCO_2 in the bulk can be assumed to be constant during its residence time. A typical solution of Eq. (5.8) is shown in Figure 5.12, where the amount of gas is represented in terms of the volume fraction $y_c = n_{CO_2}/n$, where n is the entire amount of gas in the bubble. Similar equations are valid for O_2.

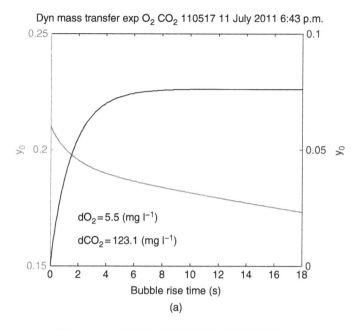

Dyn mass transfer exp O_2 CO_2 110517 11 July 2011 6:43 p.m.

$dO_2 = 5.5$ (mg l^{-1})

$dCO_2 = 123.1$ (mg l^{-1})

Bubble rise time (s)

(a)

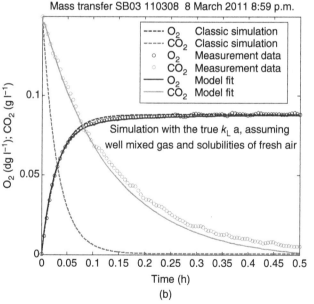

Mass transfer SB03 110308 8 March 2011 8:59 p.m.

----	O_2	Classic simulation
----	CO_2	Classic simulation
○	O_2	Measurement data
○	CO_2	Measurement data
——	O_2	Model fit
——	CO_2	Model fit

Simulation with the true $k_L a$, assuming well mixed gas and solubilities of fresh air

Time (h)

(b)

Figure 5.12 (a) For given dissolved concentrations pO_2 and pCO_2 in the liquid bulk, the volume fractions in an air bubble is shown as a function of time. While the CO_2 volume fraction rises rapidly, the volume fraction of O_2 does not drop comparably rapidly (note, the y_0-zero point is suppressed). (b) Measurement data of pO_2 and pCO_2 from a 330 l reactor (symbols) and solution of Eq. (5.10) full line. The dashed line depicts the simulation under the assumption that no saturation appears, then the time constants for O_2 and CO_2 transfer should roughly be the same.

When we assume that the mean residence time of the bubble is 18 seconds, and the bubble size is given, we can compute the amount of CO_2 the bubble can withdraw. Then, the volumetric mass transfer rate (carbon dioxide transfer rate, CTR) is the amount n_{ct} each bubble carries times the number of bubbles N_t generated per time unit divided by the total volume V_d of the culture.

$$CTR = N_t\, n_c / V_d \qquad (5.9)$$

This current CTR can then be used in the solution of the conventional mass balance equation for CO_2 around the entire culture.

$$\frac{d(CO_2)}{dt} = CPR - CTR \qquad (5.10)$$

For each time step in the solver for Eq. (5.10), we take the solution of Eq.(5.8) to compute the current CTR value with Eq. (5.9). For the case of a dynamic $k_L a$ experiment starting with a high pCO_2 in the liquid (obtained by gassing with CO_2), the solution is shown on the right side of Figure 5.12 together with experimental data.

The result depicted in Figure 5.12 shows that the volumetric CO_2 mass transfer is much smaller, although the $k_L a$ values are practically equal. If we would assume the CO_2-saturation effect negligible, the time constant of the CO_2 decay would roughly be the same as that of the pO_2 increase within the culture. The corresponding solution is plotted as the dashed curves in Figure 5.12.

Consequently, the CO_2 mass transfer in a larger cell culture does no longer depend on the $k_L a$ value. The model can be validated with experiments in which the $k_L a$ value was changed by means of the mean power input into the reactors by the impeller system. Dynamic mass transfer experiments on the 12 000 l-scale stainless steel reactor from a production system resulted in the experimental results depicted in Figure 5.13 that shows the independency of the *apparent* $k_L a$ value from the stirrer speed that definitely changes the *true* $k_L a$.

This leads to the following optimization strategy for the CO_2 stripping from and O_2 absorption into larger cell culture reactors (Sieblist et al. 2011):

The gas throughput must be increased up to the value where no CO_2 accumulation appears, i.e. pCO_2 is kept within acceptable limits. Then the impeller speed must be increased up to the value where no significant oxygen limitation appears. The latter is guaranteed by a well-performing sequential pO_2 controller that first adjusts the aeration rate and then the stirrer speed. As pCO_2 cannot be measured reliably in a production tank, an estimate can be made from the fact that, roughly, the CTR required is equal to the OUR of the cells that can easily be measured by offgas analysis. Possibilities to control animal cell cultures along predefined OUR profiles have already been published (Aehle et al. 2011b, 2012).

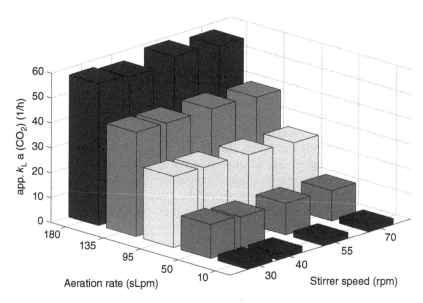

Figure 5.13 Apparent $k_L a$ for the CO_2 stripping from a two-phase dispersion as measured in a 12 000 production tank as a function of the aeration rate and the stirrer speed. The data show that the $k_L a$ does not significantly depend on the stirrer speed or the power dissipation density. It only depends on the aeration rate.

5.4 Conclusion

Practical process optimization means improving equipment and control procedures to enhance productivity and product quality. Here we focused on model-supported improvements of process operation and particularly on batch-to-batch reproducibility of microbial cultivations and cell cultures that can only be enhanced by sophisticated control techniques.

At least since FDA/EMA's process analytical technology (PAT) initiative, it became clear that there was a deficit in the development of knowledge-based design technologies as well as in techniques that can make sure that the processes will follow optimal paths. Analytical technologies are indispensable for model development as it is the only way to discriminate between model variants or ideas about the mechanism behind the production processes under consideration. And they are also prerequisite to keeping the processes tightly on track in order to guarantee high batch-to-batch reproducibility in all product characteristics.

Bioprocess optimization based on appropriate process models is not simply an exercise of numerical mathematics nor a playground for statisticians.

It is engineering disciplined that primarily needs knowledge and experience in cultivation processes on a technical scale. As an example, we presented an efficient strategy of starting the cultivations in the fed-batch mode from the early beginning, where the measurement information from the process is generally rather poor. The latter improves with time or cell proliferation. When the signals get more information content, i.e. higher signal-to-noise ratios, they can be used for feed-back control. This allows for getting closer to the true optima without running the process at too high risks for leaving the specification limits.

As there is no possibility to judge whether or not a real process is optimal, optimization is a dynamic and iterative process by itself. Hence, evolutionary process development/optimization strategies are straightforward. But we strongly emphasize to first of all make use of the a priori knowledge to formulate hypotheses and then to make experiments that can support or reject the assumptions, not the other way around.

References

Aehle, M., Simutis, R., and Lübbert, A. (2010). Comparison of viable cell concentration estimation methods for mammalian cell cultivation processes. *Cytotechnology* 62 (5): 413–422.

Aehle, M., Kuprijanov, A., Schaepe, S. et al. (2011a). Increasing batch-to-batch reproducibility of CHO cultures by robust open-loop control. *Cytotechnology* 63: 41–47.

Aehle, M., Schaepe, S., Kuprijanov, A. et al. (2011b). Simple and efficient control of CHO cell cultures. *J. Biotechnol.* 153: 56–61.

Aehle, M., Bork, K., Schaepe, S. et al. (2012). Increasing batch-to-batch reproducibility of CHO-cell cultures using a model predictive control approach. *Cytotechnology* 64 (6): 623–634.

Åström, K.J. and Hägglund, T. (2005). *Advanced PID Control*. Research Triangle Park, NC: ISA: The Instrumentation, Systems and Automation Society.

Dochain, D. (2008). *Bioprocess Control*. London: Wiley.

Galvanauskas, V., Volk, N., Simutis, R., and Lübbert, A. (2004). Design of recombinant protein production processes. *Chem. Eng. Commun.* 191: 732–748.

Glassey, J., Ignova, M., Ward, A.C. et al. (1997). Bioprocess supervision: neural networks and knowledge based systems. *J. Biotechnol.* 52: 201–205.

Gnoth, S., Kuprijanov, A., Simutis, R., and Lübbert, A. (2009). Simple adaptive pH-control in bioreactors using gain-scheduling methods. *Appl. Microbiol. Biotechnol.* 85: 955–964.

Gnoth, S., Simutis, R., and Lübbert, A. (2010). Selective expression of the soluble product fraction in *Escherichia coli* cultures employed in recombinant protein production processes. *Appl. Microbiol. Biotechnol.* 87 (6): 2047–2058.

Jenzsch, M., Gnoth, S., Beck, M. et al. (2006a). Open loop control of the biomass concentration within the growth phase of recombinant protein production processes. *J. Biotechnol.* 127: 84–94.

Jenzsch, M., Simutis, R., Eisbrenner, G. et al. (2006b). Estimation of biomass concentrations in fermentation processes for recombinant protein production. *Bioprocess. Biosyst. Eng.* 29 (1): 19–27.

Jenzsch, M., Gnoth, S., Kleinschmidt, M. et al. (2007). Improving the batch-to-batch reproducibility of microbial cultures during recombinant protein production by regulation of the total carbon dioxide production. *J. Biotechnol.* 128: 858–867.

Julier, S. and Uhlmann, J.K. (2004). Unscented filtering and nonlinear estimation. *Proc. IEEE* 92: 401–422.

Kadlec, P., Gabrys, B., and Strandt, S. (2009). Data-driven soft sensors in the process industry. *Comput. Chem. Eng.* 33: 795–814.

Kim, J.I., Song, H.S., Sunkara, S.R. et al. (2012). Exacting predictions by cybernetic model confirmed experimentally: steady state multiplicity in the chemostat. *Biotechnol. Progr.* 28: 1160–1166.

Kuprijanov, A., Gnoth, S., Simutis, R., and Lübbert, A. (2009). Advanced control of dissolved oxygen concentration in fed batch cultures during recombinant protein production. *Appl. Microbiol. Biotechnol.* 82: 221–229.

Luttmann, R., Bracewell, D.G., Cornelissen, G. et al. (2012). Soft sensors in bioprocessing: a status report and recommendations. *Biotechnol. J.* 7 (8): 1040–1048.

Mandenius, C.F. (2004). Recent developments in the monitoring, modeling and control of biological production systems. *Bioprocess. Biosyst. Eng.* 26: 347–351.

McMillan, G.K., Benton, T., Zhang, Y., and Boudreau, M.A. (2008). PAT tools for accelerated process development and improvement. *Bioprocess Int.* 6: 34–42.

Noorman, H. (2011). An industrial perspective on bioreactor scale-down: what we can learn from combined large-scale bioprocess and model fluid studies. *Biotechnol. J.* 6: 934–943.

Norgaard, M., Ravn, O., Poulsen, N.K., and Hansen, L.K. (2003). *Neural Networks for Modelling and Control of Dynamic Systems: A Practitioner's Handbook*, Advanced Textbooks in Control and Signal Processing. Berlin: Springer.

Psichogios, D.C. and Ungar, L.H. (1992). A hybrid neural network – first principles approach to process modeling. *AIChE J.* 38: 1499–1511.

Ramkrishna, D. (2003). On modeling of bioreactors for control. *J. Process Control* 13: 581–589.

Schölkopf, B. and Smola, A.J. (2002). *Learning with Kernels*. Cambridge, MA: MIT Press.

Schubert, J., Simutis, R., Dors, M. et al. (1994). Bioprocess optimization and control: application of hybrid modelling. *J. Biotechnol.* 35: 51–68.

Shuler, M.L. and Kargi, F. (2002). *Bioprocess Engineering*, 2e. Upper Saddle River, NJ: Prentice Hall.

Sieblist, C., Hägeholz, O., Aehle, M. et al. (2011). Insights into large scale cell culture reactors: II. Gas-phase mixing and CO_2-stripping. *Biotechnol. J.* 6 (12): 1547–1556.

Simutis, R. and Lübbert, A. (1997a). A comparative study on random search algorithms for biotechnical process optimization. *J. Biotechnol.* 52: 245–256.

Simutis, R. and Lübbert, A. (1997b). Exploratory analysis of bioprocesses using artificial neural network-based methods. *Biotechnol. Progr.* 13: 479–487.

Simutis, R., Havlik, I., and Lübbert, A. (1993). Fuzzy aided neural network for real time state estimation and process prediction in the alcohol formation step of production-scale beer brewing. *J. Biotechnol.* 27: 203–215.

Sonnleitner, B. and Fiechter, A. (1992). Impacts of automated bioprocess systems on modern biological research. *Adv. Biochem. Eng./Biotechnol.* 46: 143–159.

von Stosch, M., Oliveira, R., Peres, J., and Feyo de Azevedo, S. (2012). A general hybrid semi-parametric process control framework. *J. Process Control* 22: 1171–1181.

von Stosch, M., Oliveira, R., Peres, J., and Feyo de Azevedo, S. (2014). Hybrid modelling in process systems engineering: past, present and future. *Comput. Chem. Eng.* 60: 86–101.

Thompson, M.L. and Kramer, M.A. (1994). Modeling chemical processes using prior knowledge and neural networks. *AIChE J.* 40: 1328–1340.

Tipping, M.E. (2000). The relevance vector machine. *Adv. Neural. Inf. Process. Syst.* 12: 652–658.

Vapnik, V.N. (1998). *Statistical Learning Theory*. New York: Wiley.

Wiechert, W. and Noack, S. (2011). Mechanistic pathway modeling for industrial biotechnology: challenging but worthwhile. *Curr. Opin. Biotechnol.* 22: 604–610.

Zhu, M.M., Goyal, A., Rank, D.L. et al. (2005). Effects of elevated pCO_2 and osmolality on growth of CHO cells and production of antibody-fusion protein B1: a case study. *Biotechnol. Progr.* 2005 (21): 70–77.

6

Monitoring and Control of Bioreactor: Basic Concepts and Recent Advances

James Gomes[1], Viki Chopda[2], and Anurag S. Rathore[1]

[1] *Kusuma School of Biological Sciences, Indian Institute of Technology Delhi, New Delhi, India*
[2] *Department of Chemical Engineering, Indian Institute of Technology Delhi, New Delhi, India*

6.1 Introduction

Industrial biotechnology continues to be one of the dominant segments of business and economy worldwide. The variety of products manufactured by biotechnological processes (bioprocesses) ranges from antibiotics, food additives, amino acids, and vitamins to monoclonal antibodies and therapeutic proteins (Mednis et al. 2015). The quality requirements imposed on the biomanufactured products, combined with performance and productivity pressures in the context of the ever-increasing industrial competition, has led to an evolution in the way processes are controlled on an industrial scale (Rathore 2009).

The cell is the ultimate bioreactor. It is the essential, and in some cases, the only possible route to the production of complex compounds. Biotechnological manufacturing processes rely on these microorganisms for the synthesis of their products (Boudreau and McMillan 2007). Bioreactors (fermenters) are the key unit operation in biopharmaceutical, brewing, biochemical, biofuel, and waste treatment processes. Each bioreactor relies on the performance of billions of these individual "cell" reactors (Figure 6.1). It is by controlling bioreactors that the overall profitability of the process can be achieved. The goal of process control is the consistent production of the desired product at the specified quality. Sound process design with dependable process automation and control makes processes reliable (Gomes et al. 2015). This is achieved by exploiting advances in electronics and sensor development for biochemical parameters in combination with intelligent analytical systems. The analytical data should be acquired in situ, or at least at-line for continuously analyzing the desired parameter in real time (Vojinović et al. 2006).

Bioprocessing Technology for Production of Biopharmaceuticals and Bioproducts, First Edition.
Edited by Claire Komives and Weichang Zhou.
© 2019 John Wiley & Sons, Inc. Published 2019 by John Wiley & Sons, Inc.

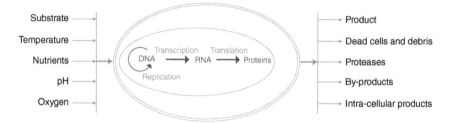

Figure 6.1 Illustration of the level of complexity involved in biomanufacturing.

Additional analysis is then carried out to cover a wide dynamic range (of activities and concentrations) for process control.

Process dynamics influences the selection of the control strategy. The volumes of bioreactors vary from few liters to several cubic meters, and they are characterized by inherent process lag. The growth of microorganisms results in the production of metabolites with changes in biomass, pH, and temperature. The strong nonlinear characteristics observed in the time profiles of measured variables have fueled design of advanced controllers based on mathematical modeling to meet process targets (Mandenius 2004). Often, supervisory systems are deployed wherein controller controlled variables together with the available information are used to dynamically change the tuning parameters of control loops. In summary, the design and implementation of control strategy require a multi-disciplinary approach to address a range of requirements and challenges and deliver solutions.

6.2 Challenges in Bioprocess Control

6.2.1 Process Dynamics and Modeling

The need for monitoring, control, and supervision systems to optimize operation and detect malfunctions in bioprocesses have become more pressing due to the changes occurring in the industry. However, in reality, very few installations are provided with such systems (Glassey et al. 2011). There are several reasons for this. First of all, biological processes are complex and involve living organisms. The complexity at the cellular level shows up due to the strong nonlinear interaction between state variables that is typically observed in bioprocesses. Thus, for most bioprocesses, using "off-the-shelf" controllers does not give the best performance because nonlinear interactions of states cannot be removed. Since microorganisms used in bioprocesses tend to switch over to a more favorable metabolic pathway depending on operating conditions in a bioreactor, variations of the state variables and their interactions should be accounted for during the design of controllers. For example, ethanol fermentation processes, the yeast metabolism depends on the relative concentrations of glucose and dissolved oxygen. Similarly in *Escherichia coli* cultivation, overflow metabolism leads to undesired acetate accumulation, which results in reduction of the recombinant protein synthesized. In such cases, a suitable feeding strategy can constrain the microorganism in a metabolic domain desirable for higher protein production. In essence, although bioprocesses are known to possess variability and a certain degree of unpredictable behavior, it is possible to guide the process along the desired trajectory and maintain it within the design space using precise controllers.

Modeling of these nonlinear dynamical systems is challenging. Further, the dearth of sensors and the wide differences in the characteristic time of measurements present hurdles not only in selection of good model structure but also in parameter estimations (Van der Schaft 1992; Vojinovic et al. 2006). Difficulties also occur during validation of these models because parametric drift caused by metabolic changes or mutation result in model prediction errors. Figure 6.2 illustrates the major hurdles impeding successful implementation of bioprocess control.

6.2.2 Limits of Hardware and Software and Their Integration

For online estimation of key variables associated with these systems (concentration of biomass, substrates, and dissolved oxygen), expensive instruments are used that require repeated calibration and regular maintenance. However, many important variables, such as the end-product recombinant protein, antibiotic, amino acid, or other intermediate metabolite concentrations are

Figure 6.2 Illustration showing major hurdles impeding in successful implementation of bioprocess control.

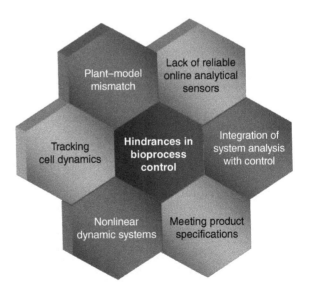

often difficult to measure online. Hence, most bioprocesses, whether in the industry or in an academic setting, make do with available measurements and complement these with advanced estimation or prediction techniques to determine the other unmeasured but important process variables. Since the bioreactor manufacturers do not provide these programs as standard options, there is limited flexibility for integration of any new third-party device (Chopda et al. 2016).

6.2.3 Regulatory Aspects

The regulatory authorities of Europe and North America laid down the guidelines of Process Analytical Technology (PAT) for the biopharmaceutical industry to bring about process consistency, ensure quality, and make biological products safer to use by people worldwide (FDA 2004). The PAT guideline takes into account the quality of the product through critical quality attributes (CQAs) and delineates how at-, in-, and online measurements may be acquired and used for quality assurance (Rathore and Winkle 2009; Rathore et al. 2010; Read et al. 2010a,b). Biopharmaceutical manufacturers must meet the stringent regulations and quality control norms that have been laid down to ensure patient safety and therefore, more of them are seeking recourse to adopting the advanced process-monitoring tools and control in manufacturing platforms for continuous improvement in quality. Recent advances include use of chemometric methods (Challa and Potumarthi 2013), building multi-PAT tool facility (Känsäkoski et al. 2006), and integrating system analysis with process control (Hinz 2006; Munson et al. 2006). The guidelines include

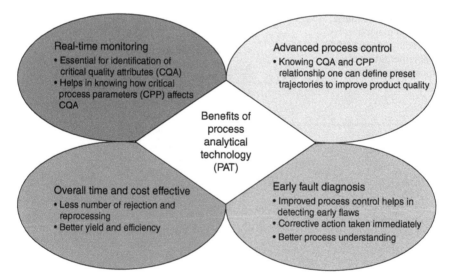

Figure 6.3 Figure depicting benefits of PAT implementation in bioprocesses.

continuous improvement, risk assessment, knowledge management, and implementation of at-line and online sensors wherever there is an opportunity to do so (FDA 2004). As a consequence, biotechnological manufacturing has resulted in improvement in productivity, consistency, and quality. Thus, both from the standpoint of process economics and quality control (Figure 6.3), it is reasonable to adopt advanced control strategies for controlling these processes.

6.3 Basic Elements of Bioprocess Control

6.3.1 Bioprocess Monitoring

Real-time process monitoring is essential for process control. Although recent development in sensor technology for bioprocesses has been significant, offline monitoring is still carried out routinely; samples are taken regularly from the bulk bioreactor medium and analyzed later. Though offline measurements are in general reliable, the obvious drawback is that it takes a long time to complete an analysis and hence not optimal for control applications and process automation. In contrast, real-time bioprocess monitoring techniques have the advantage of acquiring and providing process information as it happens and enables early warning to intervene if unprecedented process deviations occur (Lindemann et al. 1998; Clementschitsch and Bayer 2006; Mattes et al. 2007; Teixeira et al. 2009, 2011; Li et al. 2010; Chopda et al. 2013; Tiwari et al. 2013). A slew of advanced monitoring sensors for bioprocess applications have been launched in the last decade (Kornmann et al. 2003; Lee et al. 2004; Roychoudhury et al. 2006, 2007; Kandelbauer et al. 2008; Kiviharju et al. 2008; Opel et al. 2010; Petersen et al. 2010; Foley et al. 2012). There are a number of publications that describe the principles, potential advantages, limitations, and applications of these monitoring tools in depth (Crowley et al. 2000; Pollard et al. 2001; Arnold et al. 2002; Ducommun et al. 2002; Noui et al. 2002; Tamburini et al. 2003; Pons et al. 2004; Card et al. 2008; Abu-Absi et al. 2011; Justice et al. 2011; Whelan et al. 2012). Table 6.1 summarizes the literature on this topic.

6.3.2 Parameter Estimators

Despite the advances in sensor technology for biological processes in recent years, very few have been successfully implemented at manufacturing scale. Often, the challenge is to estimate the internal state of a bioreactor using a few available measurements; the unmeasured states of the process are estimated with observers designed for this purpose. These observers known as software sensors or soft-sensors are designed to be at least asymptotically convergent (Luttmann et al. 2012). The method employed for designing these soft-sensors depends on the information that may be readily acquired from the process. If the biological kinetics is not precisely known, mass balance principles are

Table 6.1 Modern PAT tools utilized widely for real-time monitoring of bio-processes.

Process analyzer	Advantages	Limitations	Monitored components	Statistical tools needed
UV spectroscopy	Noninvasive, high speed, and sensitivity	Chemo-metrics needed for interpretation	RNA, protein, cell debris, product, and nitrate	Multivariate curve resolution (MCR), PCR, PLS
Near infrared spectroscopy (NIRS)	Noninvasive, high speed, and sensitivity. No sample preparation is required. Spectral fingerprint of principle cellular constituents	Broad peaks, chemo-metrics needed for interpretation	Sugars, glycerol, lactic acid, glutamine, ammonium, acetate, viable cells, clavulanic acid, gellan gum, humidity	PCA, PLS, MSPC, MPCA, MPLS
Mid infrared spectroscopy (MIRS)	Noninvasive, high speed, and sensitivity. Higher resolution, spectral fingerprint of principle cellular constituents	Costly hardware than NIR, chemo-metrics needed for interpretation	Sugars, clavulanic acid, alcohols, antibody, acetic acid, gluconacetan, pH, phosphate	PCR, PLS, Moving Window PCA
Raman spectroscopy	Noninvasive, and no sample preparation is required	Overlaying by fluorescence activity	Pharmaceutical applications, bioprocess area yet to explore	PCR and PLS
Fluorescence spectroscopy	Noninvasive	Broad peaks, chemo-metrics needed for interpretation	Fluorophore compounds, tagged fluorescent compounds	Multiway robust PCA, NPLS-DA, and NPLS, PLS
Dielectric spectroscopy	In-line measurement of viable biomass	Signal detection limit varies with cell characteristics	Viable cells	Signal processing

used for designing parameter estimators (Paulsson et al. 2014). When accurate models can be constructed and validated, soft-sensor can have a high gain and give better performance. A soft-sensor may be designed for just monitoring a bioreactor to estimate an unmeasured variable or it may be integrated with the design of a control scheme (Luttmann et al. 2012; Krause et al. 2015).

6.3.3 Bioprocess Modeling

Historically, in mathematical modeling of bioprocesses, the cell was simply considered as a catalyst for the conversion of a biological compound to a product. The models were phenomenological and largely ignored intracellular metabolism. Mechanistic models were constructed with the expectation that the optical density of the culture and key measurements such as substrate, dissolved oxygen, and product concentration could be made (Junker and Wang 2006). These models are still useful today to obtain a coarse-grained picture of the process dynamics. In general, factors not accounted for include interactions between cells and bioreactor conditions, synthesis of metabolites not directly associated with the product of interest, and associated cellular energetics and metabolism (Chopda et al. 2016). However, if this additional information could be obtained from models, it would be useful in achieving preset process targets. Various computational software are currently being used to complement the coarse-grained models to quantify and obtain a better understanding how cellular metabolism influences productivity. From a practical perspective, the incorporation of more cellular information will increase the complexity of the model and computationally restrictive.

Unstructured models are most commonly used. These models do not account for the variation of biomass composition in response to environmental changes. However, these models are easy to write and simulate using software, such as MATLAB. Model-based controllers can be readily developed using unstructured modes and implemented online using software such as LabVIEW, which provides an interfacing platform for communicating with the bioreactor. It is well known that unstructured models give valid representation only under the conditions of balanced growth. Hence, controllers designed using these models may yield poor performance under certain metabolic conditions not adequately described by them. Another disadvantage of unstructured models is their inability to offer deeper metabolic information. In contrast, structured models are better able to provide information about the changing metabolic conditions and changing composition of the microbial biomass.

Further categorization of models is based on the degree of details in which they describe different phenotypes and the definition of internal components or "structure" of cells. For example, unsegregated models consider the microbial culture as homogeneous, whereas segregated models consider cells as individual entities and permit their classification into groups of interest. The structured and segregated models have better extrapolation capabilities than unstructured models. Hence, controls developed using these models can describe expected response of the system to perturbations. The disadvantage in incorporating more cellular information is that parameter estimation and its incorporation in controller synthesis becomes difficult and at times unpractical. Therefore, one must decide between the ease of its application and the complexity required to address the problem under consideration.

With newer methods in sequencing and improvements in mass spectrometry, it is now possible to probe deeper to understand how the cell functions (Yang et al. 2009; Zhang et al. 2014). Tools from system biology and bioinformatics can be used to determine correlations between the phenotype and its productivity under different conditions of bioreactor operation.

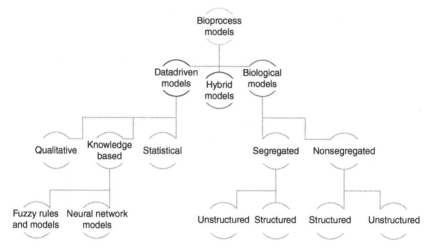

Figure 6.4 Illustration of the different bioprocess models commonly employed in monitoring and control.

The scale and scope of the model can be tuned to the required degree of graininess for elucidating a characteristic of the phenotype. The models in this category may be hybrid and employ several techniques such as combining of ordinary differential equations, network properties, and Boolean networks, tied together with heuristics. Often a bioinformatics backend is used to determine kinetics parameters of homologous proteins or to establish if new intermediates appearing in a metabolic pathway are acceptable. Further, in combination with statistical tools, these models can also complement empirical procedures, often used in process development and control or neural network based systems. Figure 6.4 illustrates the different bioprocess models currently employed in monitoring and control.

6.4 Current Practices in Bioprocess Control

6.4.1 PID Control

In industrial plants, proportional-integral-derivative (PID) controller is the most commonly used control architecture for regular process control experiments. The response generated by PID controller is based on measurement of the error value (e) which is the difference between the controlled variable and the set point at time instance t.

$$u(t) = K_C \left\{ e(t) + \frac{1}{\tau_I} \int_0^t e(t)\mathrm{d}t + \tau_D \frac{\mathrm{d}e(t)}{\mathrm{d}t} \right\} \tag{6.1}$$

Equation (6.1) is the general form of a PID controller, where K_C is the proportional gain, τ_I and τ_D are the integral and derivative time constants, and $u(t)$ is the control action. The acceptable performance of PID controller depends on the accuracy with which the tuning parameters K_C, τ_I, and τ_D is usually determined by using one of the different tuning rules. Cohen–Coon and Ziegler–Nichols tuning rules are commonly followed to make initial guesses of these values. In the cases where measurements show noisy readings, the derivative control mode can make the controller more sensitive and may result in erratic control action. This can be minimized by reducing the derivative time constant τ_d to zero so that the controller works in the proportional-integral (PI) mode.

Ideally, a tuned controller should exhibit minimum overshoot and settle quickly to the desired set point with minimum oscillation. Since the control parameters are held constant after tuning, the controller performance may deteriorate if the process kinetics changes or if there is a drift in the measurements especially in the case of bioprocesses. Also, the PID controller acts after the error has occurred and hence this "after the effect" mode can never result in perfect control.

6.4.2 Model-Based Control

In the operation of bioprocesses, laboratory or manufacturing scale, many of the delineating features of the process cannot be described satisfactorily with models. In such situations, the construction of controllers is inadequate, and this results in plant-model mismatch. It has also been widely recorded that local hydrodynamics can cause spatial variation in the composition of medium constituents within a reactor. Similarly, variation in cell population distribution can affect observed kinetics leading to parametric uncertainty in the model. Model-based controllers are more suited to address cases where the process exhibits a higher degree of complexity in terms of the state-time profiles.

Model-based control does not refer to a specific one type of control strategy. Instead, it refers to a very broad range of control methods that make explicit use of a process model. The basic idea is to drive the process to mimic the ideal response of a suitable model as illustrated in Figure 6.5. Under this broad description, innumerable variations of control structures have been developed and reported in the control literature, a complete discussion of which is beyond the scope of this chapter. We present a brief overview of model predictive control (MPC) to elucidate the principle of model-based control.

MPC addresses some of the deficiencies of the PID controller. The PID controller is derived using a linear approximation of the process. It uses the error between the measured value and the set point of the controlled variable to compute the control action. Hence, for nonlinear dynamical bioprocesses, its performance is often compromised. To address some of these issues,

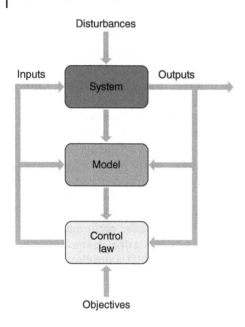

Figure 6.5 Illustrating the concept of model-based control schemes.

the MPC scheme had been developed, which uses a model response behavior as a reference to drive the control action along the desired trajectory. The inherent architecture of the MPC results is a faster and smoother control action compared to a PID controller. There are many examples in the literature claiming improved process performance using MPC. For example, Ashoori et al. (2009) presented a MPC based on a detailed unstructured kinetic model for penicillin production in fed-batch fermentation. This approach used the inverse of penicillin concentration as a cost function deployed instead of a common quadratic regulating one in the optimization block. They further benchmarked their outcomes with an auto-tuned PID controller. Using MPC, they were able to enhance penicillin concentration by 25% (Ashoori et al. 2009). Model-based controllers also found to control DO precisely (Gomes and Menawat 2000; Nayak and Gomes 2009).

The second type of control uses time series models, which represent the open-loop response of the process with a vector of impulses that are empirically determined. As these types of controller use linear models, they are refered to as linear model predictive control (LMPC). The impulse response models are formulated based on the assumption of linearity of the process. The input function is considered as an aggregate of step functions, at the respective sampling time moment. The response of the process to these aggregated inputs is the sum of the individual step responses. The process outputs are predicted over the output horizon and are a function of not only of the known past but also of not-yet-known future manipulated variables.

The future-manipulated variables are computed such as the projected outputs adapt to a desired (set point) output trajectory. As the linear model is used in the structural formulation of LMPC, it is crucial to focus on feedback stability problems.

The third type of model-based controller uses nonlinear models that offer wide range of predictability over linear models known as a nonlinear model predictive controller (NMPC). In this case, good predictability is achieved by using either mechanistic models or data-driven models such as neural network models (Kovárová-Kovar et al. 2000), and fuzzy models, etc. The nonlinear dynamic model can be used in different ways and in different phases of the control algorithm, depending on the particular NMPC approach. NMPC has two broad categories of applications (i) problem of disturbance rejection in highly complex and nonlinear processes such as bioprocesses and (ii) set point tracking problems in the cases where the metabolic/process dynamics shifting occurs frequently. However, model identification is the major challenge for successful execution of this control.

6.4.3 Adaptive Control

Besides the problems that arise because microorganisms shift metabolic regimes during the course of a process, there are other sources of uncertainties. Measurement noise electromagnetic interference is the most common. Other sources can be attributed to local hydrodynamic, temperature, and concentration gradients within the reactor at the point of measurement by the electrodes or biosensors. Calibration drift in these devices can also add to the difficulties. Further, there is a difference in the response times of these electrodes and sensors, consequently resulting in different sampling times for different measurements. It is not possible to model a process accounting for all these process uncertainties. The typical solution is to design a controller whose parameters adapt to changes in process characteristics and uncertainties. Usually, simple linear models are used to represent the process, and the parameters of the controller are adapted dynamically based on the error between the predicted and online measurement of a state variable. The parameter adaptation algorithm can be designed to be at least asymptotically (if not exponentially) convergent. For example, Ranjan and Gomes (2010) designed a decoupled adaptive controller for controlling the dissolved oxygen and glucose feeding simultaneously in the fed-batch fermentation of methionine production. They have used unstructured model for methionine production and controller derivation developed by them is as given below:

The adaptive law used such that the nonlinearities of the process were canceled and given by

$$u = (g(y))^{-1}v \qquad (6.2)$$

This adaptive law resulted in decoupling of the control system and v were defined as below:

$$v = [K_0 + K_1 y + K_2 f(y) + L y_d] \tag{6.3}$$

From this, a simplified equation between the measured inputs and desired outputs was developed

$$y = A f(x) + B(K_0 + \dot{K}_1 y + K_2 f(y) + L y_d) \tag{6.4}$$

where, $K_0, K_1, K_2,$ and L are diagonal matrices that represent unknown variable controller gains of the control law, which are determined dynamically using an adaptation law; the desired output is y_d. Error dynamics was given as

$$\dot{e} = \dot{y} - \dot{y}_m = A f(y) + B(K_0 + K_1 y + K_2 f(y) + L y_d) + A_m y_m - B_m y_d \tag{6.5}$$

After simplification, Eq. (6.5) can be reduced to

$$\dot{e} = (qI + A_m)^{-1} B(\widetilde{K}_0 + \widetilde{K}_1 y + \widetilde{K}_2 f(y) + \widetilde{L} y_d) \tag{6.6}$$

where, q denotes the Laplace operator and the elements of $(qI + A_m)^{-1} B$ are strictly positive real functions. Using the Lyapunov convergence, the following adaptation laws were chosen

$$\dot{K}_0 = -\text{sgn}(B) * \gamma_i * e \tag{6.7}$$

$$\dot{K}_1 = -\text{sgn}(B) * \gamma_i * e * y \tag{6.8}$$

$$\dot{K}_2 = -\text{sgn}(B) * \gamma_i * e * f(y) \tag{6.9}$$

$$\dot{L} = -\text{sgn}(B) * \gamma_i * e * y_d \tag{6.10}$$

where $\gamma_i > 0$, $i = 1, 2, 3,$ and 4 are the adaptation gains and sgn (B) determines the direction of search. Finally, the implementation form of the adaptive control of glucose feed rate and airflow rate are given by

$$D = \frac{1}{s_F - s}\left(K_{11} s + K_{12}\frac{\mu_1}{Y_{x/s}} x + L_1 s_d\right) \tag{6.11}$$

$$k_L a = \frac{1}{c_L^* - c_L}\left[K_{20}\frac{c_L}{s_F - s}\left(K_{11} s + K_{12}\frac{\mu_1}{Y_{x/s}} x + L_1 s_d\right)\right.$$
$$\left. + K_{21} c_L + K_{22} Y_{o/x} \mu_1 + L_2 c_{L,d}\right] \tag{6.12}$$

$$F_S = u_1 V, \quad u_1 = D \tag{6.13}$$

$$F_A = 1553.64\left(\frac{u_2 V^{0.2467}}{N^{1.14}}\right)^{2.5}, \quad u_2 = k_L a \tag{6.14}$$

In actual online control implementation, the glucose feed rate, F_S, and the airflow rate, F_A, are the true process inputs. F_S and F_A are obtained from the

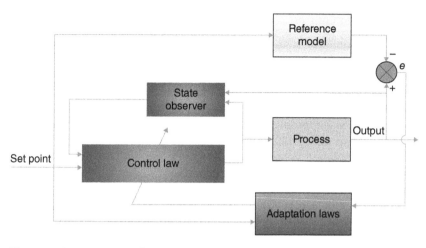

Figure 6.6 Representation of a typical adaptive control system.

relations given by Eqs. (6.13) and (6.14). Further, D is dilution rate (h^{-1}), $k_L a$ is mass transfer coefficient (h^{-1}), N is the agitation speed (rpm), and V is the bioreactor volume (l). The rate of change of substrate concentration and dissolved oxygen are denoted by $\dot{s}$ and $\dot{c}_L$, respectively. These rates were represented as v_1 and v_2 in the transformed space.

A typical adaptive control system is presented in Figure 6.6. Wu et al. (2013) proposed optimal adaptive control strategy by solving the constrained discrete-time optimization algorithm. They successfully demonstrated satisfactory output tracking performance of adaptive controller for fermentative production of PHB using *Ralstonia eutropha*. Adaptive controllers have been used in many applications such as keeping specific growth rate at desired point in fed-batch cultivation of *Bordetella pertussis* (Soons et al. 2006), for biomass maximization in yeast fermentation (Renard and Wouwer 2008), for optimal production of lactic acid (Youssef et al. 2000), and for controlling dissolved oxygen (Oliveira et al. 2005). Smets et al. (2004) described the evolution of adaptive control design and explored future opportunities for bioprocess control. However, the success of these controllers depends on the robustness of model and the selected adaptation laws.

Despite many advantages, an adaptive controller, being inherently nonlinear, is more complicated than a fixed gain controller. Before attempting to use adaptive control, it is, therefore, important to investigate whether the control problem might be solved by constant gain feedback. In the literature on adaptive control, there are many cases in which constant-gain feedback can do as well as an adaptive controller (Praly and Jiang 2004; Khalil 2008). This is one reason why there is need to search for alternatives to adaptive control.

6.4.4 Nonlinear Control

The observed nonlinear state-time profiles of biological processes exhibit the underlying interactions between different states and the inherent complexity of cellular processes. Models that best describe such behavior possess nonlinear structure. One option for controller design is to use linear approximations of the nonlinear models. This enables as simpler controller construction, but its performance is compromised since it contains truncation error. A possible remedy is to use global linearization techniques that fall in the realm of nonlinear systems theory based on differential geometry (Kanter et al. 2002; Panjapornpon and Soroush 2006). The methods of differential geometry enable the transformation of a nonlinear system into a globally linear system without incurring truncation errors. Global linearization of a nonlinear system in an input–output sense has also been suggested (Lee et al. 2008). For example, a decoupled input–output linearizing controller (DIOLC) was developed using the principles of nonlinear theory to regulate substrate and dissolved oxygen concentration in fed-batch fermentation for methionine production (Ranjan and Gomes 2009). Further, in a previously published study, we have demonstrated the implementation of the DIOLC to control Baker's yeast fermentation in the desired trajectory (Persad et al. 2013; Chopda et al. 2015). The detailed description DIOLC derivation was given in (Ranjan and Gomes 2009) and described here briefly as below:

The state-space model of a multi-input-multi-output (MIMO) nonlinear system was chosen as follows:

$$\dot{X} = f(X) + g(X)u$$
$$y = h(X) \tag{6.15}$$

where $X \in \mathfrak{R}^n$ is the vector of state variables, $u \in \mathfrak{R}^m$ is the vector of manipulated inputs, and $y \in \mathfrak{R}^p$ is the vector of measured outputs. The functions f and g are smooth vector fields and represent the kinetics and transport terms. In their case of yeast fermentation, the system was a 2×2 system. Measurements of s and c_L were performed online. The decoupling matrix was as follows:

$$E(X) = \begin{bmatrix} s_F - s & 0 \\ -c_L & c_L^* - c_L \end{bmatrix} \tag{6.16}$$

The relation between the output and input was found to be

$$\begin{bmatrix} v_1 \\ v_2 \end{bmatrix} = \begin{bmatrix} -\alpha_4 \mu_2(s)x \\ -\mu_1(s, c_L)x - \mu_3(s, e, c_L) \end{bmatrix} + E(X) \begin{bmatrix} u_1 \\ u_2 \end{bmatrix} \tag{6.17}$$

It easily follows that the decoupled input–output linearizing control laws for glucose feed and airflow rate are given by

$$u_1 = D = \frac{1}{s_F - s}(\alpha_4 \mu_2(s)x + v_1) \tag{6.18}$$

$$u_2 = k_L a = \frac{c_L}{(s_F - s)(c_L^* - c_L)}(\alpha_4 \mu_2(s)x + v_1)$$
$$+ \frac{1}{c_L^* - c_L}(\mu_1(s, c_L)x + \mu_3(s, e, c_L)x + v_2) \tag{6.19}$$

Further, v_1 and v_2 as given by the following equation

$$\begin{bmatrix} v_1 \\ v_2 \end{bmatrix} = \begin{bmatrix} \dot{s} \\ \dot{c}_L \end{bmatrix} = \begin{bmatrix} \dot{s}_d + K_{11}(s_d - s) + K_{12} \int_0^t (s_d - s)dt \\ \dot{c}_{Ld} + K_{21}(c_{Ld} + c_L) + K_{22} \int_0^t (c_{Ld} - c_L)dt \end{bmatrix} \tag{6.20}$$

where s_d and c_{Ld} are the desired values of glucose and DO concentration, and K_{ij} can be selected so that the error dynamics converges. In actual online control implementation, the glucose feed rate F_S and the airflow rate F_A, are the true process inputs. F_S and F_A are obtained from the relations given by Eqs. (6.13) and (6.14). Further, D is dilution rate (h^{-1}), $k_L a$ is mass transfer coefficient (h^{-1}), N is the agitation speed (rpm), and V is the bioreactor volume (l). The rate of change of substrate concentration and dissolved oxygen are denoted by $\dot{s}$ and $\dot{c}_L$, respectively.

The various advanced nonlinear controllers available today is the result of continual development made in this domain (Table 6.2). Although the methods of the nonlinear systems theory have been applied successfully to design better controllers, they have a few limitations. These methods are best suited for bioprocesses that can be described by ordinary differential equations using a limited number of state variables. These methods cannot capture the shift in metabolic regimes that microorganisms typically make in response to changing environmental conditions. Hence, the stability of nonlinear control system is a major concern (Niemiec and Kravaris 2003; Kravaris et al. 2004). These issues are better addressed by employing a supervisory or a heuristic decision controller that is based on proteomic or metabolic data. The volume of data that needs to be processed at this level is very high, and as a result biometric analysis and the use of multivariate statistical tools becomes indispensable. In addition, there is a growing trend toward continuous processing in which the challenge is to identify crucial parameters affecting performance (Rathore et al. 2015a). A measure of the success of nonlinear controllers is their ability to track dynamics of metabolic shifts. But, due to lack of robust online measurement devices, only a few variables can be measured online such as biomass, substrate, pH, temperature, and dissolved oxygen. Further, the time scale over which metabolic changes occur is very short and would require sophisticated devices and automated systems, which would come at a significant cost. Instead, various soft-sensors are often deployed to address this issue so that better control performance can be achieved.

Table 6.2 Advancements in nonlinear controllers: concepts and potential application related to bioprocesses.

Principle of nonlinear control	Strategy used	Potential applications
Input–output linearizing control	Exploited the connections between model predictive control and input–output linearization. The control laws were derived by solving a moving-horizon, constrained optimization problem	Applicable to stable non-linear multi-input multi-output (MIMO) systems for example, in chemical and some biological processes
Non-linear state feedback control and feedback H_∞ control	Concept of a first-order nonlinear all-pass introduced in which under appropriate coordinate transformation, a multiorder system can be truncated to linear system with lag	Applicable to first and second order nonlinear systems for example, in chemical and most biological processes, continuous stirred tank reactor (CSTR) studies, etc.
Decoupled input–output linearizing control	Interaction between state variables removed	Fed-batch fermentation
Non-linear tracking control	Introduced an inversion procedure for nonlinear systems that constructs a bounded input trajectory in the preimage of a desired output trajectory	Applicable in tracking the given trajectory in chemical and bioprocesses
Model-based geometric algorithm	An estimator that predicts the states variables of the system one-time step ahead and a controller executes action so that predicted error falls below a desired level	Accurate control of dissolved oxygen in bioreactors
Nonlinear model-state feedback control	Proposed a systemic layout for the construction of statically equivalent synthetic outputs which are further used to construct a model-state feedback controller	Method is applied to a chemical reactor control problem where a series/parallel reaction is taking place
Constrained input–output linearizing control	Implemented both input–output linearization and Lyapunov control for chemical reactor at stable and unstable steady states	Significantly useful in controlling nonlinear and unsteady bioprocesses
Advanced geometric input–output linearizing control	Uses exact input–output linearization techniques for multiple operating points and wide operating range	Applied to a nontrivial temperature control problem in a CSTR
Neural network based feedback linearization	Nonlinear control used in conjunction with artificial neural network in the area of highly nonlinear and time varying bioprocess system	Powerful synchronization for highly nonlinear and time varying uncertain system like biological processes

6.5   Intelligent Control Systems

Developing a mathematical model and implementing a robust control strategy for biological processes is a challenging area for biotechnologists and control engineers, mainly due to the intricacy of biological systems. Control systems are not only limited by the measurements or models but also by their inability to tap into the prior knowledge acquired about the system. Hence, it has been found that the knowledge or data based control structures using the human decisional factor may sometimes be a better alternative (Uraikul et al. 2007). Consequently, the intelligent techniques, such as neural nets, fuzzy structures, genetic algorithms, or hybrid systems, which are capable of replicating human expert like reasoning and decision-making when dealing with uncertainties and imprecise information have been continually developed and improved during the last two decades (Dochain 2013). However, advancements in sensor technology and improved computational power in last decade have made their pilot scale implementation possible (Ferreira et al. 2014; Borchert et al. 2015). As the human perception about the bioprocess is commonly altered by the psychological factors, the intelligent control systems that have been based on the human subjective knowledge cannot be considered robust when compared with the control systems developed using a conceptual model. Hence, it is often recommended that the intelligent control techniques be augmented with the control structure based on quantitative models. The classical control techniques and intelligent control techniques can coexist and together can bring greater benefits to the practicing engineer by expanding the repertoire of available control techniques.

6.5.1   Fuzzy Control

On various occasions, bioprocesses require human intervention to take corrective action. On such occasions, decisions are made based on the current information about the process leading to the errant event and the past knowledge and experience of the expert. The process of analysis and decision-making is complex and cannot be incorporated into the methods described in Section 6.4. Nevertheless, it is possible to represent reasoning, arguments, and imprecise statements by using a set of rules and infer logical outcomes. The fuzzy logic theory provides a mathematical platform for carrying out such reasoning processes in a quantitative manner even when available information is incomplete. In this approach, uncertain events are subjected to the systemic scientific investigation to establish a subjective correlation between variable input and disturbances to the process output.

To implement a fuzzy logic technique to a real application requires the three steps illustrated in Figure 6.7. There are mainly two types of fuzzy control. Fuzzy logic controllers that use type-1 fuzzy sets provide an effective way

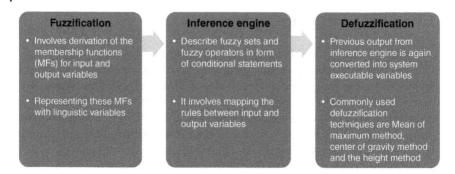

Fuzzification	Inference engine	Defuzzification
• Involves derivation of the membership functions (MFs) for input and output variables	• Describe fuzzy sets and fuzzy operators in form of conditional statements	• Previous output from inference engine is again converted into system executable variables
• Representing these MFs with linguistic variables	• It involves mapping the rules between input and output variables	• Commonly used defuzzification techniques are Mean of maximum method, center of gravity method and the height method

Figure 6.7 Illustration of conceptual stages involved in fuzzy control system.

to control systems characterized by high nonlinearities, but the presence of uncertainties within the system may deteriorate the performance of these controllers. To address this issue, the second, type-2 fuzzy systems have been proposed that use more parameters to build rules. Different fuzzy control strategies have been developed based on different classical control methods, such as PID-fuzzy control, nonlinear fuzzy control (Cosenza and Galluzzo 2012), sliding-mode fuzzy control, neural fuzzy control, adaptive fuzzy control, and phase-plan mapping fuzzy control. For example, a fuzzy model-based predictive controller (Yang et al. 2014) and fuzzy adaptive controller (Belchior et al. 2012) were used for controlling the concentration of dissolved oxygen in an activated sludge waste water treatment process. The researchers developed fuzzy rules based on a mechanistic model previously identified for the activated sludge to maintain the concentration of dissolved oxygen at the desired set point. Mokeddem and Khellaf (2012) augmented the fuzzy controller by employing genetic algorithms for selecting control parameters, which improved efficiency in predicting optimal feed profile for bioreactors. Ding et al. (2012) performed fault diagnosis and the control of *Corynebacterium glutamicum* cultures. They studied an industrial glutamate production system and proposed fuzzy control logic to capture batches having different initial biotin content. By doing so they were able to minimize the sudden changes in culture state that are often detrimental in a fermentation process and obtained final glutamate concentrations of $75–80\,\mathrm{g\,l^{-1}}$ at 34 hours.

Benefits of applying fuzzy control to such processes are many. The control strategy consists of simple if–then–else rules and is easy for a process operator to interpret. The onsite experience and knowledge of process operators and engineers can be harnessed into the design of these rules.

However, establishing rules requires a lot of experiments and its success relies heavily on either the soundness of knowledge acquisition techniques or the availability of an expert data analyst. The detailed description regarding the fuzzy system and control can be found in the literature (Karakuzu et al. 2006; Lee 2006; Jantzen 2013; Luo et al. 2015).

6.5.2 Neural Control

The concept of neural networks is inspired by the way neuronal cells work. A neuron receives signals from another neuron through its dendrites. These signals serve as inputs and are processed via complex cellular mechanisms within the neuron. When the total effect of the inputs accumulates beyond a threshold, it elicits a response and triggers an action potential that traverses down the neuronal axon and is transmitted to other neurons to which it is connected. This simple processing unit is represented as a node; several nodes are interconnected to form a network. One of the simplest architecture for such networks is the three-layered neural network consisting of the input, hidden, and an output layer. The connection between the nodes carries a weight. Based on the difference between the prediction of the network and the desired value, these weights are optimized by error-reducing algorithms. Figure 6.8 shows a typical architecture for a basic artificial neural network (ANN). This architecture consists of an input layer, one or more hidden layers, and an output layer. The number of neurons in the input and output layers are representative of the number of input and output variables. The network thus trained on existing data can be used to make predictions of the unknown variables. The performance of a neural network depends on the neural architecture, the algorithm used for determining the weights of its connections, and its activation function used for processing the inputs at a node in the hidden layer (Hussain 1999).

Figure 6.8 Conceptual representation of a typical neural network architecture.

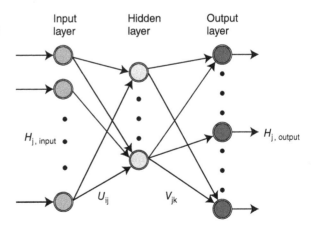

The mathematical formulation of a basic ANN is described by the equations given below. Considering hidden layer input as $H_{j,\,in}$ and output as $H_{j,\,out}$, we obtain

$$H_{j,in} = \sum_i U_{ij} I_i + \xi_j \tag{6.21}$$

$$H_{j,out} = \emptyset(H_{j,in}) \tag{6.22}$$

$$O_{k,in} = \sum_i V_{jk} H_{j,out} + W_k O_{DO,out}(n-1) + \xi_k \tag{6.23}$$

$$O_{k,out} = \psi(O_{k,in}) \tag{6.24}$$

The error equation becomes

$$\text{MSE}, E_M = \frac{1}{2} \sum_k e_k^2 \tag{6.25}$$

$$\text{Total error}, E = \sum_m E_m \tag{6.26}$$

$$e_{k,online} = d_{k,DO\ online} - O_{k,out} \tag{6.27}$$

where U_{ij}, V_{jk}, and W_k are the weights between input and hidden layer, hidden and output layer and intraconnections in output layer, respectively, d_k is the desired output, and I_i is the input to the networks. The activation function φ for input to hidden layer is a logistic sigmoidal function, and activation function ψ for hidden layer to output layer is a tan hyperbolic function. The bias is shown as ξ, α is the momentum coefficient, and η is the learning rate.

There are several advantages of using neural networks for modeling and control of complex biological processes (Montague and Morris 1994). Neural networks can learn new features of the process over time and have been applied effectively in nonlinear and multivariate systems (Boutalis 2006; Boutalis and Kosmidou 2006). Because of these characteristics, neural networks have seen successful applications in numerous chemical processes for data analysis, for optimization studies (Hussain and Ramachandran 2006; Mete et al. 2012), as a soft sensor (Kiran and Jana 2009) and in fault detection (Almeida 2002; Baruch et al. 2008; Arpornwichanop and Shomchoam 2009; Baruch and Mariaca-Gaspar 2009; Bhotmange and Shastri 2011; Yadav et al. 2013).

One of the problems with NN in process control applications is the prolonged duration required for convergence of the weights. Since the control action should be taken within the sampling period, the application of neural networks had been restricted to its role as an estimator or soft-sensor. To address this issue and extend the application of the neural network as a controller, Gadkar et al. (2005) developed an architecture for online implementation and control of biomass and ethanol through a yeast fermentation process. They applied a recurrent neural network with intra-connection within the output layer to track the dynamics of fed-batch fermentation for yeast fermentation process control. They showed a comparative study of its performance with and without online adaptation of weights and showed that online adaptations of weights gave superior performance. Guo et al. (2010) integrated the response surface methodology (RSM) combined with the artificial neural network-genetic algorithm (ANN-GA) to optimize the fermentation medium for nisin production. There are several reports in which NN was used to predict state variables in the fermentation process. For example, ANN has been extensively applied to estimate biomass (Desai et al. 2005; Jenzsch et al. 2006; Won and Yoon-Keun 2006; da Costa Albuquerque et al. 2008; Hocalar et al. 2011), substrate concentration (Jin et al. 2014), and oxygen level (Johnsson et al. 2015). Melcher et al. (2015) used neural networks as online estimators for biomass and protein concentration in recombinant *E. coli* fermentation. Recently, many researchers conducted different studies of neural network-based schemes for optimizing fed-batch bioreactors (Warnes et al. 1998; Eslamloueyan and Setoodeh 2011; Geethalakshmi et al. 2012; Peng et al. 2014). In general, neural network models were combined with basic flux balance analysis to give an improved estimate of system dynamics. This technique provided a fast and reliable way of optimizing system performance. Examples of different types of neural networks architecture and training algorithms used in bioprocess are summarized in Table 6.3. Different types of neural networks-assisted control strategies have also been constructed (Peres et al. 2001; Simon and Karim 2001; Tian et al. 2002; Patnaik 2004, 2008; Mjalli and Al-Asheh 2005; Petre et al. 2010) and successfully implemented in controlling various bioprocesses are framed under Table 6.4. Nagy (2007) had benchmarked the performance of neural network against the PID controller and MPC and demonstrated the superiority of neural network-based control for regulating yeast fermentation. These and several other examples in the literature demonstrate the power of neural networks in developing new monitoring and control tools for bioprocesses.

However, there are certain limitations in their practical application that need to be addressed to be successful. The first basic condition is the requirement of "good data set" over reasonably varied conditions, which can account for process variability and responses. Further, choice of suitable training dataset and longer duration that is required for training remains a challenging issue. Although, increased computational power of computers may solve this particular problem, their adaptation to newer operating conditions cannot be guaranteed. Their prediction capability is rather limited to trained domain area. Nevertheless, there is a tremendous potential, which is still untapped for carrying out innovative research in the field of designing and developing neural network-based bioprocess software sensors, monitoring tools, and control strategies for online implementation.

Table 6.3 Different types of neural networks commonly used for bioprocesses.

Type of neural network	Basic principle	Features
Simple feed forward neural network	Layered network in which each layer only receives inputs from previous layers. It consists of only feed-forward component with static nodes	Algorithm suffers from slow convergence times and can easily get trapped in local minima within the weight space
Radial basis function network	Uses activation functions that forms the basic building blocks for an approximation but having static nodes	Faster training rate, easy to perform online calculation of confidence limits for model estimations
Recurrent (Elman) network	It consists of both feed-forward and feedback component allowing dynamic nature to nodes	Able to estimate critical process parameters with high accuracy in real time in turn allows adaptive trajectory-tracking control
Generalized regression neural network	Combines the potential of multivariate statistical methods with numerically efficient regression neural network	Offers accurate prediction of process parameters along with efficient and fast training to neural algorithm
Probabilistic neural network	Type of feed-forward neural networks that generally implement a Gaussian distribution function in lieu of the sigmoid and hyperbolic functions commonly used in simple feed-forward neural network. Here, Bayesian decision strategies are used at the decision boundaries	Enables a reliable determination of the growth phases in bioprocesses, potential for use with other online tools for possibility of automation and control of cultivation processes
Knowledge-based modular network	Combines mechanistic, heuristic and knowledge hidden in process data records. Expectation maximization (EM) algorithm is employed to optimally utilize this records	Empowers knowledge integration for improving the quality of process modeling. However, efforts may be significant in unifying the data at all levels
Augmented neural network	Hybrid in the sense that they also utilize the available process information in the form of a model	Provides accurate and long prediction range for process parameters

6.5.3 Statistical Process Control

Systemic statistical investigation is important in bioprocesses because it exhibits behavior arising from multivariable interaction. Traditionally, statistical process control (SPC) charts have been used to monitor industrial processes for identifying batch trend and to prevent batch failures. Multivariate statistical process control (MSPC) methods are commonly employed to reduce

Table 6.4 Advancement in neural network-based control schemes for bioprocess control applications.

Neural network-assisted control strategies for bioprocesses	Features	Potential benefits and concerns for bioprocess control
Neural networks-based nonlinear approaches	NN is used as a compensator to the effects of the model uncertainties, which appear in the linearizing control law	Achieves the control target faster but with vigorous and sudden controller moves resulting in oscillatory spikes. Scarcity of reliable mathematical model
Neural networks-based model predictive control	A reliable estimation and prediction of process variables is done using historical data and heuristic rules	The control is able to react online to variations in the process and also to incorporate the new process information continuously. Scarcity of reliable mathematical model that showcase the entire bioprocess, online tuning of controller is major challenge
Neural networks-based adaptive control	Adaptation laws of neural network weights are derived from a Lyapunov stability property of the closed-loop system	Can be a good starting point for a class of nonlinear bioprocesses with incompletely known and time-varying dynamics
Expert supervisory systems	Integrates the knowledge-based systems, neural networks, and other data based modeling and pattern recognition techniques	The approach is well suited to the control of bioprocesses because of its capability of working with the fragmentary, uncertain, qualitative, and blended knowledge typically available for biological processes
Artificial intelligence and hybrid controls	Unifies various modeling, monitoring, and control approaches to deliver consistent output with minimal human interventions	Under the presence of unknown disturbances and model–plant mismatches, these controls can improve the reactor performance in terms of the amount of a desired product at the end of operation

the dimension of the large raw data set for filtering out useful information (Cimander and Mandenius 2002; Masuda et al. 2014; Rathore et al. 2014a). In nearly all applications, two basic techniques are applied: principal component analysis (PCA) and partial least squares (PLSs). PCA is a dimension reduction method that identifies a subset of uncorrelated vectors (principal components) so as to capture most variance in the data (Li et al. 2000; Russell et al. 2000; Kulkarni et al. 2004; Yoon and MacGregor 2004; Linting et al. 2007; Elshenawy et al. 2009). On the other hand, PLS is a decomposition technique that maximizes the covariance between predictor and predicted variables for each component (Chen and Liu 2002; Wang et al. 2003; Cherry and Qin 2006; De Roover et al. 2012; Xu and Goodacre 2012).

MSPC has been applied across many industries such as in the areas of semiconductors, chemicals, mining, and petrochemicals. Although the application of MSPC in the biopharmaceutical industry is in its nascent stage, it has been gaining more attention in recent years. Their application in a biological process enabled the extraction of useful data (Bouveyron and Brunet-Saumard 2014; Jacques and Preda 2014), based on which a control strategy was designed. There is rich literature available for guiding to use MSPC as a potential PAT tool (Rathore et al. 2011, 2014a,b; Rathore 2014; Lourenço et al. 2012; Stubbs et al. 2013; Mercier et al. 2014). Recently, Rathore et al. (2015b) demonstrated the use of multivariate data analysis for correlating the product quality attributes such as product glycosylation to process variables and raw materials attribute such as amino acid supplements in cell culture media. This approach was further used for process optimization to improve product expression consistency.

6.5.4 Integrated and Plant-Wide Bioprocess Control

Although the ambitious goal of establishing a fully fledged continuous process for the production of biotherapeutics by combining large-scale production and purification platforms is still far from real, but much of the needed technology is already there (Rathore et al. 2015a). The first logical step would have been the development of real-time PAT control scheme that is capable of identifying changes in key product quality attributes such as glycosylation in case mAb production, during continuous processing. In the second step, this scheme may need to be integrated with appropriate global process control strategies to ensure that the continuous process remains within defined specifications. With the systematic incorporation of PAT, it is believed that the information flow could be established in feedforward and feedback manner. This includes coordination of flow rates, management of process interruptions, and adjustments in certain operating conditions to meet certain goals. These goals may include optimizing certain quality attributes, for example to maximize product yield or to minimize performance loss over time or to minimize flux loss due to membrane fouling after multiple cycles. This type of global process control will require coordination between unit operations and between upstream and downstream processes, both of which are absent in current biomanufacturing due to the batch nature of existing processes (Figure 6.9).

Recently, many researchers have demonstrated such integrated plant-wide monitoring and control application in biomanufacturing. Borchert et al. (2015) have developed an integrated bioprocess for the production of malaria vaccine from *Pichia pastoris*. They integrated an expanded bed adsorption chromatography step into a sequential cultivation of the *Pichia* in order to link upstream

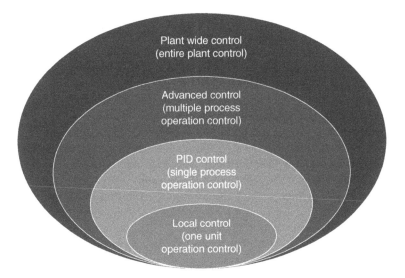

Figure 6.9 Representation of components involved in plant-wide control schemes.

production and related processing operations (such as protein expression, cell release, and product capture of the secreted protein) in a fully automated plant. They deployed a stirred conditioning vessel with a steam-sterilizable transfer line reaching from the reactor bottom valve to the vessel inlet of a purification column. This construction was designed to minimize the risk of contamination of the harvest flow. The various monitoring and control instrumentation integrated through a common program that enables automatic execution of adjustable process settings in upstream and downstream procedures. This setup provides a small, flexible, modular, and semicontinuously operated plant and is suitable to produce highly specific biopharmaceuticals. The end result was enhanced for process understanding. Resulting purities of up to 87% were achieved with recoveries of 51% in a single downstream operation. Such a type of integrated operations is the current area of research and industries are more interested in implementing such continuous processes.

6.5.5 Metabolic Control

It is well known that cells internal environment is strongly regulated by changes in local medium conditions in very complex and interactive ways. Depending on its phenotype, each cell reacts rapidly by altering subsets of signaling molecules, such as nicotine adenine dinucleotide phosphate (NAD(P)H) redox state. Hence, dynamic control strategies are essential whereby fluxes can be restored such that maximal flux will flow through the pathway leading to the synthesis of the desired end-product (Leme et al. 2014). For example in a fed-batch Chinese hamster ovary (CHO) process, lactate production is known to have a significant impact on glycolytic flux. Further, it has been shown that lactate metabolism is affected by various process parameters such as pH, temperature, humidity, and carbon dioxide. Khaparde and Roychoudhury (2014) have shown that the ratio of the lactate production rate (LPR) to the glucose uptake rate (GUR) is an important metric for monitoring cellular metabolism. They observed that the variation of cumulative urokinase produced by HT1080 cells in a hollow fiber bioreactor with the LPR/GUR ratio exhibited distinct physiological regimes. The first exhibited spindle morphology with urokinase production and the second with spherical morphology akin to metastasis and significantly reduced urokinase synthesis. Their results suggested that the modulation cycles of the hollow fiber reactor could be controlled to restrain the metabolism in the domain conducive to producing urokinase. Zalai et al. (2015) were able to control the metabolic switch to lactate consumption in CHO fed-batch cultivation via a broad range by the proper timing of pH and temperature shifts. To design dynamic control scheme, they had used time-resolved metabolic flux analysis, PLS, and design of experiments to assess the correlation of lactate metabolism and the activity of the major intracellular pathways. To build this scheme, identification, and quantification of the interactions of process parameters with specific cell metabolites is critical which then further correlated with productivity and product quality-related attributes, typically with the goal of process optimization.

6.6 Summary

It is expected that with the discussed methods of modeling, process control, and soft sensing, the dilemma of implementing PAT-based monitoring and control schemes in the biopharmaceutical industry can be significantly reduced to ensure improved and consistent product quality. In this chapter, different methods of open-loop and closed-loop control for bioprocess automation have been discussed. Since bioprocesses are inherently complex, improvement in sensor technology is imperative for monitoring and directing the behavior of cells to our advantage. However, describing the cellular function and behavior mathematically continues to present more hurdles that require effective solutions. There is no universal model or control algorithm applicable to address the range of characteristics that are observed and recorded. In a majority of cases of closed-loop control applications, on-, in-, or at-line measurements combined with soft sensors to determine the unmeasured state variables, would work. This could be further combined with a forward loop to reduce the problems generated by the dynamics of the bioprocess. Though predictive control schemes have been successfully applied in other application fields, reliable and robust process models are required to address the uncertainties related to bioprocesses. Such situations must be compensated by using knowledge databases unit with either fuzzy logic or ANN. Considering the rapid pace at which techniques and know-how are growing in this field, engineers and personnel working in the industry would need to be supported through workshops and training programs organized by experts in the field.

6.7 Future Perspectives

Major advancements in sensor technology and increasing computing power of workstations will transform the conventional manual manufacturing platform to an advanced automated PAT-based environment in the foreseeable future. Although developing models for virtual plants still remains a challenging task, working in multidisciplinary groups and forging collaborations between academia and industries will play a crucial role in translating PAT theory into its actual implementation for major benefits. Information in whatever form is very important as this will help in generating simple solutions to complex problems in bioprocesses.

Acknowledgments

The research was supported and funded by grant SB/S3/CE/093/2013 from Department of Science and Technology, India, and the DBT Centre of Excellence for Biopharmaceutical Technology (number BT/COE/34/SP15097/2015).

References

Abu-Absi, N.R., Kenty, B.M., Cuellar, M.E. et al. (2011). Real time monitoring of multiple parameters in mammalian cell culture bioreactors using an in-line Raman spectroscopy probe. *Biotechnol. Bioeng.* 108 (5): 1215–1221.

Almeida, J.S. (2002). Predictive non-linear modeling of complex data by artificial neural networks. *Curr. Opin. Biotechnol.* 13 (1): 72–76.

Arnold, S.A., Gaensakoo, R., Harvey, L.M., and McNeil, B. (2002). Use of at-line and in-situ near-infrared spectroscopy to monitor biomass in an industrial fed-batch *Escherichia coli* process. *Biotechnol. Bioeng.* 80 (4): 405–413.

Arpornwichanop, A. and Shomchoam, N. (2009). Control of fed-batch bioreactors by a hybrid on-line optimal control strategy and neural network estimator. *Neurocomputing* 72 (10): 2297–2302.

Ashoori, A., Moshiri, B., Khaki-Sedigh, A., and Bakhtiari, M.R. (2009). Optimal control of a nonlinear fed-batch fermentation process using model predictive approach. *J. Process Control* 19 (7): 1162–1173.

Baruch, I.S. and Mariaca-Gaspar, C.R. (2009). A levenberg–marquardt learning applied for recurrent neural identification and control of a wastewater treatment bioprocess. *Int. J. Intell. Syst.* 24 (11): 1094–1114.

Baruch, I., Mariaca-Gaspar, C., and Barrera-Cortes, J. (2008). *Recurrent Neural Network Identification and Adaptive Neural Control of Hydrocarbon Biodegradation Processes*. INTECH Open Access Publisher.

Belchior, C.A.C., Araújo, R.A.M., and Landeck, J.A.C. (2012). Dissolved oxygen control of the activated sludge wastewater treatment process using stable adaptive fuzzy control. *Comput. Chem. Eng.* 37: 152–162.

Bhotmange, M. and Shastri, P. (2011). *Application of Artificial Neural Networks to Food and Fermentation Technology, Artificial Neural Networks-Industrial and Control Engineering Applications*. InTech Publishing Inc. ISBN: 978-953-307-220-3.

Borchert, S.O., Voss, T., Schuetzmeier, F. et al. (2015). Development and monitoring of an integrated bioprocess for production of a potential malaria vaccine with *Pichia pastoris. J. Process Control* 35: 113–126.

Boudreau, M.A. and McMillan, G.K. (2007). *New Directions in Bioprocess Modeling and Control: Maximizing Process Analytical Technology Benefits*. ISA.

Boutalis, Y. (2006). Neural network approaches for feedback linearization. *J. Control Eng. Appl. Informatics* 6 (1): 15–26.

Boutalis, Y.S. and Kosmidou, O.I. (2006). A feedback linearization technique by using neural networks: application to bioprocess control. *J. Comput. Methods Sci. Eng.* 6 (5, 6): 269–282.

Bouveyron, C. and Brunet-Saumard, C. (2014). Model-based clustering of high-dimensional data: a review. *Comput. Stat. Data Anal.* 71: 52–78.

Card, C., Hunsaker, B., Smith, T., and Hirsch, J. (2008). Near-infrared spectroscopy for rapid, simultaneous monitoring. *BioProcess Int.* 6: 59–67.

Challa, S. and Potumarthi, R. (2013). Chemometrics-based process analytical technology (PAT) tools: applications and adaptation in pharmaceutical and biopharmaceutical industries. *Appl. Biochem. Biotechnol.* 169 (1): 66–76.

Chen, J. and Liu, K.C. (2002). On-line batch process monitoring using dynamic PCA and dynamic PLS models. *Chem. Eng. Sci.* 57 (1): 63–75.

Cherry, G. and Qin, S.J. (2006). Multiblock principal component analysis based on a combined index for semiconductor fault detection and diagnosis. *IEEE Trans. Semicond. Manuf.* 19 (2): 159–172.

Chopda, V., Rathore, A., and Gomes, J. (2013). On-line implementation of decoupled input-output linearizing controller in Baker's yeast fermentation. Dynamics and Control of Process Systems, Volume 10, Part 1, pp. 259–264.

Chopda, V.R., Rathore, A.S., and Gomes, J. (2015). Maximizing biomass concentration in Baker's yeast process by using a decoupled geometric controller for substrate and dissolved oxygen. *Bioresour. Technol.* 196: 160–168.

Chopda, V.R., Gomes, J., and Rathore, A.S. (2016). Bridging the gap between PAT concepts and implementation: an integrated software platform for fermentation. *Biotechnol. J.* 11 (1): 164–171.

Cimander, C. and Mandenius, C.F. (2002). Online monitoring of a bioprocess based on a multi-analyser system and multivariate statistical process modelling. *J. Chem. Technol. Biotechnol.* 77 (10): 1157–1168.

Clementschitsch, F. and Bayer, K. (2006). Improvement of bioprocess monitoring: development of novel concepts. *Microb. Cell Fact.* 5 (1): 19.

Cosenza, B. and Galluzzo, M. (2012). Nonlinear fuzzy control of a fed-batch reactor for penicillin production. *Comput. Chem. Eng.* 36: 273–281.

da Costa Albuquerque, C.D., de Campos-Takaki, G.M., and Fileti, A.M.F. (2008). On-line biomass estimation in biosurfactant production process by *Candida lipolytica* UCP 988. *J. Ind. Microbiol. Biotechnol.* 35 (11): 1425–1433.

Crowley, J., McCarthy, B., Nunn, N.S. et al. (2000). Monitoring a recombinant *Pichia pastoris* fed batch process using Fourier transform mid-infrared spectroscopy (FT-MIRS). *Biotechnol. Lett.* 22 (24): 1907–1912.

De Roover, K., Ceulemans, E., and Timmerman, M.E. (2012). How to perform multiblock component analysis in practice. *Behav. Res. Methods* 44 (1): 41–56.

Desai, K.M., Vaidya, B.K., Singhal, R.S., and Bhagwat, S.S. (2005). Use of an artificial neural network in modeling yeast biomass and yield of β-glucan. *Process Biochem.* 40 (5): 1617–1626.

Ding, J., Cao, Y., Mpofu, E., and Shi, Z. (2012). A hybrid support vector machine and fuzzy reasoning based fault diagnosis and rescue system for stable glutamate fermentation. *Chem. Eng. Res. Des.* 90 (9): 1197–1207.

Dochain, D. (ed.) (2013). *Automatic Control of Bioprocesses*. Wiley.

Ducommun, P., Kadouri, A., Von Stockar, U., and Marison, I.W. (2002). On-line determination of animal cell concentration in two industrial high-density culture processes by dielectric spectroscopy. *Biotechnol. Bioeng.* 77 (3): 316–323.

Elshenawy, L.M., Yin, S., Naik, A.S., and Ding, S.X. (2009). Efficient recursive principal component analysis algorithms for process monitoring. *Ind. Eng. Chem. Res.* 49 (1): 252–259.

Eslamloueyan, R. and Setoodeh, P. (2011). Optimization of fed-batch recombinant yeast fermentation for ethanol production using a reduced dynamic flux balance model based on artificial neural networks. *Chem. Eng. Commun.* 198 (11): 1309–1338.

FDA (2004). *Guidance for Industry: PAT—A Framework for Innovative Pharmaceutical Development, Manufacturing, and Quality Assurance.* Rockville, MD: FDA.

Ferreira, A.R., Dias, J.M.L., von Stosch, M. et al. (2014). Fast development of *Pichia pastoris* GS$_{115}$ Mut$^+$ cultures employing batch-to-batch control and hybrid semi-parametric modeling. *Bioprocess Biosyst. Eng.* 37 (4): 629–639.

Foley, R., Hennessy, S., and Marison, I.W. (2012). Potential of mid-infrared spectroscopy for on-line monitoring of mammalian cell culture medium components. *Appl. Spectrosc.* 66 (1): 33–39.

Gadkar, K.G., Mehra, S., and Gomes, J. (2005). On-line adaptation of neural networks for bioprocess control. *Comput. Chem. Eng.* 29 (5): 1047–1057.

Geethalakshmi, S., Narendran, S., Pappa, N., and Ramalingam, S. (2012). Development of a hybrid neural network model to predict feeding method in fed-batch cultivation for enhanced recombinant streptokinase productivity in *Escherichia coli. J. Chem. Technol. Biotechnol.* 87 (2): 280–285.

Glassey, J., Gernaey, K.V., Clemens, C. et al. (2011). Process analytical technology (PAT) for biopharmaceuticals. *Biotechnol. J.* 6 (4): 369–377.

Gomes, J. and Menawat, A.S. (2000). Precise control of dissolved oxygen in bioreactors – a model-based geometric algorithm. *Chem. Eng. Sci.* 55 (1): 67–78.

Gomes, J., Chopda, V.R., and Rathore, A.S. (2015). Integrating systems analysis and control for implementing process analytical technology in bioprocess development. *J. Chem. Technol. Biotechnol.* 90 (4): 583–589.

Guo, W.L., Zhang, Y.B., Lu, J.H. et al. (2010). Optimization of fermentation medium for nisin production from *Lactococcus lactis* subsp. *lactis* using response surface methodology (RSM) combined with artificial neural network-genetic algorithm (ANN-GA). *Afr. J. Biotechnol.* 9 (38): 6264–6272.

Hinz, D.C. (2006). Process analytical technologies in the pharmaceutical industry: the FDA's PAT initiative. *Anal. Bioanal. Chem.* 384 (5): 1036–1042.

Hocalar, A., Türker, M., Karakuzu, C., and Yüzgeç, U. (2011). Comparison of different estimation techniques for biomass concentration in large scale yeast fermentation. *ISA Trans.* 50 (2): 303–314.

Hussain, M.A. (1999). Review of the applications of neural networks in chemical process control—simulation and online implementation. *Artif. Intell. Eng.* 13 (1): 55–68.

Hussain, M.A. and Ramachandran, K.B. (2006). The study of neural network-based controller for controlling dissolved oxygen concentration in a sequencing batch reactor. *Bioprocess Biosyst. Eng.* 28 (4): 251–265.

Jacques, J. and Preda, C. (2014). Model-based clustering for multivariate functional data. *Comput. Stat. Data Anal.* 71: 92–106.

Jantzen, J. (2013). *Foundations of Fuzzy Control: A Practical Approach.* Wiley.

Jenzsch, M., Simutis, R., Eisbrenner, G. et al. (2006). Estimation of biomass concentrations in fermentation processes for recombinant protein production. *Bioprocess Biosyst. Eng.* 29 (1): 19–27.

Jin, H., Chen, X., Yang, J. et al. (2014). Hybrid intelligent control of substrate feeding for industrial fed-batch chlortetracycline fermentation process. *ISA Trans.* 53 (6): 1822–1837.

Johnsson, O., Andersson, J., Lidén, G. et al. (2015). Modelling of the oxygen level response to feed rate perturbations in an industrial scale fermentation process. *Process Biochem.* 50 (4): 507–516.

Junker, B.H. and Wang, H.Y. (2006). Bioprocess monitoring and computer control: key roots of the current PAT initiative. *Biotechnol. Bioeng.* 95 (2): 226–261.

Justice, C., Brix, A., Freimark, D. et al. (2011). Process control in cell culture technology using dielectric spectroscopy. *Biotechnol. Adv.* 29 (4): 391–401.

Kandelbauer, A., Kessler, W., and Kessler, R.W. (2008). Online UV–visible spectroscopy and multivariate curve resolution as powerful tool for model-free investigation of laccase-catalysed oxidation. *Anal. Bioanal. Chem.* 390 (5): 1303–1315.

Känsäkoski, M., Kurkinen, M., von Weymarn, N. et al. (2006). Process analytical technology (PAT) needs and applications in the bioprocess industry. VTT Technical Research Centre of Finland, 60, p. 99.

Kanter, J.M., Soroush, M., and Seider, W.D. (2002). Nonlinear controller design for input-constrained, multivariable processes. *Ind. Eng. Chem. Res.* 41 (16): 3735–3744.

Karakuzu, C., Türker, M., and Öztürk, S. (2006). Modelling, on-line state estimation and fuzzy control of production scale fed-batch Baker's yeast fermentation. *Control. Eng. Pract.* 14 (8): 959–974.

Khalil, H.K. (2008). High-gain observers in nonlinear feedback control. Control, Automation and Systems, 2008. ICCAS 2008. International Conference on October 2008. IEEE.

Khaparde, S.S. and Roychoudhury, P.K. (2014). Amino acid supplementation enhances urokinase production by HT-1080 cells. *J. Ind. Microbiol. Biotechnol.* 41 (6): 1035–1038.

Kiran, A.U.M. and Jana, A.K. (2009). Control of continuous fed-batch fermentation process using neural network based model predictive controller. *Bioprocess Biosyst. Eng.* 32 (6): 801–808.

Kiviharju, K., Salonen, K., Moilanen, U., and Eerikäinen, T. (2008). Biomass measurement online: the performance of in situ measurements and software sensors. *J. Ind. Microbiol. Biotechnol.* 35 (7): 657–665.

Kornmann, H., Rhiel, M., Cannizzaro, C. et al. (2003). Methodology for real-time, multianalyte monitoring of fermentations using an in-situ mid-infrared sensor. *Biotechnol. Bioeng.* 82 (6): 702–709.

Kovárová-Kovar, K., Gehlen, S., Kunze, A. et al. (2000). Application of model-predictive control based on artificial neural networks to optimize the fed-batch process for riboflavin production. *J. Biotechnol.* 79 (1): 39–52.

Krause, D., Hussein, M.A., and Becker, T. (2015). Online monitoring of bioprocesses via multivariate sensor prediction within swarm intelligence decision making. *Chemom. Intell. Lab. Syst.* 145: 48–59.

Kravaris, C., Niemiec, M., and Kazantzis, N. (2004). Singular PDEs and the assignment of zero dynamics in nonlinear systems. *Syst. Control Lett.* 51 (1): 67–77.

Kulkarni, S.G., Chaudhary, A.K., Nandi, S. et al. (2004). Modeling and monitoring of batch processes using principal component analysis (PCA) assisted generalized regression neural networks (GRNN). *Biochem. Eng. J.* 18 (3): 193–210.

Lee, K.H. (2006). *First Course on Fuzzy Theory and Applications*, vol. 27. Springer Science & Business Media.

Lee, H.L., Boccazzi, P., Gorret, N. et al. (2004). In situ bioprocess monitoring of Escherichia coli bioreactions using Raman spectroscopy. *Vib. Spectrosc.* 35 (1): 131–137.

Lee, Y.I., Kouvaritakis, B., and Cannon, M. (2008). Input–output feedback linearization for nonminimum phase nonlinear systems through periodic use of synthetic outputs. *Syst. Control Lett.* 57 (8): 626–630.

Leme, J., Fernández Núñez, E.G., de Almeida Parizotto, L. et al. (2014). A multivariate calibration procedure for UV/VIS spectrometric monitoring of BHK-21 cell metabolism and growth. *Biotechnol. Prog.* 30 (1): 241–248.

Li, W., Yue, H.H., Valle-Cervantes, S., and Qin, S.J. (2000). Recursive PCA for adaptive process monitoring. *J. Process Control* 10 (5): 471–486.

Li, B., Ryan, P.W., Ray, B.H. et al. (2010). Rapid characterization and quality control of complex cell culture media solutions using Raman spectroscopy and chemometrics. *Biotechnol. Bioeng.* 107 (2): 290–301.

Lindemann, C., Marose, S., Nielsen, H.O., and Scheper, T. (1998). 2-Dimensional fluorescence spectroscopy for on-line bioprocess monitoring. *Sens. Actuators, B* 51 (1): 273–277.

Linting, M., Meulman, J.J., Groenen, P.J., and van der Kooij, A.J. (2007). Nonlinear principal components analysis: introduction and application. *Psychol. Methods* 12 (3): 336.

Lourenço, N.D., Lopes, J.A., Almeida, C.F. et al. (2012). Bioreactor monitoring with spectroscopy and chemometrics: a review. *Anal. Bioanal. Chem.* 404 (4): 1211–1237.

Luo, Y., Wang, Z., Wei, G. et al. (2015). Fuzzy-logic-based control, filtering, and fault detection for networked systems: a survey. *Math. Prob. Eng.* 2015: 1, 543725–11.

Luttmann, R., Bracewell, D.G., Cornelissen, G. et al. (2012). Soft sensors in bioprocessing: a status report and recommendations. *Biotechnol. J.* 7 (8): 1040–1048.

Mandenius, C.F. (2004). Recent developments in the monitoring, modeling and control of biological production systems. *Bioprocess Biosyst. Eng.* 26 (6): 347–351.

Masuda, Y., Kaneko, H., and Funatsu, K. (2014). Multivariate statistical process control method including soft sensors for both early and accurate fault detection. *Ind. Eng. Chem. Res.* 53 (20): 8553–8564.

Mattes, R.A., Root, D., Chang, D. et al. (2007). In situ monitoring of CHO cell culture medium using near-infrared spectroscopy. *BioProcess Int.* 5 (1).

Mednis, M., Meitalovs, J., Viļums, S. et al. (2015). Bioprocess monitoring and control using mobile devices. *Inf. Technol. Control* 39 (3): 195–201.

Melcher, M., Scharl, T., Spangl, B. et al. (2015). The potential of random forest and neural networks for biomass and recombinant protein modeling in *Escherichia coli* fed-batch fermentations. *Biotechnol. J.* 10 (11): 1770–1782.

Mercier, S.M., Diepenbroek, B., Wijffels, R.H., and Streefland, M. (2014). Multivariate PAT solutions for biopharmaceutical cultivation: current progress and limitations. *Trends Biotechnol.* 32 (6): 329–336.

Mete, T., Ozkan, G., Hapoglu, H., and Alpbaz, M. (2012). Control of dissolved oxygen concentration using neural network in a batch bioreactor. *Comput. Appl. Eng. Educ.* 20 (4): 619–628.

Mjalli, F.S. and Al-Asheh, S. (2005). Neural-networks-based feedback linearization versus model predictive control of continuous alcoholic fermentation process. *Chem. Eng. Technol.* 28 (10): 1191–1200.

Mokeddem, D. and Khellaf, A. (2012). Optimal feeding profile for a fuzzy logic controller in a bioreactors using genetic algorithm. *Nonlinear Dyn.* 67 (4): 2835–2845.

Montague, G. and Morris, J. (1994). Neural-network contributions in biotechnology. *Trends Biotechnol.* 12 (8): 312–324.

Munson, J., Freeman Stanfield, C., and Gujral, B. (2006). A review of process analytical technology (PAT) in the US pharmaceutical industry. *Curr. Pharm. Anal.* 2 (4): 405–414.

Nagy, Z.K. (2007). Model based control of a yeast fermentation bioreactor using optimally designed artificial neural networks. *Chem. Eng. J.* 127 (1): 95–109.

Nayak, R. and Gomes, J. (2009). Sequential adaptive networks: an ensemble of neural networks for feed forward control of L-methionine production. *Chem. Eng. Sci.* 64 (10): 2401–2412.

Niemiec, M.P. and Kravaris, C. (2003). Nonlinear model-state feedback control for nonminimum-phase processes. *Automatica* 39 (7): 1295–1302.

Noui, L., Hill, J., Keay, P.J. et al. (2002). Development of a high resolution UV spectrophotometer for at-line monitoring of bioprocesses. *Chem. Eng. Process. Process Intensif.* 41 (2): 107–114.

Oliveira, R., Clemente, J.J., Cunha, A.E., and Carrondo, M.J.T. (2005). Adaptive dissolved oxygen control through the glycerol feeding in a recombinant *Pichia pastoris* cultivation in conditions of oxygen transfer limitation. *J. Biotechnol.* 116 (1): 35–50.

Opel, C.F., Li, J., and Amanullah, A. (2010). Quantitative modeling of viable cell density, cell size, intracellular conductivity, and membrane capacitance in batch and fed-batch CHO processes using dielectric spectroscopy. *Biotechnol. Prog.* 26 (4): 1187–1199.

Panjapornpon, C. and Soroush, M. (2006). Control of non-minimum-phase nonlinear systems through constrained input-output linearization. *2006 American Control Conference*. IEEE, p. 6.

Patnaik, P.R. (2004). Neural and hybrid neural modeling and control of fed-batch fermentation for streptokinase: comparative evaluation under non-ideal conditions. *Can. J. Chem. Eng.* 82 (3): 599–606.

Patnaik, P.R. (2008). Neural and hybrid optimizations of the fed-batch synthesis of poly-β-hydroxybutyrate by *Ralstonia eutropha* in a nonideal bioreactor. *Biorem. J.* 12 (3): 117–130.

Paulsson, D., Gustavsson, R., and Mandenius, C.F. (2014). A soft sensor for bioprocess control based on sequential filtering of metabolic heat signals. *Sensors* 14 (10): 17864–17882.

Peng, W., Zhong, J., Yang, J. et al. (2014). The artificial neural network approach based on uniform design to optimize the fed-batch fermentation condition: application to the production of iturin A. *Microb. Cell Fact.* 13 (1): 54.

Peres, J., Oliveira, R., and De Azevedo, S.F. (2001). Knowledge based modular networks for process modelling and control. *Comput. Chem. Eng.* 25 (4): 783–791.

Persad, A., Chopda, V.R., Rathore, A.S., and Gomes, J. (2013). Comparative performance of decoupled input–output linearizing controller and linear interpolation PID controller: enhancing biomass and ethanol production in saccharomyces cerevisiae. *Appl. Biochem. Biotechnol.* 169 (4): 1219–1240.

Petersen, N., Ödman, P., Padrell, A.E.C. et al. (2010). In situ near infrared spectroscopy for analyte-specific monitoring of glucose and ammonium in streptomyces coelicolor fermentations. *Biotechnol. Prog.* 26 (1): 263–271.

Petre, E., Selişteanu, D., Şendrescu, D., and Ionete, C. (2010). Neural networks-based adaptive control for a class of nonlinear bioprocesses. *Neural Comput. Appl.* 19 (2): 169–178.

Pollard, D.J., Buccino, R., Connors, N. et al. (2001). Real-time analyte monitoring of a fungal fermentation, at pilot scale, using in situ mid-infrared spectroscopy. *Bioprocess Biosyst. Eng.* 24 (1): 13–24.

Pons, M.N., Le Bonté, S., and Potier, O. (2004). Spectral analysis and fingerprinting for biomedia characterisation. *J. Biotechnol.* 113 (1): 211–230.

Praly, L. and Jiang, Z.P. (2004). Linear output feedback with dynamic high gain for nonlinear systems. *Syst. Control Lett.* 53 (2): 107–116.

Ranjan, A.P. and Gomes, J. (2009). Simultaneous dissolved oxygen and glucose regulation in fed-batch methionine production using decoupled input–output linearizing control. *J. Process Control* 19 (4): 664–677.

Ranjan, A.P. and Gomes, J. (2010). Decoupled adaptive control of glucose and dissolved oxygen for fed-batch methionine production using linear reference model. In: *Proceedings of the 2010 American Control Conference*, 5862–5867.

Rathore, A.S. (2009). Roadmap for implementation of quality by design (QbD) for biotechnology products. *Trends Biotechnol.* 27 (9): 546–553.

Rathore, A.S. (2014). QbD/PAT for bioprocessing: moving from theory to implementation. *Curr. Opin. Chem. Eng.* 6: 1–8.

Rathore, A.S. and Winkle, H. (2009). Quality by design for biopharmaceuticals. *Nat. Biotechnol.* 27 (1): 26–34.

Rathore, A.S., Bhambure, R., and Ghare, V. (2010). Process analytical technology (PAT) for biopharmaceutical products. *Anal. Bioanal. Chem.* 398 (1): 137–154.

Rathore, A.S., Bhushan, N., and Hadpe, S. (2011). Chemometrics applications in biotech processes: a review. *Biotechnol. Prog.* 27 (2): 307–315.

Rathore, A.S., Mittal, S., Pathak, M., and Arora, A. (2014a). Guidance for performing multivariate data analysis of bioprocessing data: pitfalls and recommendations. *Biotechnol. Prog.* 30 (4): 967–973.

Rathore, A.S., Mittal, S., Pathak, M., and Mahalingam, V. (2014b). Chemometrics application in biotech processes: assessing comparability across processes and scales. *J. Chem. Technol. Biotechnol.* 89 (9): 1311–1316.

Rathore, A.S., Agarwal, H., Sharma, A.K. et al. (2015a). Continuous processing for production of biopharmaceuticals. *Prep. Biochem. Biotechnol.* 45: 836–849.

Rathore, A.S., Singh, S.K., Pathak, M. et al. (2015b). Fermentanomics: relating quality attributes of a monoclonal antibody to cell culture process variables and raw materials using multivariate data analysis. *Biotechnol. Prog.* 31: 1586–1599.

Read, E.K., Park, J.T., Shah, R.B. et al. (2010a). Process analytical technology (PAT) for biopharmaceutical products: Part I. Concepts and applications. *Biotechnol. Bioeng.* 105 (2): 276–284.

Read, E.K., Shah, R.B., Riley, B.S. et al. (2010b). Process analytical technology (PAT) for biopharmaceutical products: Part II. Concepts and applications. *Biotechnol. Bioeng.* 105 (2): 285–295.

Renard, F. and Wouwer, A.V. (2008). Robust adaptive control of yeast fed-batch cultures. *Comput. Chem. Eng.* 32 (6): 1238–1248.

Roychoudhury, P., Harvey, L.M., and McNeil, B. (2006). At-line monitoring of ammonium, glucose, methyl oleate and biomass in a complex antibiotic fermentation process using attenuated total reflectance-mid-infrared (ATR-MIR) spectroscopy. *Anal. Chim. Acta* 561 (1): 218–224.

Roychoudhury, P., O'Kennedy, R., McNeil, B., and Harvey, L.M. (2007). Multiplexing fibre optic near infrared (NIR) spectroscopy as an emerging technology to monitor industrial bioprocesses. *Anal. Chim. Acta* 590 (1): 110–117.

Russell, E.L., Chiang, L.H., and Braatz, R.D. (2000). Fault detection in industrial processes using canonical variate analysis and dynamic principal component analysis. *Chemom. Intell. Lab. Syst.* 51 (1): 81–93.

Simon, L. and Karim, M.N. (2001). Probabilistic neural networks using Bayesian decision strategies and a modified Gompertz model for growth phase classification in the batch culture of *Bacillus subtilis*. *Biochem. Eng. J.* 7 (1): 41–48.

Smets, I.Y., Claes, J.E., November, E.J. et al. (2004). Optimal adaptive control of (bio)chemical reactors: past, present and future. *J. Process Control* 14 (7): 795–805.

Soons, Z.I.T.A., Voogt, J.A., Van Straten, G., and Van Boxtel, A.J.B. (2006). Constant specific growth rate in fed-batch cultivation of *Bordetella pertussis* using adaptive control. *J. Biotechnol.* 125 (2): 252–268.

Stubbs, S., Zhang, J., and Morris, J. (2013). Multiway interval partial least squares for batch process performance monitoring. *Ind. Eng. Chem. Res.* 52 (35): 12399–12407.

Tamburini, E., Vaccari, G., Tosi, S., and Trilli, A. (2003). Near-infrared spectroscopy: a tool for monitoring submerged fermentation processes using an immersion optical-fiber probe. *Appl. Spectrosc.* 57 (2): 132–138.

Teixeira, A.P., Oliveira, R., Alves, P.M., and Carrondo, M.J.T. (2009). Advances in on-line monitoring and control of mammalian cell cultures: supporting the PAT initiative. *Biotechnol. Adv.* 27 (6): 726–732.

Teixeira, A.P., Duarte, T.M., Carrondo, M.J.T., and Alves, P.M. (2011). Synchronous fluorescence spectroscopy as a novel tool to enable PAT applications in bioprocesses. *Biotechnol. Bioeng.* 108 (8): 1852–1861.

Tian, Y., Zhang, J., and Morris, J. (2002). Optimal control of a fed-batch bioreactor based upon an augmented recurrent neural network model. *Neurocomputing* 48 (1): 919–936.

Tiwari, S., Suraishkumar, G.K., and Chandavarkar, A. (2013). Robust near-infra-red spectroscopic probe for dynamic monitoring of critical nutrient ratio in microbial fermentation processes. *Biochem. Eng. J.* 71: 47–56.

Uraikul, V., Chan, C.W., and Tontiwachwuthikul, P. (2007). Artificial intelligence for monitoring and supervisory control of process systems. *Eng. Appl. Artif. Intell.* 20 (2): 115–131.

Van der Schaft, A.J. (1992). L 2-gain analysis of nonlinear systems and nonlinear state-feedback H∞ control. *IEEE Trans. Autom. Control* 37 (6): 770–784.

Vojinović, V., Cabral, J.M.S., and Fonseca, L.P. (2006). Real-time bioprocess monitoring: Part I: In situ sensors. *Sens. Actuators, B* 114 (2): 1083–1091.

Wang, X., Kruger, U., and Lennox, B. (2003). Recursive partial least squares algorithms for monitoring complex industrial processes. *Control. Eng. Pract.* 11 (6): 613–632.

Warnes, M.R., Glassey, J., Montague, G.A., and Kara, B. (1998). Application of radial basis function and feedforward artificial neural networks to the *Escherichia coli* fermentation process. *Neurocomputing* 20 (1): 67–82.

Whelan, J., Craven, S., and Glennon, B. (2012). In situ Raman spectroscopy for simultaneous monitoring of multiple process parameters in mammalian cell culture bioreactors. *Biotechnol. Prog.* 28 (5): 1355–1362.

Won, H. and Yoon-Keun, C. (2006). An artificial neural network for biomass estimation from automatic pH control signal. *Biotechnol. Bioprocess Eng.* 11 (4): 351–356.

Wu, W., Lai, S.Y., Jang, M.F., and Chou, Y.S. (2013). Optimal adaptive control schemes for PHB production in fed-batch fermentation of *Ralstonia eutropha*. *J. Process Control* 23 (8): 1159–1168.

Xu, Y. and Goodacre, R. (2012). Multiblock principal component analysis: an efficient tool for analyzing metabolomics data which contain two influential factors. *Metabolomics* 8 (1): 37–51.

Yadav, M., Sehrawat, N., Sangwan, A. et al. (2013). Artificial neural network (ANN): application in media optimization for industrial microbiology and comparison with response surface methodology (RSM). *Adv. Appl. Sci. Res.* 4 (4): 457–460.

Yang, M.Q., Athey, B.D., Arabnia, H.R. et al. (2009). High-throughput next-generation sequencing technologies foster new cutting-edge computing techniques in bioinformatics. *BMC Genomics* 10 (Suppl. 1): I1.

Yang, T., Qiu, W., Ma, Y. et al. (2014). Fuzzy model-based predictive control of dissolved oxygen in activated sludge processes. *Neurocomputing* 136: 88–95.

Yoon, S. and MacGregor, J.F. (2004). Principal-component analysis of multiscale data for process monitoring and fault diagnosis. *AIChE J.* 50 (11): 2891–2903.

Youssef, C.B., Guillou, V., and Olmos-Dichara, A. (2000). Modelling and adaptive control strategy in a lactic fermentation process. *Control. Eng. Pract.* 8 (11): 1297–1307.

Zalai, D., Koczka, K., Párta, L. et al. (2015). Combining mechanistic and data-driven approaches to gain process knowledge on the control of the metabolic shift to lactate uptake in a fed-batch CHO process. *Biotechnol. Prog.* 31 (6): 1657–1668.

Zhang, Z., Wu, S., Stenoien, D.L., and Paša-Tolic, L. (2014). High-throughput proteomics. *Annu. Rev. Anal. Chem.* 7: 427–454.

Part III

Host Strain Technologies

7

Metabolic Engineering for Biocatalyst Robustness to Organic Inhibitors

Liam Royce and Laura R. Jarboe

4134 Biorenewables Research Laboratory, Department of Chemical and Biological Engineering, Iowa State University, Ames, IA, USA

7.1   Introduction

The field of metabolic engineering has grown enormously since its definition in 1991 (Bailey 1991; Stephanopoulos and Vallino 1991). Since then, enormous progress has been made in the use of engineered microbes for production of biorenewable fuels and chemicals. Two years after its initial definition, Cameron and Tong classified five types of metabolic engineering: enhanced production of native metabolites; heterologous production of foreign metabolites; utilization of new substrates for metabolism; improved or new metabolic pathways for chemical degradation; and modification of cell properties that facilitate bioprocessing (Cameron and Tong 1993). Many landmark metabolic engineering projects have dealt with the first three strategies, including but not limited to (Ohta et al. 1991; Zhang et al. 1995; Steinbuchel 2001; Nakamura and Whited 2003; Jeffries and Jin 2004; Lutke-Eversloh and Stephanopoulos 2007; Atsumi et al. 2008), and can be thought of as relating mainly to metabolic functionality. Here we focus on the modification of cell properties that facilitate bioprocessing. Such modifications have focused on a wide variety of biocatalyst properties; here we focus on the property of biocatalyst robustness, particularly in regards to organic inhibitor tolerance. This improved robustness can be geared toward increasing product tolerance, possibly enabling increased product titers, and increased ability to use "dirty" biomass-derived sugars.

While industrial (Chotani et al. 2000) and academic researchers (Zeng and Biebl 2002; Demain 2006; Jarboe et al. 2007; Yan and Liao 2009) have reported attainment of high titers, yields, and productivities for some products, this is something that remains difficult to achieve, especially with next-generation biofuels and biochemicals. High product titers are desirable

Bioprocessing Technology for Production of Biopharmaceuticals and Bioproducts, First Edition.
Edited by Claire Komives and Weichang Zhou.
© 2019 John Wiley & Sons, Inc. Published 2019 by John Wiley & Sons, Inc.

to make downstream separation steps cost-effective, but can be toxic to the microbial biocatalyst. Biocatalyst inhibition is also problematic when biomass-derived sugars are used as the fermentation substrate, as biomass hydrolysate or pyrolysate ("dirty sugars") contains trace contaminants that are inhibitory to the microbe (Mills et al. 2009; Lian et al. 2010; Jarboe et al. 2011a). This toxicity means that the amount of sugars that can be utilized by the organism is limited, and thus the amount of product formed is decreased relative to fermentations using "clean" sugars. One approach for dealing with this product or feedstock toxicity is to remove the inhibitory compounds (Martinez et al. 2000; Lennen et al. 2010; Lian et al. 2010; Lakshmanaswamy et al. 2011; Chi et al. 2013). However, here we focus on the complementary approach of improving microbial robustness to organic inhibitors. Note that there are also reports of inhibitory side products; these can often be addressed by re-distribution of the metabolic flux (Agrawal et al. 2012).

Biocatalyst robustness as it relates to the economically viable production of fuels and chemicals has been discussed in a variety of recent reviews (Warnecke and Gill 2005; Fischer et al. 2008; Mills et al. 2009; Nicolaou et al. 2010; Dunlop 2011; Jarboe et al. 2011b). In this chapter, we highlight some of the key concepts in this field and cover some of the recent findings. The discussion includes metabolic evolution coupled with reverse engineering, a valuable tool for gaining insight into the mechanism of inhibition and strategies for increasing tolerance. Figure 7.1 highlights the possible mechanisms of inhibition by organic compounds, while Table 7.1 lists successful strategies to increase tolerance. Most of our discussion involves either *Escherichia coli* or *Saccharomyces cerevisiae*, since they are the two most commonly used organisms both for metabolic engineering and for understanding tolerance.

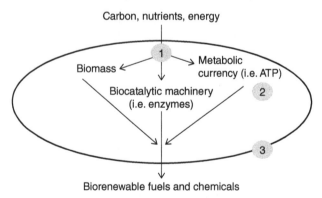

Figure 7.1 Reported mechanisms of inhibition often take the form of (1) inhibition reactions that produce metabolic precursors, limiting the formation of biomass and/or the enzymes needed for metabolism; (2) depletion of general metabolic precursors, such as ATP and NADPH; (3) damage to the cell membrane, interfering with membrane-associated reactions, and retention of valuable metabolites.

Table 7.1 Common types of microbial inhibition by organic inhibitors.

General mechanism	Example	Source
Inhibition of production of essential metabolites	Depletion of NADPH by aldehyde reduction limits production of cysteine	Miller et al. (2009a)
	Inhibition of methionine biosynthesis during acetate challenge	Whitfield et al. (1970), Roe et al. (2002)
	Inhibition of isoleucine and leucine biosynthesis during valine challenge	Manten and Rowley (1953)
	Inhibition of chorismate and threonine production by 3-HP	Warnecke et al. (2010)
Membrane damage	Membrane leakage induced during carboxylic acid challenge and production	Zaldivar and Ingram (1999), Lennen and Pfleger (2013), Liu et al. (2013), Royce et al. (2013)
	Reduced function of membrane-associated processes during carboxylic acid challenge	Teixeira et al. (2004), Cipak et al. (2008), Ruenwai et al. (2011)
Intracellular acidification	Low intracellular pH during acetate challenge	Trcek et al. (2015)

7.2 Mechanisms of Inhibition

The mechanism by which different compounds limit biocatalyst growth and metabolism vary according to the chemistry of the molecule. Discussed in this section are inhibition of production of essential metabolites, such as amino acids and nucleotides, membrane damage, and perturbation of intracellular pH (Table 7.1), but it should be noted that there are a variety of known pathogenesis-related inhibition mechanisms, such as antibiotics and reactive oxygen or reactive nitrogen species (Wright 2003; Dzidic et al. 2008; De Pascale and Wright 2010). Omics analysis, such as transcriptome analysis, has often played a key role in identification of these mechanisms of inhibition, as reviewed elsewhere (Jarboe et al. 2011b).

Given the necessity of amino acids and nucleotides for production of proteins, transcripts, and genome replication, disturbance of their production can lead to biocatalyst inhibition. This can be especially problematic when the desired growth condition is defined minimal media, relative to metabolite-containing rich media. Depletion of essential metabolic building blocks can trigger the global stringent response (Jain et al. 2006; Durfee et al. 2008). For example, it was shown that the biomass-derived inhibitor furfural indirectly inhibits cysteine production in *E. coli* by depleting the NADPH needed for conversion of sulfate to hydrogen sulfide (Miller et al. 2009a). The accumulation of carboxylate ions in *E. coli*, such as during challenge with carboxylic acids, can increase the intracellular ionic strength to levels sufficient to inhibit homocysteine transmethylase (MetE), an enzyme required for methionine biosynthesis (Whitfield et al. 1970; Roe et al. 2002). The feedback loop that regulates valine production in *E. coli* results in valine-mediated inhibition of production of leucine and isoleucine. Thus, valine can be inhibitory to some *E. coli* strains when leucine and isoleucine are not exogenously supplied (Manten and Rowley 1953). 3-Hydroxypropionic acid (3-HP) can limit amino acid synthesis via inhibition of the chorismate and threonine superpathway (Warnecke et al. 2010). This limitation of biosynthesis pathways is distinct from situations where the product compound inhibits its own production; enzyme improvement for alleviating this problem has been reviewed elsewhere (Jarboe et al. 2012).

Another commonly noted mechanism of inhibition is membrane damage. This is especially problematic when dealing with hydrophobic compounds, such as carboxylic acids and butanol (Lennen et al. 2011; Jarboe et al. 2013; Liu et al. 2013; Royce et al. 2013). Overton's Rule provides a general rule of thumb for predicting the membrane permeability of a compound, whereby compounds with increased lipid solubility (i.e. hydrophobicity) have greater membrane permeability (Al-Awqati 1999). Membrane damage can manifest as a failure to maintain appropriate fluidity and/or leakage of valuable metabolites. Leakage is often monitored using Mg^{2+} as a reporter molecule (Zaldivar and Ingram 1999; Liu et al. 2013; Royce et al. 2013) or permeability to nucleic acid stains (Lennen and Pfleger 2013). Fluidity can be measured via the diffusivity of a fluorescent reporter molecule (Beney et al. 2004; Mykytczuk et al. 2007). Recent studies have visualized membrane pores, and eventual disintegration, caused by an antimicrobial peptide (Rakowska et al. 2013), though such studies have not yet been performed with the type of inhibitors of interest here. Membrane damage can result in improper functioning of membrane-associated reactions, such as the electron transport chain, sometimes resulting in increased production of reactive oxygen species (Teixeira et al. 2004; Cipak et al. 2008; Ruenwai et al. 2011).

Exogenous challenge with carboxylic acids, especially acetate, has been shown to disrupt intracellular pH (Viegas et al. 1998; Ricke 2003; Royce et al. 2014; Trcek et al. 2015). This drop in intracellular pH is due to dissociation of the carboxylic acid in the cell interior. The acidification can inhibit cellular processes and impose a hefty ATP burden as ATPase is used to remove the excess protons (Viegas et al. 1998).

Thus, biocatalyst inhibition can occur at a variety of levels and scales. Inhibition of a single biosynthesis-related enzyme can completely inhibit biosynthesis of the machinery needed for growth and metabolism. An increase in the harshness of the intracellular environment, such as increased osmotic stress or a drop in pH can stress multiple enzymes. Depletion of generic cofactors, such as NAD(P)H and ATP, can have a general slowing effect on metabolism. Membrane damage can impact membrane-associated processes and result in leakage of valuable metabolites. A variety of metabolic engineering strategies have been demonstrated for dealing with these various types of inhibition.

7.3 Mechanisms of Tolerance

Strategies for increasing biocatalyst robustness of organic inhibitors have arisen both as a deliberate response to known mechanisms of inhibition and from analysis of evolved strains. Here we highlight a few of these known mechanisms (Table 7.2), while evolutionary strategies to increase tolerance are described later in this chapter.

A design scheme that has arisen somewhat independently of known inhibition mechanisms or analysis of evolved strains is the provision of appropriate transporter proteins. This strategy asserts that product toxicity can be alleviated, at least in part, by providing cells with the opportunity to expel the problematic compound. This strategy has proven effective in improving both tolerance and production of limonene (Dunlop et al. 2011) and valine (Park et al. 2007) by *E. coli*, as well as alkane tolerance in *S. cerevisiae* (Ling et al. 2013), geraniol tolerance in *E. coli* (Shah et al. 2013) and decanoate tolerance in *E. coli* (Lennen et al. 2013).

Problems with biosynthesis due to inhibition of enzyme activity can be addressed by replacing the sensitive enzyme with an isozyme or mutant enzyme that is resistant to the inhibitory effect. This approach has been successfully demonstrated with both valine and 3-HP (Warnecke et al. 2010; Park et al. 2011). Depletion of generic cofactors, such as furfural-mediated depletion of NADPH, can be addressed by either deleting the enzyme responsible for this depletion (Miller et al. 2009b) or implementing metabolic changes to increase cofactor availability (Wang et al. 2011, 2013; Zheng et al. 2013). For instance, provision of an NADH-dependent furfural reductase enabled *E. coli* to reduce furfural to the less toxic furfuryl alcohol, while not depleting the NADPH needed for biosynthesis (Wang et al. 2011; Zheng et al. 2013).

Acetate and other organic acids are known to decrease the intracellular pH of microorganisms, as previously discussed. The balance between extracellular and intracellular pH greatly affects the proton motive force (PMF), which is important for ATP production and transport processes. *E. coli* can adapt to changes in pH by accommodating the electrochemical potential (the balance of chemical charges) in order to maintain the appropriate PMF (Royce et al. 2014). Other adaptations of organic acid tolerance include addition of cyclopropane fatty acids in the cell membrane, as discussed below.

Table 7.2 Common methods for improving tolerance.

Method	Example	Source
Provision of exporters for the inhibitory compound	Production of limonene	Dunlop et al. (2011)
	Production of valine	Park et al. (2007)
	Alkane tolerance	Ling et al. (2013)
Membrane engineering	Alteration of the distribution of saturated and unsaturated fatty acids improved ethanol tolerance	Luo et al. (2009)
	Alteration of the distribution of saturated and unsaturated fatty acids improved carboxylic acid tolerance	Lennen and Pfleger (2013)
Directed evolution	Improved isobutanol tolerance	Atsumi et al. (2010), Minty et al. (2011)
	Improved tolerance of biomass hydrolysate	Huang et al. (2009), Mills et al. (2009), Geddes et al. (2011)
Enrichment of expression libraries	multi-SCalar Analysis of Library Enrichments (SCALEs)	Gall et al. (2008), Warnecke et al. (2010, 2012), Sandoval et al. (2011), Woodruff et al. (2013)
	Global transcription machinery engineering (gTME)	Alper and Stephanopoulos (2007, Pan et al. (2009), Zhang et al. (2010), Chen et al. (2011), Liu et al. (2011), Ma et al. (2011), Yang et al. (2011), Lanza and Alper (2012), Ma and Yu (2012), Wang et al. (2012), Chong et al. (2013)
Combinatorial expression of protective genes	Ethanol tolerance	Nicolaou et al. (2012)

As mentioned above, membrane damage is a frequently cited mechanism of inhibition and membrane-associated genes have been identified in the analysis of many strains evolved for various tolerance phenotypes (Sandoval et al. 2011; Woodruff et al. 2013). For this reason, we have designated a distinct section for describing membrane engineering efforts.

7.4 Membrane Engineering

Modulation of the membrane composition enables microbes to respond to environmental challenges (Zhang and Rock 2008). The lipid-rich cell membrane is often targeted at the molecular level by hydrophobic compounds. As described above, this can result in decreased ability of the membrane to retain valuable metabolites and decreased function of membrane-associated metabolic processes. For this reason, many of the most interesting recent breakthroughs in metabolic engineering for inhibitor tolerance are related to membrane engineering. We describe these results here, after a brief summary of the physiological role of the different bacterial membrane components.

Bacterial membranes are a mixture of many different lipid entities (Table 7.3). Each lipid species plays a distinct physiological role. The main lipid component of the *E. coli* membrane is the straight-chain, unsaturated palmitic acid (C16:0).

Table 7.3 Physiological significance of selected types of bacterial membrane lipids.

C14:0 *not shown: C12:0, C18:0*	*E. coli* membranes contain minor amounts of these lipids. *E. coli* incorporates C14:0 into lipid A (Raetz et al. 2007). Fluidity is dependent on chain length and degree of saturation; short lipids increase fluidity and long lipids decrease fluidity
3-hydroxy-C14:0	C12:0 or C14:0 fatty acids bind to the hydroxyl group of 3-hydroxy-C14:0 in lipid A of *E. coli*
C16:0 (palmitic acid)	*E. coli* bacteria membranes are predominantly C16:0
Cyclopropane C17cyc *not shown: C19cyc*	Gram-negative bacteria incorporate more cyclopropane fatty acids during stationary phase as a stress response (Chang and Cronan 1999). C17cyc helps mediate acid stress
Monounsaturated *cis*-C16:1 *not shown: C18:1*	An increase in unsaturated lipids increases membrane fluidity

(Continued)

Table 7.3 (Continued)

trans C18:1	Trans fatty acids occur naturally in *Pseudomonas putida*, which is solvent-tolerant. The trans species acts like saturated fatty acids by decreasing membrane fluidity (Zhang and Rock 2008)
Polyunsaturated C18:3 *not shown: C18:2*	Generally present in gram-positive bacteria (e.g. *Lactobacillus plantarum*)
Branched iso-C17:0 Branched anteiso-C17:0	Membrane fluidity depends on iso or anteiso forms. These lipids are usually observed in gram-positive bacteria. Combined substituents are also possible- iso-C17:1 and anteiso-C17:1
Docosahexaenoic acid (DHA) 22:6(*n*−3) *not shown: EPA 20:5(n−3)*	DHA and other long chain polyunsaturated fatty acids are unique to marine bacteria. PUFA increase membrane fluidity in cold, high pressure environments (Urakawa et al. 1998)

Typical metrics for describing membrane composition include average chain length and the unsaturated/saturated ratio. A variety of studies have reported changes in both of these metrics in response to challenge with inhibitory compounds (Lennen et al. 2011; Wu et al. 2012; Liu et al. 2013; Royce et al. 2013). Palmitic acid can by modified by either changing the length of the carbon backbone or adding a substituent such as a double bond or cyclopropane group (Table 7.2). An important note is that cyclopropane lipids are formed as an intramembrane modification of a *cis*-unsaturated fatty acid and not from *de novo* fatty acid biosynthesis (Zhang and Rock 2008).

The available modifications to standard membrane lipids and relative distribution are dependent on the bacterial species and the environment. For example, branched lipids are a major component in the membrane of gram-positive bacteria (Cronan and Thomas 2009). Branched fatty acids can increase or decrease membrane fluidity, dependent upon the ante- or iso-conformation (Zhang and Rock 2008). The difference comes from the packing efficiency of the fatty acid within the lipid bilayer. The methyl group in the anteiso form is further away from the end, which increases the bulk. Another intriguing variable is the presence of cyclopropane fatty acids. These occur when the double bond in unsaturated lipids is methylated, such as by the *E. coli* Cfa enzyme (Grogan and Cronan 1997). Teichoic acid is a major component in the cell wall of gram-positive bacteria and provides structural rigidity to the cell wall by cross-linking peptidoglycan layers (Moat et al. 2002). Sterols increase the membrane rigidity, decreasing passive transport of lactic acid and its resulting accumulation (Vanderrest et al. 1995). Peptidoglycan is a three-dimensional mesh outside the plasma membrane and can serve as a structural barrier against osmotic pressure and toxic compounds (Heidrich et al. 2002; Typas et al. 2012). The outer layer of polymers on the surface of a bacteria cell, known as lipopolysaccharides (LPS), can also be a protection against inhibitors. The hydrophobicity and surface charge of the cell is attributed to LPS and may affect the interaction of the cell with inhibitors (Aono and Kobayashi 1997; Lee et al. 2009).

Each lipid component has an overall effect on the fluidity of the membrane through hydrophobic interactions between the lipids and proteins in the membrane (Zhang and Rock 2008). Membrane fluidity is an important part of cellular physiology and affects the function of the membrane, including respiration, passive and active transport, and protein function. Typically, as the length of the fatty acid increases the membrane fluidity decreases due to a higher packing efficiency (Mykytczuk et al. 2007). Fatty acid length has also an effect on the membrane thickness (Lewis and Engelman 1983; In't Veld et al. 1991) and curvature (Xu et al. 2008). The membrane thickness is important for proper membrane function such as protein conformation and nutrient transport (In't Veld et al. 1991; Engelman 2005). Membrane fluidity is temperature-dependent; at low temperatures, bacterial membranes are generally less fluid (Zhang and Rock 2008). *Moritella marina*, a bacterial species found in deep, cold ocean waters (Kautharapu and Jarboe 2012) copes with this by producing polyunsaturated fatty acids that then increase the fluidity of the membrane.

Membrane fluidity needs to be in an optimal range for proper membrane function. Solvents, such as alcohols and aromatics, have a fluidizing effect on the membrane (Bernal et al. 2007; Huffer et al. 2011), and thus a change in the membrane fluidity is needed to counterbalance fluidizing agents. Bacteria respond to solvent stress in a number of ways, most notably a change in the membrane lipid composition. The ethanol producer *Zymomonas mobilis* contains relatively more octadecenoic fatty acid (C18:1) than *E. coli* (Liu and Qureshi 2009). It has been shown that *E. coli* exposure to ethanol increases the relative amount of C18:1 compared to C16:0. This change is accompanied by an altered growth rate and membrane fluidity (Ingram 1982; Liu and Qureshi 2009). This difference in membrane composition could explain why *Z. mobilis* has increased ethanol tolerance. *Bacillus cereus* predominantly uses branched lipids in its membrane; the relative abundance of saturated and iso-branched lipids increases during anaerobiosis, presumably in response to ethanol production (de Sarrau et al. 2012). Biosynthesis of teichoic acid and capsular polysaccharides by *Lactobacillus plantarum* is stimulated by the presence of ethanol (van Bokhorst-van de Veen et al. 2011), leading to the suggestion that an increase in the cell wall thickness can act as a barrier against solvents.

Cyclopropane fatty acids have been demonstrated as useful (Chang and Cronan 1999; Shabala and Ross 2008), but not essential (To et al. 2011) for microbial acid tolerance. Motivated by an observed increase in both saturated and cyclopropane fatty acids in *Clostridium acetobutylicum* during butanol challenge (Vollherbstschneck et al. 1984), Zhao et al. overexpressed the native *cfa* gene in *C. acetobutylicum*. This increased Cfa expression did enable increased butanol tolerance, though it was unfortunately accompanied by decreased butanol production (Zhao et al. 2003). It may be that a change in the membrane properties of *C. acetobutylicum* upon overexpression of *cfa* decreased solventogenesis. Liu et al. enabled cyclopropane fatty acid production in *S. cerevisiae*, an organism that does not inherently produce these compounds, in order to mitigate membrane leakage during octanoic acid challenge (Liu et al. 2013). While the cyclopropane fatty acids were successfully produced, there was no change in octanoic acid tolerance.

Luo et al. increased expression of the native *fabA* desaturase in *E. coli* with the goal of improving ethanol tolerance (Luo et al. 2009). This *fabA* overexpression increased the saturated lipid content and improved ethanol tolerance. The physiological effect of *fabA* overexpression is not clear at this time, but an increase in saturated fatty acids inherently decreases the fluidity of the membrane, possibly either affecting membrane functions or the transport of ethanol into the cell.

Similarly, Lennen and Pfleger noted a decrease in saturated fatty acid content in the cell membrane during production of carboxylic acids and proposed that this change in membrane composition was actually a mechanism of carboxylic acid toxicity. They successfully engineered their strain to restore closer to normal saturated fatty acid content, though levels were still lower than that observed for the nonproducing control strain (Lennen and Pfleger 2013). The viability of the modified carboxylic-acid producing strains was significantly increased in the strain engineered for increased saturated fatty acid content (Lennen and Pfleger 2013). Other membrane engineering strategies inspired by microbial evolution experiments are discussed below.

7.5 Evolutionary and Metagenomic Strategies for Increasing Tolerance

Evolution is essentially a selection of random mutations that confer increased fitness (Elena and Lenski 2003; Demain 2006; Beaume et al. 2013; Kussell 2013). These mutations can take a variety of forms and influence protein function via changes in amino acid sequence or increased, decreased, or even total lack of gene expression. This evolutionary process has been leveraged in the development of strains with useful metabolic behaviors (Yomano et al. 2008; Zhang et al. 2009; Cobb et al. 2012; Reyes et al. 2013; Jin et al. 2016), but here we are more interested in its use in the development of strains with increased robustness to organic inhibitors. While these robust strains are inherently useful, identifying and understanding the mutations that confer increased robustness can enable development of strain engineering strategies. The idea that we can distill metabolic engineering strategies from the evolved strains follows the proverb commonly known as Orgel's Second Rule: "evolution is cleverer than you are." The results of reverse engineering studies are included in a distinct section below. Table 7.4 lists selected reports of evolutionary studies for ethanol, acetate, and butanol.

The most basic evolutionary studies take advantage of the natural "background" mutation rate. For wild-type *E. coli*, the most recent estimate for the rate of point mutations is 8.9×10^{-11} events per base pair per generation (Wielgoss et al. 2011). The mutation rate can be increased by the use of mutagens, such as ethyl methane sulfonate (EMS) (Kim et al. 2007), or deletion of proofreading enzymes, such as the methyl-directed mismatch repair system (Shaver et al. 2002). Such studies have successfully evolved tolerance to inhibitory compounds, such as furfural (Lin et al. 2009; Miller et al. 2009b), isobutanol (Atsumi et al. 2010; Minty et al. 2011), 2-butanol (Ghiaci et al. 2013), acetic acid (Steiner and Sauer 2003), biomass pyrolysate (Liang et al. 2013), biomass hydrolysate (Huang et al. 2009; Mills et al. 2009; Geddes et al. 2011), and others. In this traditional evolutionary scheme, cells are typically subdiluted in batch cultures on a regular basis (i.e. every 24 or 48 hours) or upon attainment of a certain cell density. Inhibitor concentration is typically increased in a stepwise fashion as cells gain tolerance. An interesting variation of this traditional approach is the Visualizing Evolution in Real Time (VERT) method, in which fluorescence-based cell sorting is used to track distinct gains in fitness (Reyes et al. 2012a; Winkler and Kao 2012; Winkler et al. 2013); the utility of this

Table 7.4 Evolution for inhibitor tolerance. It is important to note that many mutations are synergistic in development of the final tolerance phenotype.

Organism	Evolutionary method	Result	Reference
Inhibitor: Ethanol			
E. coli	SCALEs	Enriched populations in 15 and 30 g l^{-1} ethanol in minimal media	Woodruff et al. (2013)
	Transcription Machinery Engineering of Sigma 70	Final strain showed growth (6-h doubling time) in the presence of 60 g l^{-1} ethanol in rich media; control strain had no growth	Alper and Stephanopoulos (2007)
	Transcription Machinery Engineering of CRP	Final strain had a growth rate of 0.08 h^{-1} in the presence of 62 g l^{-1} ethanol, relative to 0.06 h^{-1} of original strain	Chong et al. (2013)
	Transcription Machinery Engineering of IrrE	Best strain showed a 10-fold increase in the number of cells surviving 1-h shock with 12.5% ethanol	Chen et al. (2011)
S. cerevisiae	Transcription Machinery Engineering of Spt15	Significantly improved viability during 30 h of culturing with 20% ethanol	Alper and Stephanopoulos (2007)
	Transcription Machinery Engineering of Spt15	Evolved strains grew in the presence of 15% ethanol; control strain did not tolerance concentrations above 10%	Yang et al. (2011)
Inhibitor: Acetate			
E. coli	SCALEs	Enriched population in 1.75 g l^{-1} acetate in minimal media at neutral pH	Sandoval et al. (2011)
	Transcription Machinery Engineering of IrrE	Best strain showed increased growth in the presence of 0.05% acetate in rich medium	Chen et al. (2011)
Inhibitor: Butanol			
E. coli	Sequential transfers	Increased growth and viability in the presence of 6 and 8 g l^{-1} isobutanol in rich medium	Atsumi et al. (2010)
	Sequential transfers	Increased growth in the presence of 1% (w/v) isobutanol	Minty et al. (2011)
	VERT	At least 10-fold increased survival to shock with 2 vol% *n*-butanol for 1 h in minimal medium	Reyes et al. (2012b)
	Transcription Machinery Engineering of IrrE	Best strain showed a 100-fold increase in the number of cells surviving 1-h shock with 2.1% butanol	Chen et al. (2011)

method was demonstrated during evolution of *n*-butanol tolerance in *E. coli* (Reyes et al. 2012b).

Other evolutionary-type studies rely on growth-based selection for improved tolerance, but do not only rely on the natural background mutation rate. Historically, transposon mutagenesis to randomly inactivate genes has been an effective method for identifying genes involved in a particular phenotype (Suzuki et al. 2008). The multi-SCalar Analysis of Library Enrichments (SCALEs) method uses a plasmid library containing *E. coli* genomic fragments of various lengths to comprehensively increase expression of individual genes and gene clusters (Gill et al. 2002; Lynch et al. 2007; Bonomo et al. 2008; Warner et al. 2009). DNA fragments and genes conferring increased tolerance to the focal compound are identified and interpreted. This approach has been used with ethanol (Woodruff et al. 2013), acetate (Sandoval et al. 2011), 3-HP (Warnecke et al. 2010, 2012), and 1-naphthol (Gall et al. 2008).

Another opportunity for biological discovery is the use of metagenomic screens, in which a plasmid library is generated from a mixture of various genomes and screened for the desired activity in an inhibitory condition, are most often used to find enzymes that are tolerant to specific stresses (Lu et al. 2013; Shi et al. 2013). However, a search of a DNA library generated from the human gut microbiota not only identified multiple salt-tolerant *Collinsella* genes, they also found that expression of these genes in *E. coli* resulted in increased salt tolerance (Culligan et al. 2012). These metagenomic screens provide the opportunity to discover and utilize tolerance-conferring genes from organisms or pathways that have not yet been characterized.

Expression libraries typically only enable expression of one gene or multiple genes that are co-located within the genome. It has been shown that a stepwise, combinatorial approach to expression libraries can further improve tolerance (Nicolaou et al. 2011). This approach has been demonstrated, for example, in regards to oxidative stress (Nicolaou et al. 2013), acid stress (Gaida et al. 2013), and ethanol (Nicolaou et al. 2012).

Instead of relying on altered expression of a few plasmid-associated genes, global transcription machinery engineering (gTME) uses variants of genetic regulators to perturb the expression of multiple genes (Alper and Stephanopoulos 2007; Lanza and Alper 2012). For example, ethanol tolerance in *E. coli* has been addressed by enriching library of cAMP receptor protein mutants (Chong et al. 2013), *Rhodococcus ruber* tolerance of acrylamide and acrylonirtrile through mutation of sigma 70 (Ma and Yu 2012), ethanol and biomass hydrolysate tolerance in *S. cerevisiae* through mutation of the TATA-binding protein (Liu et al. 2011; Yang et al. 2011). gTME can also be applied to a foreign regulator. Specifically, a mutation library of IrrE global regulator from radiation-resistant *Deinococcus radiodurans* has been introduced into *E. coli* and screened for resistance to a variety of inhibitors, such as ethanol, butanol, acetate, osmotic stress, biomass hydrolysate, and salt stress (Pan et al. 2009; Zhang et al. 2010; Chen et al. 2011; Ma et al. 2011; Wang et al. 2012).

7.6 Reverse Engineering of Improved Strains

One of the major benefits of reverse engineering-evolved strains is the opportunity for biological discovery. For example, reverse engineering of furfural-tolerant *E. coli* identified YqhD as the major *E. coli* furfural reductase (Miller et al. 2009b). However, not all reverse engineering results are so clear-cut. Analysis of the results of SCALEs for 3-HP tolerance identified the 21-amino acid peptide IroK as important to 3-HP tolerance, independent of 3-HP transport, but the biological function of this peptide remains unknown at this time (Warnecke et al. 2012).

Several analyses of *E. coli* strains evolved for isobutanol or *n*-butanol tolerance have noted possible changes in LPS content or composition (Atsumi et al. 2010; Minty et al. 2011; Reyes et al. 2013). One evolved isobutanol-tolerant mutant had an insertion mutation that essentially inactivated the gene *yhbJ*, a mutation that can lead to increased production of glucosamine-6-phosphate, a major component of peptidoglycan and LPS synthesis (Atsumi et al. 2010). Another reverse engineering of isobutanol-tolerant *E. coli* found mutations in *fepE* and *yjgQ*, which contribute to LPS synthesis as well as *eptB*, which adds a phophoethanolamine group (a type of phospholipid) to LPS (Minty et al. 2011). Minty et al. found that a mutated form, designated *hfq**, can increase isobutanol tolerance (Minty et al. 2011). Hfq is an RNA-binding protein that affects RNA processing and regulation.

Each of these different evolutionary methods and subsequent reverse engineering efforts provide the opportunity to gain insight into mechanisms of inhibition and methods for increasing tolerance. These methods usually highlight multiple genes and traits that lead to a combinatorial form of tolerance. Note that it has been challenging to distill clear design strategies from some of the more advanced evolutionary schemes due to the large number of genes with perturbed expression.

It is important to note that many mutations found in evolution studies are synergistic in development of the final tolerance phenotype; rarely does one mutation dominate. It was suggested that global regulators at the transcriptional and post-transcriptional level may be key components in metabolic engineering for tolerance due to their vast reach over the genome. Interestingly, stress responses may either overlap (van Bokhorst-van de Veen et al. 2011) or be antagonistic (Reyes et al. 2013). This result is consistent with the idea that genes of overlapping function have a negative epistasis (He et al. 2010). Epistasis can be quantified by a multiplicative fitness model (Minty et al. 2011) that can analyze how particular mutations are affected by subsequent mutations.

7.7 Concluding Remarks

Production of biorenewable fuels and chemicals from biomass is desirable as a means of potentially increasing energy security, stabilizing fuels costs and addressing climate change and increasing carbon dioxide levels. Microbes are an appealing biocatalyst for this production and excellent strategies and techniques for rationally and predictably engineering biocatalyst metabolism have been developed. However, biocatalyst inhibition by either the product compound or inhibitors in the cheap, "dirty," biomass-derived sugars remain problematic in the attainment of economically viable yields, titers, and productivities. Multiple approaches can be used to increase titers and productivities: removal of the inhibitory organic compounds, redistribution of metabolic flux to limit production of inhibitory side products, and improving microbial robustness. Here we have summarized representative mechanisms of inhibition and evolutionary and engineering strategies for increasing tolerance. Evolutionary strategies have proven quite adept at increasing tolerance (Table 7.1), but additional efforts on reverse engineering these improved strains can provide guidance for new design strategies.

Acknowledgments

This work was supported by the NSF Engineering Research Center for Biorenewable Chemicals (CBiRC), NSF award number EEC-0813570.

References

Agrawal, M., Wang, Y., and Chen, R.R. (2012). Engineering efficient xylose metabolism into an acetic acid-tolerant *Zymomonas mobilis* strain by introducing adaptation-induced mutations. *Biotechnol. Lett.* 34 (10): 1825–1832.

Al-Awqati, Q. (1999). One hundred years of membrane permeability: does Overton still rule? *Nat. Cell Biol.* 1 (8): E201–E202.

Alper, H. and Stephanopoulos, G. (2007). Global transcription machinery engineering: a new approach for improving cellular phenotype. *Metab. Eng.* 9 (3): 258–267.

Aono, R. and Kobayashi, H. (1997). Cell surface properties of organic solvent-tolerant mutants of *Escherichia coli* K-12. *Appl. Environ. Microbiol.* 63 (9): 3637–3642.

Atsumi, S., Hanai, T., and Liao, J.C. (2008). Non-fermentative pathways for synthesis of branched-chain higher alcohols as biofuels. *Nature* 451 (7174): 86–89.

Atsumi, S., Wu, T.-Y., Machado, I.M.P. et al. (2010). Evolution, genomic analysis, and reconstruction of isobutanol tolerance in *Escherichia coli*. *Mol. Syst. Biol.* 6: doi: 10.1038/msb.2010.98.

Bailey, J.E. (1991). Toward a science of metabolic engineering. *Science* 252 (5013): 1668–1675.

Beaume, M., Monina, N., Schrenzel, J., and Francois, P. (2013). Bacterial genome evolution within a clonal population: from in vitro investigations to in vivo observations. *Future Microbiol* 8 (5): 661–674.

Beney, L., Mille, Y., and Gervais, P. (2004). Death of *Escherichia coli* during rapid and severe dehydration is related to lipid phase transition. *Appl. Microbiol. Biotechnol.* 65 (4): 457–464.

Bernal, P., Segura, A., and Ramos, J.-L. (2007). Compensatory role of the *cis–trans*-isomerase and cardiolipin synthase in the membrane fluidity of *Pseudomonas putida* DOT-T1E. *Environ. Microbiol.* 9 (7): 1658–1664.

van Bokhorst-van de Veen, H., Abee, T., Tempelaars, M. et al. (2011). Short- and long-term adaptation to ethanol stress and its cross-protective consequences in *Lactobacillus plantarum*. *Appl. Environ. Microbiol.* 77 (15): 5247–5256.

Bonomo, J., Lynch, M.D., Warnecke, T. et al. (2008). Genome-scale analysis of anti-metabolite directed strain engineering. *Metab. Eng.* 10 (2): 109–120.

Cameron, D. and Tong, I.T. (1993). Cellular and metabolic engineering. *Appl. Biochem. Biotechnol.* 38 (1): 105–140.

Chang, Y.Y. and Cronan, J.E. (1999). Membrane cyclopropane fatty acid content is a major factor in acid resistance of *Escherichia coli*. *Mol. Microbiol.* 33 (2): 249–259.

Chen, T., Wang, J., Yang, R. et al. (2011). Laboratory-evolved mutants of an exogenous global regulator, IrrE from *Deinococcus radiodurans*, enhance stress tolerances of *Escherichia coli*. *PLoS ONE* 6 (1): doi: 10.1371/journal.pone.0016228.

Chi, Z., Rover, M., Jun, E. et al. (2013). Overliming detoxification of pyrolytic sugar syrup for direct fermentation of levoglucosan to ethanol. *Bioresour. Technol.* 150: 220–227.

Chong, H., Huang, L., Yeow, J. et al. (2013). Improving ethanol tolerance of *Escherichia coli* by rewiring its global regulator cAMP receptor protein (CRP). *PLoS ONE* 8 (2): doi: 10.1371/journal.pone.0057628.

Chotani, G., Dodge, T., Hsu, A. et al. (2000). The commercial production of chemicals using pathway engineering. *Biochim. Biophys. Acta, Protein Struct. Mol. Enzymol.* 1543 (2): 434–455.

Cipak, A., Jaganjac, M., Tehlivets, O. et al. (2008). Adaptation to oxidative stress induced by polyunsaturated fatty acids in yeast. *Biochim. Biophys. Acta, Mol. Cell. Biol. Lipids* 1781 (6–7): 283–287.

Cobb, R.E., Si, T., and Zhao, H. (2012). Directed evolution: an evolving and enabling synthetic biology tool. *Curr. Opin. Chem. Biol.* 16 (3–4): 285–291.

Cronan, J.E. and Thomas, J. (2009). Bacterial fatty acid synthesis and its relationships with polyketide synthetic pathways. In: *Complex Enzymes in Microbial Natural Product Biosynthesis, Part B: Polyketides, Aminocoumarins and Carbohydrates*, vol. 459 (ed. D.A. Hopwood), 395–433. Elsevier.

Culligan, E.P., Sleator, R.D., Marchesi, J.R., and Hill, C. (2012). Functional metagenomics reveals novel salt tolerance loci from the human gut microbiome. *ISME J.* 6 (10): 1916–1925.

De Pascale, G. and Wright, G.D. (2010). Antibiotic resistance by enzyme inactivation: from mechanisms to solutions. *ChemBioChem* 11 (10): 1325–1334.

Demain, A.L. (2006). From natural products discovery to commercialization: a success story. *J. Ind. Microbiol. Biotechnol.* 33 (7): 486–495.

Dunlop, M.J. (2011). Engineering microbes for tolerance to next-generation biofuels. *Biotechnol. Biofuels* 4.

Dunlop, M.J., Dossani, Z.Y., Szmidt, H.L. et al. (2011). Engineering microbial biofuel tolerance and export using efflux pumps. *Mol. Syst. Biol.* 7: 487.

Durfee, T., Hansen, A.-M., Zhi, H. et al. (2008). Transcription profiling of the stringent response in *Escherichia coli*. *J. Bacteriol.* 190 (3): 1084–1096.

Dzidic, S., Suskovic, J., and Kos, B. (2008). Antibiotic resistance mechanisms in bacteria: biochemical and genetic aspects. *Food Technol. Biotechnol.* 46 (1): 11–21.

Elena, S.F. and Lenski, R.E. (2003). Evolution experiments with microorganisms: the dynamics and genetic bases of adaptation. *Nat. Rev. Genet.* 4 (6): 457–469.

Engelman, D.M. (2005). Membranes are more mosaic than fluid. *Nature* 438 (7068): 578–580.

Fischer, C.R., Klein-Marcuschamer, D., and Stephanopoulos, G. (2008). Selection and optimization of microbial hosts for biofuels production. *Metab. Eng.* 10 (6): 295–304.

Gaida, S.M., Al-Hinai, M.A., Indurthi, D.C. et al. (2013). Synthetic tolerance: three noncoding small RNAs, DsrA, ArcZ and RprA, acting supra-additively against acid stress. *Nucleic Acids Res.* 41 (18): 8726–8737.

Gall, S., Lynch, M.D., Sandoval, N.R., and Gill, R.T. (2008). Parallel mapping of genotypes to phenotypes contributing to overall biological fitness. *Metab. Eng.* 10 (6): 382–393.

Geddes, C.C., Mullinnix, M.T., Nieves, I.U. et al. (2011). Simplified process for ethanol production from sugarcane bagasse using hydrolysate-resistant *Escherichia coli* strain MM160. *Bioresour. Technol.* 102 (3): 2702–2711.

Ghiaci, P., Norbeck, J., and Larsson, C. (2013). Physiological adaptations of *Saccharomyces cerevisiae* evolved for improved butanol tolerance. *Biotechnol. Biofuels* 6 (1): 101.

Gill, R.T., Wildt, S., Yang, Y.T. et al. (2002). Genome-wide screening for trait conferring genes using DNA microarrays. *Proc. Natl. Acad. Sci. U.S.A.* 99 (10): 7033–7038.

Grogan, D.W. and Cronan, J.E. (1997). Cyclopropane ring formation in membrane lipids of bacteria. *Microbiol. Mol. Biol. Rev.* 61 (4): 429–441.

He, X., Qian, W., Wang, Z. et al. (2010). Prevalent positive epistasis in *Escherichia coli* and *Saccharomyces cerevisiae* metabolic networks. *Nat. Genet.* 42 (3): 272–U120.

Heidrich, C., Ursinus, A., Berger, J. et al. (2002). Effects of multiple deletions of murein hydrolases on viability, septum cleavage, and sensitivity to large toxic molecules in *Escherichia coli*. *J. Bacteriol.* 184 (22): 6093–6099.

Huang, C.-F., Lin, T.-H., Guo, G.-L., and Hwang, W.-S. (2009). Enhanced ethanol production by fermentation of rice straw hydrolysate without detoxification using a newly adapted strain of *Pichia stipitis*. *Bioresour. Technol.* 100 (17): 3914–3920.

Huffer, S., Clark, M.E., Ning, J.C. et al. (2011). Role of alcohols in growth, lipid composition, and membrane fluidity of yeasts, bacteria, and archaea. *Appl. Environ. Microbiol.* 77 (18): 6400–6408.

Ingram, L.O. (1982). Regulation of fatty acid composition in *Escherichia coli* - a proposed common mechanism for changes induced by ethanol, chaotropic agents, and a reduction of growth temperature. *J. Bacteriol.* 149 (1): 166–172.

In't Veld, G., Driessen, A.J., Op den Kamp, J.A., and Konings, W.N. (1991). Hydrophobic membrane thickness and lipid-protein interactions of the leucine transport system of *Lactococcus lactis*. *Biochim. Biophys. Acta* 1065 (2): 203–212.

Jain, V., Kumar, M., and Chatterji, D. (2006). ppGpp: stringent response and survival. *J. Microbiol.* 44 (1): 1–10.

Jarboe, L.R., Grabar, T.B., Yomano, L.P. et al. (2007). Development of ethanologenic bacteria. In: *Biofuels*, Advances in Biochemical Engineering/Biotechnology, vol. 108 (ed. L. Olsson), 237–261. Berlin, Heidelberg: Springer-Verlag.

Jarboe, L.R., Wen, Z., Choi, D., and Brown, R.C. (2011a). Hybrid thermochemical processing: fermentation of pyrolysis-derived bio-oil. *Appl. Microbiol. Biotechnol.* 91 (6): 1519–1523.

Jarboe, L.R., Liu, P., and Royce, L.A. (2011b). Engineering inhibitor tolerance for the production of biorenewable fuels and chemicals. *Curr. Opin. Chem. Eng.* 1: 38–42.

Jarboe, L.R., Liu, P., Kautharapu, K.B., and Ingram, L.O. (2012). Optimization of enzyme parameters for fermentative production of biorenewable fuels and chemicals. *Comput. Struct. Biotechnol. J.* 3: e201210005.

Jarboe, L.R., Royce, L.A., and Liu, P. (2013). Understanding biocatalyst inhibition by carboxylic acids. *Front. Microbiol.* 4: doi: 10.3389/fmicb.2013.00272.

Jeffries, T.W. and Jin, Y.S. (2004). Metabolic engineering for improved fermentation of pentoses by yeasts. *Appl. Microbiol. Biotechnol.* 63 (5): 495–509.

Jin, T., Chen, Y., and Jarboe, L.R. (2016). Evolutionary methods for improving the prodution of biorenewable fuels and chemicals. In: *Biotechnology for Biofuel Production and Optimization* (ed. C. Trinh and C. Eckert), 265–290. Elsevier.

Kautharapu, K.B. and Jarboe, L.R. (2012). Genome sequence of the psychrophilic deep-sea bacterium *Moritella marina* MP-1 (ATCC 15381). *J. Bacteriol.* 194 (22): 6296–6297.

Kim, Y., Ingram, L.O., and Shanmugam, K.T. (2007). Construction of an *Escherichia coli* K-12 mutant for homoethanologenic fermentation of glucose or xylose without foreign genes. *Appl. Environ. Microbiol.* 73 (6): 1766–1771.

Kussell, E. (2013). Evolution in microbes. *Annu. Rev. Biophys.* 42: 493–514.

Lakshmanaswamy, A., Rajaraman, E., Eiteman, M.A., and Altman, E. (2011). Microbial removal of acetate selectively from sugar mixtures. *J. Ind. Microbiol. Biotechnol.* 38 (9): 1477–1484.

Lanza, A.M. and Alper, H.S. (2012). Using transcription machinery engineering to elicit complex cellular phenotypes. *Methods Mol. Biol. (Clifton, N.J.)* 813: 229–248.

Lee, J.-H., Lee, K.-L., Yeo, W.-S. et al. (2009). SoxRS-mediated lipopolysaccharide modification enhances resistance against multiple drugs in *Escherichia coli*. *J. Bacteriol.* 191 (13): 4441–4450.

Lennen, R.M. and Pfleger, B.F. (2013). Modulating membrane composition alters free fatty acid tolerance in *Escherichia coli*. *PLoS ONE* 8 (1): doi: 10.1371/journal.pone.0054031.

Lennen, R.M., Braden, D.J., West, R.M. et al. (2010). A process for microbial hydrocarbon synthesis: overproduction of fatty acids in *Escherichia coli* and catalytic conversion to alkanes. *Biotechnol. Bioeng.* 106 (2): 193–202.

Lennen, R.M., Kruziki, M.A., Kumar, K. et al. (2011). Membrane stresses induced by overproduction of free fatty acids in *Escherichia coli*. *Appl. Environ. Microbiol.* 77 (22): 8114–8128.

Lennen, R.M., Politz, M.G., Kruziki, M.A., and Pfleger, B.F. (2013). Identification of transport proteins involved in free fatty acid efflux in *Escherichia coli*. *J. Bacteriol.* 195 (1): 135–144.

Lewis, B.A. and Engelman, D.M. (1983). Lipid bilayer thickness varies linearly with acyl chain length in fluid phosphatidylcholine vesicles. *J. Mol. Biol.* 166 (2): 211–217.

Lian, J., Chen, S., Zhou, S. et al. (2010). Separation, hydrolysis and fermentation of pyrolytic sugars to produce ethanol and lipids. *Bioresour. Technol.* 101 (24): 9688–9699.

Liang, Y., Zhao, X., Chi, Z. et al. (2013). Utilization of acetic acid-rich pyrolytic bio-oil by microalga *Chlamydomonas reinhardtii*: reducing bio-oil toxicity and enhancing algal toxicity tolerance. *Bioresour. Technol.* 133: 500–506.

Lin, F.-M., Qiao, B., and Yuan, Y.-J. (2009). Comparative proteomic analysis of tolerance and adaptation of ethanologenic *Saccharomyces cerevisiae* to furfural, a lignocellulosic inhibitory compound. *Appl. Environ. Microbiol.* 75 (11): 3765–3776.

Ling, H., Chen, B., Kang, A. et al. (2013). Transcriptome response to alkane biofuels in *Saccharomyces cerevisiae*: identification of efflux pumps involved in alkane tolerance. *Biotechnol. Biofuels* 6 (1): 95.

Liu, S. and Qureshi, N. (2009). How microbes tolerate ethanol and butanol. *New Biotechnol.* 26 (3–4): 117–121.

Liu, H., Liu, K., Yan, M. et al. (2011). gTME for improved adaptation of *Saccharomyces cerevisiae* to Corn Cob acid hydrolysate. *Appl. Biochem. Biotechnol.* 164 (7): 1150–1159.

Liu, P., Chernyshov, A., Najdi, T. et al. (2013). Membrane stress caused by octanoic acid in *Saccharomyces cerevisiae*. *Appl. Microbiol. Biotechnol.* 97 (7): 3239–3251.

Lu, J., Du, L., Wei, Y. et al. (2013). Expression and characterization of a novel highly glucose-tolerant beta-glucosidase from a soil metagenome. *Acta Biochim. Biophys. Sin.* 45 (8): 664–673.

Luo, L.H., Seo, P.-S., Seo, J.-W. et al. (2009). Improved ethanol tolerance in *Escherichia coli* by changing the cellular fatty acids composition through genetic manipulation. *Biotechnol. Lett.* 31 (12): 1867–1871.

Lutke-Eversloh, T. and Stephanopoulos, G. (2007). L-tyrosine production by deregulated strains of *Escherichia coli*. *Appl. Microbiol. Biotechnol.* 75 (1): 103–110.

Lynch, M.D., Warnecke, T., and Gill, R.T. (2007). SCALEs: multiscale analysis of library enrichment. *Nat. Methods* 4 (1): 87–93.

Ma, Y. and Yu, H. (2012). Engineering of *Rhodococcus* cell catalysts for tolerance improvement by sigma factor mutation and active plasmid partition. *J. Ind. Microbiol. Biotechnol.* 39 (10): 1421–1430.

Ma, R., Zhang, Y., Hong, H. et al. (2011). Improved osmotic tolerance and ethanol production of ethanologenic *Escherichia coli* by IrrE, a global regulator of radiation-resistance of *Deinococcus radiodurans*. *Curr. Microbiol.* 62 (2): 659–664.

Manten, A. and Rowley, D. (1953). Genetic analysis of valine inhibition in the K-12 strain of *Bacterium coli*. *J. Gen. Microbiol.* 9 (2): 226–233.

Martinez, A., Rodriguez, M.E., York, S.W. et al. (2000). Effects of Ca(OH)$_2$ treatments ("overliming") on the composition and toxicity of bagasse hemicellulose hydrolysates. *Biotechnol. Bioeng.* 69 (5): 526–536.

Miller, E.N., Jarboe, L.R., Turner, P.C. et al. (2009a). Furfural inhibits growth by limiting sulfur assimilation in ethanologenic *Escherichia coli* strain LY180. *Appl. Environ. Microbiol.* 75 (19): 6132–6141.

Miller, E.N., Jarboe, L.R., Yomano, L.P. et al. (2009b). Silencing of NADPH-dependent oxidoreductase genes (yqhD and dkgA) in furfural-resistant ethanologenic *Escherichia coli*. *Appl. Environ. Microbiol.* 75 (13): 4315–4323.

Mills, T.Y., Sandoval, N.R., and Gill, R.T. (2009). Cellulosic hydrolysate toxicity and tolerance mechanisms in *Escherichia coli*. *Biotechnol. Biofuels* 2 (1): 26.

Minty, J.J., Lesnefsky, A.A., Lin, F. et al. (2011). Evolution combined with genomic study elucidates genetic bases of isobutanol tolerance in *Escherichia coli*. *Microb. Cell Fact.* 10 (1): 18.

Moat, A.G., Foster, J.W., and Spector, M.P. (2002). *Microbial Physiology*, 4e, i–xx, 1-715.

Mykytczuk, N.C.S., Trevors, J.T., Leduc, L.G., and Ferroni, G.D. (2007). Fluorescence polarization in studies of bacterial cytoplasmic membrane fluidity under environmental stress. *Prog. Biophys. Mol. Biol.* 95 (1–3): 60–82.

Nakamura, C.E. and Whited, G.M. (2003). Metabolic engineering for the microbial production of 1,3-propanediol. *Curr. Opin. Biotechnol.* 14 (5): 454–459.

Nicolaou, S.A., Gaida, S.M., and Papoutsakis, E.T. (2010). A comparative view of metabolite and substrate stress and tolerance in microbial bioprocessing: from biofuels and chemicals, to biocatalysis and bioremediation. *Metab. Eng.* 12 (4): 307–331.

Nicolaou, S.A., Gaida, S.M., and Papoutsakis, E.T. (2011). Coexisting/coexpressing genomic libraries (CoGeL) identify interactions among distantly located genetic loci for developing complex microbial phenotypes. *Nucleic Acids Res.* 39 (22): e152.

Nicolaou, S.A., Gaida, S.M., and Papoutsakis, E.T. (2012). Exploring the combinatorial genomic space in *Escherichia coli* for ethanol tolerance. *Biotechnol. J.* 7 (11): 1337–1345.

Nicolaou, S.A., Fast, A.G., Nakamaru-Ogiso, E., and Papoutsakis, E.T. (2013). Overexpression of fetA (ybbL) and fetB (ybbM), encoding an iron exporter, enhances resistance to oxidative stress in *Escherichia coli*. *Appl. Environ. Microbiol.* 79 (23): 7210–7219.

Ohta, K., Beall, D.S., Mejia, J.P. et al. (1991). Genetic improvement of *Escherichia coli* for ethanol production - chromosomal integration of *Zymomonas mobilis* genes encoding pyruvate decarboxylase and alcohol dehydrogenase II. *Appl. Environ. Microbiol.* 57 (4): 893–900.

Pan, J., Wang, J., Zhou, Z. et al. (2009). IrrE, a global regulator of extreme radiation resistance in *Deinococcus radiodurans*, enhances salt tolerance in *Escherichia coli* and *Brassica napus*. *PLoS ONE* 4 (2): doi: 10.1371/journal.pone.0004422.

Park, J.H., Lee, K.H., Kim, T.Y., and Lee, S.Y. (2007). Metabolic engineering of *Escherichia coli* for the production of L-valine based on transcriptome analysis and in silico gene knockout simulation. *Proc. Natl. Acad. Sci. U.S.A.* 104 (19): 7797–7802.

Park, J.H., Jang, Y.-S., Lee, J.W., and Lee, S.Y. (2011). *Escherichia coli* W as a new platform strain for the enhanced production of L-valine by systems metabolic engineering. *Biotechnol. Bioeng.* 108 (5): 1140–1147.

Raetz, C.R.H., Reynolds, C.M., Trent, M.S., and Bishop, R.E. (2007). Lipid A modification systems in gram-negative bacteria. *Annu. Rev. Biochem.* 295–329.

Rakowska, P.D., Jiang, H., Ray, S. et al. (2013). Nanoscale imaging reveals laterally expanding antimicrobial pores in lipid bilayers. *Proc. Natl. Acad. Sci. U.S.A.* 110 (22): 8918–8923.

Reyes, L.H., Winkler, J., and Kao, K.C. (2012a). Visualizing evolution in real-time method for strain engineering. *Front. Microbiol.* 3: doi: 10.3389/fmicb.2012.00198.

Reyes, L.H., Almario, M.P., Winkler, J. et al. (2012b). Visualizing evolution in real time to determine the molecular mechanisms of *n*-butanol tolerance in *Escherichia coli*. *Metab. Eng.* 14 (5): 579–590.

Reyes, L.H., Abdelaal, A.S., and Kao, K.C. (2013). Genetic determinants for *n*-butanol tolerance in evolved *Escherichia coli* mutants: cross adaptation and antagonistic pleiotropy between n-butanol and other stressors. *Appl. Environ. Microbiol.* 79 (17): 5313–5320.

Ricke, S.C. (2003). Perspectives on the use of organic acids and short chain fatty acids as antimicrobials. *Poult. Sci.* 82 (4): 632–639.

Roe, A.J., O'Byrne, C., McLaggan, D., and Booth, I.R. (2002). Inhibition of *Escherichia coli* growth by acetic acid: a problem with methionine biosynthesis and homocysteine toxicity. *Microbiology* 148: 2215–2222.

Royce, L.A., Liu, P., Stebbins, M.J. et al. (2013). The damaging effects of short chain fatty acids on *Escherichia coli* membranes. *Appl. Microbiol. Biotechnol.* 97 (18): 8317–8327.

Royce, L.A., Boggess, E., Fu, Y. et al. (2014). Transcriptomic analysis of carboxylic acid challenge in *Escherichia coli*: beyond membrane damage. *PLoS ONE* 9 (2): e89580.

Ruenwai, R., Neiss, A., Laoteng, K. et al. (2011). Heterologous production of polyunsaturated fatty acids in *Saccharomyces cerevisiae* causes a global transcriptional response resulting in reduced proteasomal activity and increased oxidative stress. *Biotechnol. J.* 6 (3): 343–356.

Sandoval, N.R., Mills, T.Y., Zhang, M., and Gill, R.T. (2011). Elucidating acetate tolerance in *E. coli* using a genome-wide approach. *Metab. Eng.* 13 (2): 214–224.

de Sarrau, B., Clavel, T., Clerte, C. et al. (2012). Influence of anaerobiosis and low temperature on *Bacillus cereus* growth, metabolism, and membrane properties. *Appl. Environ. Microbiol.* 78 (6): 1715–1723.

Shabala, L. and Ross, T. (2008). Cyclopropane fatty acids improve *Escherichia coli* survival in acidified minimal media by reducing membrane permeability to H^+ and enhanced ability to extrude H^+. *Res. Microbiol.* 159 (6): 458–461.

Shah, A.A., Wang, C., Chung, Y.-R. et al. (2013). Enhancement of geraniol resistance of *Escherichia coli* by MarA overexpression. *J. Biosci. Bioeng.* 115 (3): 253–258.

Shaver, A.C., Dombrowski, P.G., Sweeney, J.Y. et al. (2002). Fitness evolution and the rise of mutator alleles in experimental *Escherichia coli* populations. *Genetics* 162 (2): 557–566.

Shi, Y., Pan, Y., Li, B. et al. (2013). Molecular cloning of a novel bioH gene from an environmental metagenome encoding a carboxylesterase with exceptional tolerance to organic solvents. *BMC Biotechnol.* 13 (1): doi: 10.1186/1472-6750-13-13.

Steinbuchel, A. (2001). Perspectives for biotechnological production and utilization of biopolymers: metabolic engineering of polyhydroxyalkanoate biosynthesis pathways as a successful example. *Macromol. Biosci.* 1 (1): 1–24.

Steiner, P. and Sauer, U. (2003). Long-term continuous evolution of acetate resistant *Acetobacter aceti*. *Biotechnol. Bioeng.* 84 (1): 40–44.

Stephanopoulos, G. and Vallino, J.J. (1991). Network rigidity and metabolic engineering in metabolite overproduction. *Science* 252 (5013): 1675–1681.

Suzuki, N., Inui, M., and Yukawa, H. (2008). Random genome deletion methods applicable to prokaryotes. *Appl. Microbiol. Biotechnol.* 79 (4): 519–526.

Teixeira, M.C., Telo, J.P., Duarte, N.F., and Sa-Correia, I. (2004). The herbicide 2,4-dichlorophenoxyacetic acid induces the generation of free-radicals and associated oxidative stress responses in yeast. *Biochem. Biophys. Res. Commun.* 324 (3): 1101–1107.

To, T.M.H., Grandvalet, C., and Tourdot-Marechal, R. (2011). Cyclopropanation of membrane unsaturated fatty acids is not essential to the acid stress response of *Lactococcus lactis* subsp. cremoris. *Appl. Environ. Microbiol.* 77 (10): 3327–3334.

Trcek, J., Mira, N.P., and Jarboe, L.R. (2015). Adaptation and tolerance of bacteria against acetic acid. *Appl. Microbiol. Biotechnol.* 99 (15): 6215–6229.

Typas, A., Banzhaf, M., Gross, C.A., and Vollmer, W. (2012). From the regulation of peptidoglycan synthesis to bacterial growth and morphology. *Nat. Rev. Microbiol.* 10 (2): 123–136.

Urakawa, H., Kita-Tsukamoto, K., Steven, S.E. et al. (1998). A proposal to transfer *Vibrio marinus* (Russell 1891) to a new genus *Moritella* gen. nov. as *Moritella marina* comb. nov. *FEMS Microbiol. Lett.* 165 (2): 373–378.

Vanderrest, M.E., Kamminga, A.H., Nakano, A. et al. (1995). The plasma membrane of *Saccharomyces cerevisiae* – structure, function and biogenesis. *Microbiol. Rev.* 59 (2): 304–322.

Viegas, C.A., Almeida, P.F., Cavaco, M., and Sa-Correia, I. (1998). The H^+-ATPase in the plasma membrane of *Saccharomyces cerevisiae* is activated during growth latency in octanoic acid-supplemented medium accompanying the decrease in intracellular pH and cell viability. *Appl. Environ. Microbiol.* 64 (2): 779–783.

Vollherbstschneck, K., Sands, J.A., and Montenecourt, B.S. (1984). Effect of butanol on lipid composition and fluidity of *Clostridium acetobutylicum* ATCC 824. *Appl. Environ. Microbiol.* 47 (1): 193–194.

Wang, X., Miller, E.N., Yomano, L.P. et al. (2011). Increased furfural tolerance due to overexpression of NADH-dependent oxidoreductase FucO in *Escherichia coli* strains engineered for the production of ethanol and lactate. *Appl. Environ. Microbiol.* 77 (15): 5132–5140.

Wang, J., Zhang, Y., Chen, Y. et al. (2012). Global regulator engineering significantly improved *Escherichia coli* tolerances toward inhibitors of lignocellulosic hydrolysates. *Biotechnol. Bioeng.* 109 (12): 3133–3142.

Wang, X., Yomano, L.P., Lee, J.Y. et al. (2013). Engineering furfural tolerance in *Escherichia coli* improves the fermentation of lignocellulosic sugars into renewable chemicals. *Proc. Natl. Acad. Sci. U.S.A.* 110 (10): 4021–4026.

Warnecke, T. and Gill, R.T. (2005). Organic acid toxicity, tolerance, and production in *Escherichia coli* biorefining applications. *Microb. Cell Fact.* 4: 25.

Warnecke, T.E., Lynch, M.D., Karimpour-Fard, A. et al. (2010). Rapid dissection of a complex phenotype through genomic-scale mapping of fitness altering genes. *Metab. Eng.* 12 (3): 241–250.

Warnecke, T.E., Lynch, M.D., Lipscomb, M.L., and Gill, R.T. (2012). Identification of a 21 amino acid peptide conferring 3-hydroxypropionic acid stress-tolerance to *Escherichia coli*. *Biotechnol. Bioeng.* 109 (5): 1347–1352.

Warner, J.R., Patnaik, R., and Gill, R.T. (2009). Genomics enabled approaches in strain engineering. *Curr. Opin. Microbiol.* 12 (3): 223–230.

Whitfield, C.D., Steers, E.J., and Weissbac, H. (1970). Purification and properties of 5-methyltetrahydropteroyltri-glutamate-homocysteine-transmethylase. *J. Biol. Chem.* 245 (2): 390–401.

Wielgoss, S., Barrick, J.E., Tenaillon, O. et al. (2011). Mutation rate inferred from synonymous substitutions in a long-term evolution experiment with *Escherichia coli*. *G3-Genes Genomes Genet.* 1 (3): 183–186.

Winkler, J. and Kao, K.C. (2012). Computational identification of adaptive mutants using the VERT system. *J. Biol. Eng.* 6 (1): doi: 10.1186/1754-1611-6-3.

Winkler, J., Reyes, L.H., and Kao, K.C. (2013). Adaptive laboratory evolution for strain engineering. *Methods Mol. Biol. (Clifton, N.J.)* 985: 211–222.

Woodruff, L.B.A., Pandhal, J., Ow, S.Y. et al. (2013). Genome-scale identification and characterization of ethanol tolerance genes in *Escherichia coli*. *Metab. Eng.* 15: 124–133.

Wright, G.D. (2003). Mechanisms of resistance to antibiotics. *Curr. Opin. Chem. Biol.* 7 (5): 563–569.

Wu, C., Zhang, J., Wang, M. et al. (2012). *Lactobacillus casei* combats acid stress by maintaining cell membrane functionality. *J. Ind. Microbiol. Biotechnol.* 39 (7): 1031–1039.

Xu, Q., Kim, M., Ho, K.W.D. et al. (2008). Membrane hydrocarbon thickness modulates the dynamics of a membrane transport protein. *Biophys. J.* 95 (6): 2849–2858.

Yan, Y. and Liao, J.C. (2009). Engineering metabolic systems for production of advanced fuels. *J. Ind. Microbiol. Biotechnol.* 36 (4): 471–479.

Yang, J., Bae, J.Y., Lee, Y.M. et al. (2011). Construction of *Saccharomyces cerevisiae* strains with enhanced ethanol tolerance by mutagenesis of the TATA-binding protein gene and identification of novel genes associated with ethanol tolerance. *Biotechnol. Bioeng.* 108 (8): 1776–1787.

Yomano, L.P., York, S.W., Zhou, S. et al. (2008). Re-engineering *Escherichia coli* for ethanol production. *Biotechnol. Lett.* 30 (12): 2097–2103.

Zaldivar, J. and Ingram, L.O. (1999). Effect of organic acids on the growth and fermentation of ethanologenic *Escherichia coli* LY01. *Biotechnol. Bioeng.* 66 (4): 203–210.

Zeng, A.-P. and Biebl, H. (2002). Bulk chemicals from biotechnology: the case of 1,3-propanediol production and the new trends. *Adv. Biochem. Eng./Biotechnol.* 74: 239–259.

Zhang, Y.-M. and Rock, C.O. (2008). Membrane lipid homeostasis in bacteria. *Nat. Rev. Microbiol.* 6 (3): 222–233.

Zhang, M., Eddy, C., Deanda, K. et al. (1995). Metabolic engineering of a pentose metabolism pathway in ethanologenic *Zymomonas mobilis*. *Science* 267 (5195): 240–243.

Zhang, X., Jantama, K., Shanmugam, K.T., and Ingram, L.O. (2009). Reengineering *Escherichia coli* for succinate production in mineral salts medium. *Appl. Environ. Microbiol.* 75 (24): 7807–7813.

Zhang, Y., Ma, R., Zhao, Z. et al. (2010). irrE, an exogenous gene from *Deinococcus radiodurans*, improves the growth of and ethanol production by a *Zymomonas mobilis* strain under ethanol and acid stresses. *J. Microbiol. Biotechnol.* 20 (7): 1156–1162.

Zhao, Y.S., Hindorff, L.A., Chuang, A. et al. (2003). Expression of a cloned cyclopropane fatty acid synthase gene reduces solvent formation in *Clostridium acetobutylicum* ATCC 824. *Appl. Environ. Microbiol.* 69 (5): 2831–2841.

Zheng, H., Wang, X., Yomano, L.P. et al. (2013). Improving *Escherichia coli* FucO for furfural tolerance by saturation mutagenesis of individual amino acid positions. *Appl. Environ. Microbiol.* 79 (10): 3202–3208.

Index

a

acetate 245
adaptive control system 211–213
adaptive law 211–212
agitation/shear stress 156–157
airlift reactor (ALR) 135, 151–153
ammonia 82
ANNs. *see* artificial neural networks
    (ANNs)
artificial neural network-genetic
    algorithm (ANN-GA) 221
artificial neural networks (ANNs) 177
  bioprocess control 219–220
  in bioprocess engineering
    181–182, 190
asialoglycoprotein receptor
    (ASGPR) 76

b

Bacillus, 3–5
Bacillus cereus, 250
Bacillus subtilis
  cell transformation 7–8
  genetics by electroporation 9
  and heterologous protein production
    9–21
  production strain 6–7
  promoters 5–6
  protoplasts-mediated manipulations
    9

secretion system
  downstream processing 17–21
  fermentation and recovery
    11–12
  fermentation stoichiometry
    12–14
  fermentor kinetics and outputs
    14–17
bacterial membrane lipids 247–248
batch culture mode 145
bench-scale culture systems,
    incubators
  dynamic culture systems 149–151
  static culture systems 147, 148,
    150
biocatalyst inhibition 242, 244
biological license approval (BLA) 136,
    137
biopharmaceutical manufacturers
    204
bioprocess control
  adaptive control 211–213
  implementation of 203
  intelligent control systems
    fuzzy control 217–219
    integrated bioprocess 224–225
    metabolic control 225–226
    neural control 219–223
    plant-wide bioprocess control
      224–225

bioprocess control (*contd.*)
   statistical process control
     222–224
   mathematical modeling of
     206–208
   model-based control 209–211
   nonlinear control 214–216
   online estimation of 203–204
   parameter estimators 205–206
   PID control 208–209
   process analytical technology (PAT)
     in 204, 206
   process dynamics and modeling
     202–203
   proportional-integral-derivative
     (PID) 208–209
   real-time process monitoring 205,
     206
   regulatory aspects 204–205
black-box models 182, 183
Bordetella pertussis, 213
branched fatty acids 249
branched lipids 249

c
calibration drift 211
capillary electrophoresis 97–99
capillary gel electrophoresis coupled to
   LIF (CGE–LIF) detection
   99–100
capsular polysaccharides 250
carbon dioxide evolution rate (CER)
   16
carbon dioxide production rates
   (CPRs) 180–181, 190, 191
carbon dioxide transfer rate (CTR)
   196
CellCube module 147, 148
cell culture systems
   airlift bioreactor 151–153
   fixed/fluidized-bed
     bioreactor 152
   roller bottles 149, 150

   rotating-wall vessel bioreactor 153,
     154
   spinner flask 149, 150–151
   static culture systems 147, 148, 150
   stirred tank bioreactors (STRs)
     153, 155–157
   wave bioreactor 152–154
cell expression systems 84
cell-free systems 56–57
cell growth 77, 78
cell mass formation 13–14
cell metabolism 141–143
cell reactors 201, 202
CellStack culture chamber 147, 148
cell transformation 7–8
cellular therapies 133
central cannabinoid receptor 1 (CB1)
   44
chemoenzymatic modifications 89
chemometric methods 204
Chinese hamster ovary (CHO) 136,
   137, 226
Clostridium acetobutylicum, 250
Cohen–Coon tuning rules 209
continuous culture 145, 146
controlled-fed perfusion 159
Corynebacterium glutamicum, 218
CPRs. *see* carbon dioxide production
   rates (CPRs)
critical quality attributes (CQAs) 204
culture operating modes 145–146
current good manufacturing practice
   (cGMP) standards 134
cyclopropane fatty acids 250
cyclopropane lipids 248

d
Darcy's law 18
decoupled input–output linearizing
   controller (DIOLC) 214
Deinococcus radiodurans, 253
discrete-time optimization algorithm
   213

disposable bioreactors   135
dissolved oxygen (DO)   83
dolichol (Dol) phosphate   72
dry cell weight (DCW)   17
DSA-FACE   100
dynamic differential equations   184,
        185

e

Enbrel™   134
endogenous hexose transporters
        49
engineered expression systems,
        GPCRs
    bacteria   42–48
    cell-free systems   56–57
    insect cells   51–53
    mammalian cells   54
    transgenic animals   54–56
    yeasts   48–51
error-reducing algorithms   219
Escherichia coli, 242, 254
    cultivation   203
ethanol fermentation processes   203
ethyl methane sulfonate (EMS)   251
extended Kalman filter (EKF)   184

f

fed-batch culture   79–80, 145, 146
fed-batch feeding strategy   134
fed-batch fermentation process   14
fed-batch mode   146
fermentation stoichiometry   12–14
fetal bovine serum   143
Fick's first law   194
filtration processes   18
fixed/fluidized-bed bioreactor   149,
        152
fluorescence-activated cell sorting
        (FACS)   45
fluorophore-assisted carbohydrate
        electrophoresis (FACE)
        99–100

follicle stimulating hormone (FSH)   83
fuzzy control system   217–219
fuzzy logic controllers   217–218
fuzzy logic technique   217

g

galactosyltransferase   78
gene knockout/knockin   86–88
genes/gene clusters   46
genetics, electroporation   9
global transcription machinery
        engineering (gTME)   253
glucose catabolic pathways   142
glucose uptake rate (GUR)   226
glutamine   142
glycans, derivatization of   91
glycoengineering   78
glycoprotein processing
    inhibitors   88–89
    reactions   73
glycosylation   71
    ammonia   82
    analysis
        capillary electrophoresis   97–99
        and CGE-LIF   99–100
        fluorophore-assisted carbohydrate
            electrophoresis (FACE)
            99–100
        glycans, derivatization of   91
        high pH anion exchange
            chromatography with pulsed
            amperometric detection
            (HPAEC-PAD)   96–97
        hydrophobic interaction liquid
            chromatography (HILIC)
            93–95
        lectin microarrays   91–93
        mass spectrometry (MS)
            100–108
        and porous graphitic carbon
            (PGC) chromatography   95–96
        release of glycans from
            glycoproteins   90–91

glycosylation (*contd.*)
  reversed phase (RP) 95–96
  weak anion exchange (WAX) 96
  dissolved oxygen (DO) 83
  fed-batch cultures and supplements
    79–80
  host cell systems 83–85
  modification
    processing inhibitors and *in vitro*,
      88–89
    siRNA and gene
      knockout/knockin 86–88
  N-linked glycans 72–74
  nutrient depletion 76–78
  O-linked glycans 74–76
  pH 82–83
  quality by design (QbD) 71
  specific culture supplements 80–82
GnTI 87
G-protein coupled receptors (GPCRs)
    29
  druggability 31
  engineered expression systems
    42–57
  functional expression of 52
  predicted/actual topologies of 30
  recombinant GPCR production
    39–42
  structures of 33–35
green fluorescent protein (GFP) 45

h
hematopoietic stem cell (HSC) 137,
    138–139
Herceptin™ 134
heterologous protein production
    9–21
high pH anion exchange
    chromatography with pulsed
    amperometric detection
    (HPAEC-PAD) 96–97
host cell systems 83–85
human neurokinin receptor 1, 46

hybridoma cells 145
hybridoma technology 133
hydrophobic interaction liquid
    chromatography (HILIC)
    93–95
3-hydroxypropionic acid (3-HP) 244
HyperFlask vessel 147, 148
hypogonadotropic hypogonadism
    (HH) 31

i
inactivated hepatitis A vaccine 134
industrial bioprocess optimization
  balances 176–177
  CO_2-removal in large-scale cell
    cultures 194–197
  dynamic alternatives 183–186
  goals of 179–180
  model-aided feedback control
    controlled process operation 190
    soluble product 190, 192–193
  model identification 177–178
  open loop-controlled cultivations
    evolutionary modeling approach
      188, 189
    robust cultivation profiles 184,
      187–188
  static models 180–183
intelligent control systems
  fuzzy control 217–219
  integrated bioprocess 224–225
  metabolic control 225–226
  neural control 219–223
  plant-wide bioprocess control
    224–225
  statistical process control 222–224
in vitro amplification 8
in vitro modification, of glycans
    88–89

k
Kalman filter 183, 184
kernel functions 182, 183

l

lactate production rate (LPR)   226
Lactobacillus plantarum, 250
lectin microarrays   91–93
Leudeking–Piret equation   14
Leukemia inhibitory factor (LIF)   144
ligand-binding-competent receptor
     45
linear model predictive control
     (LMPC)   210–211
lipid component   247–248
lipid-oligosaccharide   72
lipid-rich cell membrane   246
lipopolysaccharides (LPS)   249
Lyapunov convergence   212

m

mammalian cell culture technology
   adherent cells   141
   case studies
      antibody production in   159–161
      mouse ESC culture   161–162
   cell culture systems (*see* cell culture
      systems)
   cell metabolism   141–143
   cell products, therapeutics
      134–137
   culture medium design   143
   fetal bovine serum   143
   kinetic models for   158, 159
   microcarriers   140–141
   physicochemical parameters
      144–145
   viral vaccines   133
mammalian expression systems   134,
   135
mannose-binding receptor   76
mass spectrometry (MS)
   derivatization techniques   102–103
   fragmentation of carbohydrates
      103–108
   ionization   100–102
medium exchange culture   145, 146

membrane damage   244
membrane engineering   246–251
membrane fluidity   249
mesenchymal stem cell (MSC)   137,
   138–139
metabolic control   225–226
metabolic engineering
   for biocatalyst robustness, organic
      inhibitors
      Escherichia coli or *Saccharomyces
      cerevisiae,* 242
   evolutionary and metagenomic
      strategies   251–254
   mechanisms of inhibition   242,
      243–245
   mechanisms of tolerance
      245–246
   membrane engineering   246–251
   reverse engineering-evolved
      strains   254
   types of   241
microbes   241, 242, 246
microcarriers   136, 140–141
model predictive control (MPC)
   209–211
monoclonal antibody (mAb)   133, 134
Monod-type expressions   177
mouse embryonic stem cells (mESC)
   145, 161–162
multi-input-multi-output (MIMO)
   nonlinear system   214
multivariate statistical process control
   (MSPC) methods
   bioprocess control   222, 224
murine myeloma lines SP2/0   136

n

National Aeronautics and Space
   Administration (NASA)   154
Navier–Stokes equation   177
nephrogenic diabetes insipidus (NDI)
   31
neural control   219–223

neural stem cell (NSC) 137, 138–139
nicotine adenine dinucleotide
    phosphate (NAD(P)H) redox
    state 225
N-linked glycans 72–74
nonlinearmodel predictive controller
    (NMPC) 211
nonlinear systems theory 215
nonsecreting null (NS0) 136
nutrient depletion 76–78

o
oligosaccharide chains 72
O-linked glycans 74–76
Omics analysis 243
Orgel's Second Rule 251
orphan receptors 31
Overton's Rule 244
oxygen uptake rate (OUR) 16, 181,
    190, 191, 196

p
partial least squares (PLSs) 223
pathogenesis-related inhibition
    mechanisms 243
peptidoglycan 249
perfused bioreactors 137
perfusion culture 145, 146
Pichia pastoris, 224
plant-wide bioprocess control
    224–225
plasmid-associated genes 253
porous graphitic carbon
    (PGC) chromatography
    95–96
principal component analysis (PCA)
    223
process analytical technology (PAT)
    204, 206
process dynamics 202–203
production host, *Bacillus,* 3–5

proportional-integral-derivative (PID)
    190
    bioprocess control 208–209
proton motive force (PMF) 245
pyruvate 142

q
quality by design (QbD) 71

r
Ralstonia eutropha, 213
Regenerative Medicine applications
    136
release of glycans from glycoproteins
    90–91
relevance vector machines (RVMs)
    182, 183
repeated batch culture 145, 146
response surface methodology (RSM)
    221
reversed phase (RP) 95–96
reverse engineering-evolved strains
    254
Rhodococcus ruber, 253
roller bottles (RB) 150
rotary vacuum drum filters (RVDFs)
    18
rotating-wall vessel bioreactor 154
rotating-wall vessel reactor (RWVR)
    153, 154

s
Saccharomyces cerevisiae, 242
SCalar Analysis of Library Enrichments
    (SCALEs) method 253
secretion system 9–21
    downstream processing 17–21
    fermentation and recovery 11–12
    fermentation stoichiometry 12–14
    fermentor kinetics and outputs
    14–17

serum substitutes 143
signal recognition particle (SRP) 43
siRNA 86–88
sound process design 201
stable isotope tracers 80
static culture systems 147, 148, 150
stem cell culture
  cell culture systems (*see* cell culture
    systems)
  cell metabolism 141–143
  culture medium design 143
  culture operating modes 145–147
  culture parameters 144–145
  ESCs and HSC 137, 138–139
  MSC and NSC 137, 138–139
  Regenerative Medicine applications
    136
  stirred and perfused bioreactors
    136
stirred bioreactors 137
stirred tank bioreactors (STRs)
  cultivation of mammalian cells
    153, 155–157
  mESC expansion on microcarriers in
    161–162
support vector machines (SVMs) 182,
  183
suspension cell culture processes 135

t
teichoic acid 249, 250
tissue engineering 133
tissue-plasminogen activator (t-PA)
  136, 160
tricarboxylic acid (TCA) 142
TripleFlask 147, 148
tumor necrosis factor (TNF) 134
type-2 fuzzy systems 218

u
unscented Kalman filters (UKFs) 184,
  185

v
Vaqta™ 134
viral vaccines 133
visualizing evolution in real time
  (VERT) method 251

w
wave bioreactor 136, 152–154
weak anion exchange (WAX) 96

z
Ziegler–Nichols tuning rules 209
Zymomonas mobilis, 250